Theôria as Cognition in Plato

Brill's Plato Studies Series

VOLUME 22

The titles published in this series are listed at *brill.com/bpss*

Theôria as Cognition in Plato

Edited by

Refik Güremen, Mauro Bonazzi and Elisabetta Cattanei

BRILL

ISSN 2452-2945
ISBN 978-90-04-74711-1
ISBN 978-90-04-74712-8 (PDF)
DOI https://doi.org/10.1163/9789004747128

Library of Congress Control Number: 2025051192

Brill and Koninklijke Brill BV are part of De Gruyter Brill.
www.degruyterbrill.com

Questions about General Product Safety Regulation: productsafety@degruyterbrill.com.

Contents

Notes on Contributors VII

Introduction 1
Refik Güremen, Mauro Bonazzi and Elisabetta Cattanei

PART 1
Setting the Problem

1 The Psychology of Contemplation 7
Mauro Bonazzi

2 The Dark Side of *Theôria* as Cognition in Plato: Mathematics 25
Elisabetta Cattanei

PART 2
***From Sense Perception to* Theôria**

3 From Mathematics to Dialectic in Plato's *Republic* 43
Damien Storey

4 Plato on Noticing 60
Andy German

5 Plato on the Ascent to Knowledge 78
Béatrice Lienemann

6 Plato and Aristotle on Knowledge 98
Hikmet Unlu

7 *Theôria* as Understanding of First Causes 116
Abida Malik

PART 3
Image Perception, Imagery and Theôria *as Higher Cognition*

8 Imaging Cognition in Plato's *Republic* 141
Marco Picciafuochi

9 Seeing the Divine through Its Visible Images: Plato's Philosopher and the *Theôros* 163
Cesare Simone Astorino

10 Vision, Blindness, and the Eleusinian Mysteries in Plato's *Phaedrus* 183
Giulia Cervato

PART 4
The Lived Experience of Theôria *as Higher Cognition*

11 Plato on Euripides and *Baccheia* as a Metaphor of Higher Cognition 207
Eva Anagnostou-Laoutides

12 *Theôria* in *Laws* 12 230
Joseph Bjelde

13 Understanding the Epistemic Journey of Plato's Philosophical *Theôros* as Uprootedness 240
Shane Fair

14 *Theôria* in Islamic Platonism 256
Victoria Rowe Holbrook

Index 279

Notes on Contributors

Eva Anagnostou-Laoutides
is Associate Professor at the Department of History and Archaeology, Macquarie University, Sydney, Australia. She works on ancient leadership legacies, political and intellectual, and the role of ritual in shaping key conceptual metaphors about legitimate leaders. She has published on the role of regeneration narratives in managing political crises during the Hellenistic and Augustan periods. She also works on metaphors about philosophical inspiration in Plato and their reception by Christian authors from the time of the early Church to the Quattrocento. Her research has been funded by the Australian Research Council and more recently by the Gerda Henkel Foundation.

Cesare Simone Astorino
graduated from the University of Milan with a thesis supervised by Prof. Franco Trabattoni, focused on the ways Plato presents his main metaphysical ideas and the reasons behind these stylistic choices. He is currently a Ph.D. candidate at the University of Pavia (Northwestern Italian Philosophy Convention, FINO, supervisor: Prof. Franco Ferrari). His dissertation delves into Plato's epistemology, aiming to establish a constructive role for sensible bodies by highlighting their nature as faithful images of the forms and their description as cult statues. He has spent research stays first at Trier Universität and then at the Cambridge Faculty of Classics.

Joseph Bjelde
received his Ph.D. from UC Berkeley in 2012 and has since 2013 been Wissenschaftlicher Mitarbeiter at the Humboldt Universität zu Berlin. He works primarily on epistemology in ancient Greek philosophy, and is especially interested in the epistemic and practical significance of dialectic and teaching.

Mauro Bonazzi
is Professor of History of Ancient Philosophy at the University of Bologna since 2023. Before he taught at the University of Milan and at Utrecht University. He also held teaching positions at Paris, Lille, Bordeaux, Clermont-Ferrand and Leuven. He is a specialist of the sophists, Plato and the Platonist tradition. His recent publications include *The Sophists* (Cambridge University Press, 2020) and *Platonism. A Concise History from the Early Academy to Late Antiquity* (Cambridge University Press, 2023). For the Italian publisher Einaudi he trans-

lated and commented the *Meno* and the *Phaedrus*. He is currently working to a new commented translation of Plato's *Republic*.

Elisabetta Cattanei
is Full Professor of History of Ancient Philosophy at the Catholic University in Milan, Italy. After graduating (1988) and getting a Ph.D. in Philosophy (1993) at the Catholic University in Milan, she did her research at the Universität Tübingen (WS 1993/94-SS 1995) and a post-Doc program (1996/97–98/99) in the field of History of Ancient Mathematics (Regensburg-Paris). In 1999 she became Research Professor at the University of Cagliari (Italy) where she worked from 2006 to 2015 as Associate Professor, and from 2016 until 2018 as Full Professor of History of Ancient Philosophy. From October 2018 until 2022, she was Full Professor of History of Ancient Philosophy at the University of Genoa, Italy. In the last 15 years she has been directing "local units" of three different National Research Projects (PRIN) about Aristotle and Aristoteleanism. She is Vice President (2025–28) and Next President (2028–31) of the International Plato Society (IPS) and former president of the Italian Society of History of Ancient Philosophy (SISFA – Società Italiana di Storia della Filosofia Antica). She takes part to the scientific and editorial committees of several series and reviews pertaining ancient philosophy and science. Her publications – in national and international context – are focused on the relations between mathematics and philosophy in the Antiquity, with special regard to Plato, Aristotle, Alexander of Aphrodisias and Proclus.

Giulia Cervato
is a Ph.D. student at the University of Padua. She previously studied at Alma Mater Studiorum – University of Bologna and Université Catholique de Louvain and attended several conferences both on a national and an international level. Her research interests span aesthetics and hermeneutics in Platonism and Neoplatonism (with particular reference to Hermias of Alexandria's *Commentary on the* Phaedrus) and revolve around the tension between philosophical thinking and poetic inspiration, mimesis, and philosophy of image. Moreover, she examines the reception of Platonism in contemporary philosophy, especially concerning iconography and aniconism, and how these ideas intersect with contemporary Jewish philosophical traditions, with particular reference to Emmanuel Levinas' thought.

Shane Fair
is an independent scholar and a recent Ph.D. graduate of the University of Ottawa (2021). His doctoral dissertation examined the role of wonder (θαῦμα,

θαυμάζειν) in Plato's dialogues. The thesis sought to distinguish two distinct-but-related types of philosophical wonder in Plato (aporetic and contemplative wonder), which both shed light on what Plato may have meant by his claim that "wonder is the *pathos* of the philosopher" (*Tht.* 155d2–4). His research interests lie predominantly in examining the cross sections between Greek philosophy and other contemporary cultural and scientific practices (e.g. literature/poetry, medicine, and geometry); in particular, how Greek philosophy is infused with, and in some sense can only be understood in light of these practices.

Andy German
is a faculty member and Chair of the Philosophy Department at Ben-Gurion University, in Israel. His research and teaching focus on Plato and Aristotle, as well as on the German philosophical tradition of the 19th and early 20th Century – principally, Hegel. He was co-editor of the volume of essays titled *Knowledge and Ignorance of Self in Platonic Philosophy* (Cambridge University Press, 2019) and has published on the Greeks and the Germans in various journals such as the *Review of Metaphysics*, the *British Journal of the History of Philosophy*, *Ancient Philosophy*, and the *International Philosophical Quarterly*. His current project centers on the Aristotelian account of the relation between *physis* and *logos* and how it has been assimilated and transformed in the thought of Rousseau, Kant and Hegel.

Refik Güremen
is Associate Professor at the Middle East Technical University (METU). His work in ancient philosophy appeared in journals such as the *Australasian Journal of Philosophy*, *Metaphilosophy*, *Polis*, *Methexis*, and the *International Journal for the Study of Skepticism*. He co-edited (with P.-M. Morel and J. Hammerstaedt) *Diogenes of Oinoanda. Epicureanism and Philosophical Debates* (Leuven University Press, 2017) and *Aristote. L'animal politique* (Publications de la Sorbonne, 2017, with A. Jaulin).

Victoria Rowe Holbrook
is a Senior Research Fellow at the Orient Institute Istanbul. She is a graduate of Harvard and Princeton and a member of the Society of Fellows and the American Philosophical Society. She has taught at Columbia, Ohio State, Bosphorus, Koç, Bilkent and Istanbul Bilgi universities. She is the author of *The Unreadable Shores of Love: Turkish Modernity and Mystic Romance*, translated into Turkish as *Aşkın Okunmaz Kıyıları: Türk Modernitesi ve Mistik Romans*; writings bridging literary, philosophical, and religious thought; and transla-

tions from Turkish and Persian, including *The White Castle* by Nobel Prize-winner Orhan Pamuk; *The New Cultural Climate in Turkey: Living in a Shop Window* by Nurdan Gürbilek; the Ottoman Turkish philosophical verse romance *Beauty and Love* by Galip; *Listen: Commentary on the Spiritual Couplets of Mevlana Rumi* by Kenan Rifai; and *O Humankind: Surah Ya-Sin*, a collection of Quranic commentary spanning a thousand years, compiled by Cemalnur Sargut. She is now working on a book about Platonism in the rest of the world.

Béatrice Lienemann
is Professor of Ancient Philosophy at the Université de Fribourg, in Switzerland. Her main research interests are ancient philosophy (especially metaphysics, epistemology and ethics) and their relations to the modern debates of practical and theoretical philosophy.

Abida Malik
is Assistant Professor at the Institute of Business and Vocational Education at Johannes Kepler University Linz, Austria. Her area of specialization is in ancient philosophy, contemporary epistemology, and virtue theory. She did her Ph.D. at the University of Bonn, Germany (with a research stay at Universidad Panamericana in Mexico City). Her dissertation *Seelen im Wandel. Eine Studie zum Charakterbegriff bei Platon* (Verlag Karl Alber, 2020) is the first comprehensive study on Plato's concept of character. The book analyses Plato's views of the evolvement of personality traits, explores the possibility of character change for adults, and looks at the educational and political implications of Plato's views on character. In contemporary epistemology, Abida Malik focuses particularly on the concept of tacit knowledge, on which she published the papers "Tacit Knowing: What it is and Why it Matters" (*Episteme*, 2023) and "Can Tacit Know-How be Acquired via Testimony?" (*Grazer Philosophische Studien*, 2023). Currently, she is developing a new project on epistemic environments in educational settings. This project seeks to describe current epistemic environments at educational institutions with the aim to develop measures for improving and enhancing these environments. In this study, she combines measures for environment transformation with a virtue epistemological approach to education.

Damien Storey
is Assistant Professor of Philosophy at Koç University, Istanbul. He works in ancient philosophy and has interests in contemporary ethical and political philosophy. His research explores questions in Plato's ethics, psychology,

and epistemology, and in particular Plato's views on the nature of belief and perception and how these relate to non-rational cognition. He is currently working on the images of the Sun, Line, and Cave in Plato's *Republic*.

Marco Picciafuochi
obtained his master's degree in 2021 from the University of Roma Tre – Tor Vergata and the Martin-Luther-Universität Halle-Wittenberg, under the supervision of Francesco Aronadio and Johannes Hübner, with a thesis on the conceptual system underlying Plato's use of figurative language in the *Gorgias*. In 2023 he spent six months at Durham University as a visiting research student, under the supervision of Phillip S. Horky. He is currently a Ph.D. student at the University of Roma Tre – Tor Vergata, working on Plato's philosophical use of images in the *Republic*, under the supervision of Francesco Aronadio. His research interests include ancient philosophy, Plato's writing style in relation to his epistemology, and the use of methods from cognitive linguistics and lexicology to study ancient texts.

Hikmet Unlu
is Assistant Professor of Philosophy at Hacettepe University, Turkey. His research focuses primarily on the intersection of ancient Greek philosophy and classical phenomenology.

Introduction

Refik Güremen, Mauro Bonazzi and Elisabetta Cattanei

Andrea Nightingale's *Spectacles of Truth in Classical Greek Philosophy: Theoria in its Cultural Context* (Cambridge University Press, 2004) has been a landmark in our understanding of the Ancient Greek notion of philosophical *theôria*. In her book, Nightingale provides a detailed historical and philosophical analysis of how this concept evolved from its origins in religious pilgrimages to its more intellectualized version in the works of Plato and Aristotle. She argues that *theôria* served as a crucial metaphor for philosophical inquiry and knowledge in the classical Greek world, transitioning from a religious practice to a philosophical one. Originally, *theôria* referred to a journey undertaken to witness religious festivals or spectacles, but Plato and Aristotle transformed the idea into a metaphor for the intellectual "journey" of contemplating eternal truths. She claims that the concept of *theôria* bridges the gap between religion, ethics, and epistemology. Nightingale's book was groundbreaking in that it showed, in compelling ways, how the philosophical notion of *theôria* is deeply embedded in the cultural and social context of ancient Greece.

Like all groundbreaking research, Nightingale's book has inspired new questions to explore and offered fresh approaches to traditional problems. This is the motivation behind the articles collected in this volume. The central aim of the volume is to investigate and analyse Plato's notion of *theôria* in terms of its cognitive aspects, which is reflected in the title: *Theôria as Cognition in Plato*. More specifically, this book seeks to provide detailed analyses of the psychological and epistemic components of *theôria* as discussed in various Platonic dialogues. As readers will see, this exploration has proven highly fruitful. Nightingale's work reveals that Plato's concept of *theôria* is deeply embedded in its historical and cultural context, shaped by metaphors, allegories, figurative language, and dialogical nuances. By paying close attention to the theoretical function and significance of these elements, we can discern Plato's effort to develop a profound understanding of the cognitive experience of acquiring knowledge and the means by which it is achieved. However, the true novelty of this book lies not in reiterating this point. Instead, it emerges from its focus on the *specific* processes by which *theôria*, as a form of cognition with its own proper psychology, interacts with other cognitive processes like sensation, memory, recollection and image perception.

The papers in this book can be read individually. However, to assist readers, we have grouped them into subsections based on thematic unity. Conse-

 | HTTPS://DOI.ORG/10.1163/9789004747128-002

quently, the book is divided into four main parts, each varying in length. In the first two papers, Mauro Bonazzi and Elisabetta Cattanei establish the framework of the problem. Bonazzi proposes an intellectualist interpretation of Plato's notion of *theôria* as discussed in the *Republic*. He argues that Plato recognizes a distinct, autonomous intellectual space for philosophical *theôria*, separate from the practical (i.e., political and ethical) aspects of knowledge. Bonazzi's argument identifies the object of inquiry for the rest of the book. Elisabetta Cattanei, through a discussion of mathematics in Plato, demonstrates that when *theôria* is viewed in terms of its own theoretical dynamics, it also reveals "dark sides," leading to certain *aporiai*. Her paper highlights a set of issues that are central to many of the subsequent discussions in the book.

The papers in the second part of the book focus on two central issues. Plato often treats sense perception as a mere "trigger" for a cognitive process that ascends to knowledge. But how exactly does this process function? What specific and positive contribution does sensation make toward attaining knowledge? From the analyses in this section, one might even argue that Plato's epistemology reveals an empiricist dimension. The second issue discussed in this section concerns the nature of *theôria*, understood as the cognition of first principles. Plato appears to envision a "leap" at a key stage in the cognitive process, transitioning from perceptual experience to knowledge of these principles. The papers in this section also address this cognitive "blind spot."

The third section of the book explores how Plato models and draws on certain cognitive experiences (e.g., image perception) that are tied to religious practices of his time. It is argued that Plato's use of metaphors culminates in an imagery that equips him with the conceptual "machinery" to explain the progression from sensation to higher forms of cognition, such as *dianoia*, *epistêmê*, and *noêsis*. Papers in the fourth section introduce existential dimensions to this cognitive ascent from sense perception to *theôria*, emphasizing how these existential aspects are deeply embedded in a cultural, and predominantly religious, context.

This book includes some of the papers presented at the first regional meeting of the International Plato Society (IPS) in the Middle East, which took place in May 2023. The conference was held in Ankara, at Middle East Technical University (METU). The event would not have been possible without the generous support of IPS. Similarly, the organization of the conference benefited from the generous support of The Scientific and Technological Research Council of Türkiye (TÜBITAK).

We owe special thanks to Edward Halper, who was serving as IPS president at the time. His contributions were invaluable, not only in the administrative processes of organizing the conference but also in the intellectual development of the event. We also extend our gratitude to our colleagues from the METU Philosophy Department for their various forms of support. But we owe a special debt of gratitude to our colleagues Büşra Akkökler, Berk Yaylım, and Selma Aydın Bayram for their unwavering assistance throughout the organization of this event.

We would like to thank our colleagues Mehmet Erginel, James Griffith, Tonguç Seferoğlu, Ömer Aygün and Daniel Wolt, who participated as speakers in the IPS Middle East Conference but whose papers are not included in this book, for their valuable contributions.

Bibliography

Andrea Nightingale. (2004). *Spectacles of Truth in Classical Greek Philosophy: Theoria in its Cultural Context.* Cambridge: Cambridge University Press.

PART 1

Setting the Problem

CHAPTER 1

The Psychology of Contemplation

Mauro Bonazzi

1 Plato's Psychology: The Standard Account

The story is well known. One of the central questions, in Plato's dialogues, pertains to "what we ourselves are", as quoted in the *Alcibiades I* (128e10–11). The answer is clear, to a certain extent. We are our soul, *psyche*, rather than our body, *soma*, or the composite of body and soul (*Alc. I* 129a–130e). In some dialogues, notably the *Phaedo*, this assertion is confirmed by a stark dichotomy between the soul and the body. However, in later dialogues, the picture becomes more intricate. We are our soul – this point is uncontested –, but our soul seems no longer to be a simple, unified entity. Instead, the soul is fragmented in three different "parts" (whatever "part" might imply concerning a non-corporeal entity),[1] designated as *epithymetikon*, *thymoeides*, *logistikon* – appetite, spirit and reason in standard English translations. The *Republic*, the *Phaedrus* and the *Timaeus* are the most eloquent witness to this doctrine.[2] So far so good.

Nonetheless much remains to be considered. The precise reason for Plato's development of the tripartite theory of the soul is not clearly known. It is argued that this shift occurred because Plato recognized that a strong soul/body dualism would not sufficiently explain the complexity of the human self. As it emerges from the *Phaedo*, a dualistic perspective implies an overly rigid opposition between a pure and an impure, between a good and an evil component. In this view the body becomes the source of all evil, yet this seems to oversimplify our nature and actions. As aptly noted by Julia Annas, "the various kinds of complexity to be found in human behavior cannot be adequately accounted for without allowing that the person is himself or herself complex" (Annas 1981: 124). With this change what we may term the dark side of the human self is incorporated into the soul; in contrast to the early dialogues echoing Socrates' views, desire is not solely oriented toward the good (see for instance *Men.* 77b–78a) because irrational and blind desires, not all related

1 In the *Republic* Plato uses *eide* (435c1 and 5, e1, 439e1), *gene* (441c6, 443d3) and *mere* (442b10, 442c4, 444b3).

2 The standard account is Robinson (1972).

HTTPS://DOI.ORG/10.1163/9789004747128-003

to bodily needs, are acknowledged. Scholars have delved competently and insightfully into the tripartition, focusing on its most intriguing innovation. In traditional Greek accounts of the soul or human self the opposition between reason and passion was commonplace. Plato's innovation lies in the further subdivision of what opposes reason in two distinct parts (*thymoeides* and *epithymetikon*). Over recent years, several scholars have significantly contributed to a better understanding of the philosophical significance of this new reconfiguration of the soul.[3] This is especially crucial due to the practical and political implications of such a doctrine, as it appears in the *Republic*. As Socrates famously claims in Book 2 (*Rep.* 368c–369b), a proper comprehension of the human self is essential for explaining what justice is and how a just city should be ruled – two issues of paramount importance.

To sum up. The soul/body dualism can likely be traced back to the historical Socrates, but it proved to be inadequate or incomplete. Tripartition, constituting a comprehensive account of our soul, is the Platonic theory that philosophers and politicians need when reentering the cave. Certainly, there are still numerous other philosophical and exegetical issues to be addressed. However, or so it is often assumed, it is within these conceptual coordinates that a thorough exploration of Plato's psychology should be undertaken.

In the present paper I advocate for a distinct approach by centering on the rational part. Undoubtedly, among the three components, reason appears to be the less contentious. In fact, reason holds a central position across all the dialogues, spanning from the *Alcibiades 1* and the *Phaedo* to the *Republic*, the *Phaedrus* or the *Timaeus*, irrespective of the theoretical framework in question, be it monistic or tripartite. We assert ourselves rational animals, don't we? However, the precise meaning of "rational" is less transparent than one might assume, as it will become evident through an in-depth examination of the *Republic*. Indeed, it could even be questioned whether a unified, consistent rational part exists within our soul. To scrutinize this issue, I will initially delve into the sections of Plato's *Republic* that introduce the soul, particularly the rational part, and argue that the overall depiction lacks consistency. Subsequently, I will explore the implications for the relationship between philosophy and politics in Plato. There is a widespread, albeit implicit, assumption that Plato's philosophy, particularly in the *Republic*, possesses a decisive practical, political dimension. However, an analysis of the tripartition in the *Republic* will show that this is controversial because the ultimate goal of the rational part appears to be theoretical contemplation. Interestingly, it will also turn out

3 See for instance Lorenz (2006) and Renaut (2014).

that there is more continuity in the nature of the soul throughout Plato's dialogues than it is conventionally assumed.

2 Two Contrasting Accounts of the Rational Part in the *Republic*

The structure and objectives of the *Republic* are anything but straightforward. The dialogue between Socrates and his interlocutors explores a multitude of diverse topics, making it challenging to integrate them into a cohesive interpretation. Moreover, as is often the case, these topics are frequently left incompletely addressed, awaiting further exploration that never materializes (with specific reference to the soul see *Rep.* 435c9–d4, referred to at *Rep.* 504b1–c2). Admittedly, Socrates will conclude at the end of Book 9 that he has successfully demonstrated the greater happiness of the just person compared to the unjust person – the demonstration requested by his companions in the dialogue. To achieve this, he has introduced a multitude of doctrines; essentially, all the fundamental tenets (Forms, soul, dialectics …) of what we may call Plato's philosophy are present in the *Republic*. However, these doctrines are introduced and discussed solely as tools or means to the ultimate objective of proving the superiority of the just life, rather than as distinct and independent subjects of investigation. Consequently, it is not surprising that several difficulties and unclarities arise when delving into these topics in greater detail.

The case of the soul, particularly regarding its tripartition and the rational part is probably the most evident one. The discussion of the (rational) soul unfolds into distinct sections: first in Book 4, representing one of the most detailed expositions of the tripartition in the dialogues; and then in Books 8 and 9, constituting one of the less lucid expositions of the doctrine in the dialogues, to the extent that it is often left aside, without substantial justification.[4] Let's explore both instances.

The aim of Book 4 is to illustrate the structural parallel between the city and the soul. When "constructing in words" the *kallipolis*, Socrates first delineates a city divided in three parts, and then he investigates whether this tripartition is also applicable to the soul. To be sure, that Socrates is trying to prove such parallel seems to expose him to the charge of circular reasoning. Particularly among those advocating for a political interpretation of the dialogue, the tri-

4 With two remarkable exceptions: Lear (1992), and especially Roochnik (2003). I don't agree however with Roochnik's (2003: 109–110) reading of Books 8–9 as the longer way referred to at *Rep.* 435d.

partition has been frequently dismissed as an artificial classification: a contrived construct that merely serves to confirm the parallel between the soul and the city, and therefore the validity of the tripartite *kallipolis*. However, against such critiques, it can be rightly pointed out that the analysis of the soul is grounded in a robust foundation – an empirical, phenomenological examination of human behaviors. The emergence of tripartition is not the result of idle speculation (which Socrates himself is wary of) but rather stems from a meticulous analysis of human behavior. In fact, the *Republic* provides a detailed account of a psychological doctrine that will be further expounded upon in other dialogues.[5] Dismissing it as mere fictional construct is a debatable choice.

Considering the dialogical context, in Book 4 the emphasis lies on the two irrational parts, the *epithymetikon* (opposed to reason in the example of the thirsty man) and the *thymoeides* (opposed to reason in the example of Odysseus). This emphasis is easily explained because Socrates must primarily establish the existence of two non-rational components, and not just one. As for the rational part, what insights do we gain about it in Book 4?

In the text this part is described in different ways: as that by which we learn (*Rep.* 436a9) or as the lover of learning (*philomathes*: *Rep.* 435e7); as that which possesses wisdom (*sophon*: *Rep.* 441e3, 442c4) or science (*Rep.* 442c5); and especially and repeatedly as that by which we calculate (439d5). In Book 4 this function is by far the most important one (the desire to learn occurs only one time and it is never put to work, see Ferrari [2007: 176]). This part is the part that calculates about what is better and worse (*to analogisamenon peri tou beltionos te kai cheironos*: *Rep.* 441c1–2; *ek logismou*: *Rep.* 439d1). In Book 4 what counts is the strategic and practical calculation of the best course of action. The most common name for this part is calculative, *logistikon* (*Rep.* 439d5; 440e4, 441a5 *et passim*; Annas [1981] translates “the reasoning part”, but such translation leaves this aspect in the shadow). The task of this part is to govern, *archein*, because it is wise and can take care of the other parts and the whole soul (*Rep.* 441e4–5, see also *kratoun*: *Rep.* 439c8).

> Does it not belong to the rational part to rule, being wise and exercising forethought on behalf of the entire soul […]? – Assuredly.
>
> *Rep.* 441e

5 Regarding the city/soul analogy it could even be argued, inverting the order of the exposition in the dialogue, that it is precisely this psychological foundation that grounds the tripartition of the *kallipolis*, see for instance *Rep.* 435d–436a.

Its wisdom is *euboulia* (*Rep.* 428d). Wisdom consists in this part's ruling the whole person and giving orders based on "the knowledge of what is advantageous (*to sympheron*) for each of the parts and for the whole which is common to all three" (*Rep.* 442c5–8).[6]

In short, it appears that this part has a cognitive component – the soul learns and knows what is good or bad – and a practical function: its knowledge is essentially practical. Interestingly, this part does not seem to have a proper motivation, as John Ferrari among others has observed: "it had been unclear whether this element [reason] even had a goal of its own or was merely a supervisor that placed limits on the interests of the other elements in the interest of the individual as a whole."[7] Since it somehow knows the best course of action, it is entitled to rule. This incidentally explains the importance of spirit. In order to rule, the calculative part needs the support of this other part, spirit, with its positive drive towards goodness, countering the impulses of the desiderative part (*epithymetikon*).

Upon revisiting the exploration of the soul in Books 8 and 9, following the extensive digression of Books 5–7, a discernible shift becomes apparent. At this juncture, Socrates has further deepened the analysis of the (many paradoxes of the) *kallipolis*. However, the connection between this revised city, as delineated in Books 5–7, and the ideal city envisioned in Book 4 remains somewhat elusive. The same ambiguity extends to the understanding of the soul. In books 8 and 9, as agreed at the end of Book 4, Socrates proceeds to explain how the perfect governance of the *kallipolis* can deteriorate into various other forms, resembling the political regimes encountered in our everyday life. Since Book 4 has established the structural affinity between the city and the soul, the analysis here draws on this parallel. Each subsequent city – timocratic, oligarchic, democratic, tyrannical – corresponds to a specific configuration of the citizens' soul. The structural analogy, rooted in the relationship between the city and the soul, persists from Book 4 to Book 8. However, a notable challenge arises in the transition from clear tripartition of Book 4 to the more intricate and complex psychological models introduced in Books 8 and 9. The configurations of the soul in these later books appear to be more convoluted and nuanced than the well-defined divisions outlined in Book 4 (Roochnik 2003: 111–132). This complexity represents a first difficulty.

In the context of our current inquiry, a second issue warrants consideration. In these degenerated cities, the pivotal role once assigned to the rational part no longer persists. In other words, a proper treatment of the rational part

6 See Kahn (1987: 84).

7 Ferrari (2007: 165–166).

should have taken place in Books 5–7. But this did not happen, at least explicitly. The narrative in Books 8 and 9 instead places emphasis on the two other components. Notably, explicit references to the rational part only emerge towards the conclusion of Book 9, during the comparison between the just individual and the tyrant. And what we learn here is different from Book 4.

To be sure, also in this book the rational part is called (with an explicit reference to Book 4, see *Rep.* 580d9) "that by which we learn" (*to hoi manthanei*: *Rep.* 580d9) and "the lover of learning" (*philomathes*: *Rep.* 581d10). But the differences are more striking:

1) First, it is remarkable that this discussion takes place in the context of an analysis of desires, with Socrates explicitly attributing the rational part its own kind of desire:

> As there are three parts, there are also three kinds of pleasure, one peculiar to each part, and so with desires.
>
> *Rep.* 580d7–8

This is a striking formulation, as John Cooper, has rightly remarked. Before, the word for desires, *epithymiai*, was used as the generic name for the urgent bodily appetites (thirst, hunger, and sexual desire) that serve as paradigms for the third part of the soul, *to epithymetikon*, which is so named after them. Now, desires are also attributed to reason. "The desires of reason are thus implied to be strong impulses of some kind which we experience simply and directly because we possess the power of reason" (Cooper 1984: 5). It is now explicitly stated that also the rational part has its own desire, a proper object of desire and a considerable motivational force of its own, as other passages confirm:[8]

> It is obvious to anyone that the part by which we learn is always wholly straining to know where the truth lies.
>
> *Rep.* 581b 5–6,

With this reference to truth, *philomathia* is equated to *philosophia*:

> Then would it be appropriate for us to call it learning-loving and wisdom-loving?
>
> *Rep.* 581b10

8 On the importance of desires, passions, love in contrast to Book 4, see Roochnik (2003: 4).

2) But if this is the case, as the above quotes make clear, another important difference from Book 4 emerges. True, also in Book 4 we had found an occurrence of *philomathes*, "lover of learning". But in Book 9 this desire seems to take a very specific direction, which is not easy to reconcile with Book 4. *Philomathes* is now qualified as *philosophon*, desire of wisdom (and of truth). This does not seem to implicate any practical involvement of the rational part. The rational part is now interested (it is dragged: *Rep.* 533d2, 521d4) only in knowing and wisdom. The philosopher is described as a *lover* of wisdom, dedicated to "contemplate the very things themselves in each case, ever remaining the same and unchanged (*Rep.* 479e6–480a4)". Clearly, this is not the type of practical wisdom described in Book 4. The desire of the rational part is to know not to rule; reason is not calculative anymore but it is philosophical (Ferrari 2007: 191–194):

> But the kind of pleasure connected with the vision of what is (*thes tou ontos theas*) cannot be tasted by anyone except the lover of wisdom.
>
> *Rep.* 582c7–9

> Therefore, of the three pleasures, the most pleasant would belong to that part of the soul with which we learn; and the man among us in whom this part rules has the most pleasant life.
>
> *Rep.* 583a1–3

3 Two Attempts: Kahn and Cooper

In the scholarly literature two notable attempts at reconciling this tension between a practical and a contemplative reason can be found in the seminal articles by Charles Kahn (1987) and John Cooper (1984), both addressing Plato's theory of soul, desire and motivation. Since it is evident in Book 9 that reason has its own desire, this must be the case already in Book 4 – as a matter of fact, it is typical of Plato to proleptically introduce views that will become clear only at later stages of the argument. Assuming that this is the case – in other words, that the *logistikon* of Book 4 has also some kind of desire – the problem is therefore to understand what this desire is.[9] Now, it is doubtful that

9 See Kahn (1987) and Cooper (1984). One alternative option, proposed by Cooper (see also Cross and Woozley 1964: 118–19), is that the rational part has actually two different goals – to rule (Book 4) and to know the truth (Book 9). But such a solution is debatable given the central claim of the *Republic* that each part (of the soul and of the city) must perform its own task and function – the principle of *oikeiopragia*, in other words (or the functional theory, as

this desire can be directed to concrete objects such as the ones of the two examples in Book 4. In the case of thirst it is not clear what reason wants, apart from stressing that it is not good to drink in that given situation; in Odysseus' example the soul simply shows spirit how to be successful in achieving its (that is, spirit's) goal (see for instance Ferrari 2007: 170–174). Therefore, the two scholars conclude, if there is a desire, it must be of another, higher, order. It is the desire of the good. Already in Book 4 the rational part is moved by its own desire of the good. It is this desire which gives the soul the strength/motivation to oppose the other desires.[10] Given the dialogical context this desire remains obscure in Book 4 and will become apparent only in Book 9, after the digression on the Form of the Good and the description of the philosopher in Books 5–7.[11]

To be sure, this is a very interesting solution, which serves to bridge the gap between Books 4 and 9. It is debatable, however, that it solves all the problems. For that to be the case this desire of the good should also reconcile the practical component of Book 4 with the contemplative dimension of Book 9. This is precisely what both Cooper and Kahn assume, arguing that this desire of the good (that is, to know the good) also involves a practical commitment. This is controversial, though. The importance of the (Form of the) Good cannot be denied. A closer reading of the *Republic*, however, shows that the emphasis on the Good does not solve all the problems – on the contrary.

First, it should be remarked that the rational part of Books 8 and 9 desires wisdom and not so much the good.[12] That wisdom and the good are the same should be proven. For the sake of the argument, we can however admit that wisdom and good are strictly dependent – they are certainly linked by the common reference to the Form of the Good. Once this is admitted, the problem is to understand what the desire for the (Form of the) Good means. This leads to some complicated issues. The precise nature and role of this Form

Santas [2010: 63–67] argues). As a matter of fact, it is never said that the rational soul has the desire to rule. Besides, if the rational part were to have two different desires (to rule, to know), a strong conflict would emerge within the rational part!

10 Incidentally, one may observe that this reading works better in the first case than in the second, because the example of Odysseus seems to imply that the desire of reason does correspond with that of spirit (vengeance).

11 See for instance Kahn (1987: 82): "the knowledge of truth must also be a knowledge of value, of what is worth pursuing, so that the desire to know the truth will ultimately be a desire to know *and* possess the good"; and see also (1987: 89): "So reason's desire to rule in the soul is not some third thing, over and above its judgment and desire for what is beneficial: its desire to rule is just the expression or the spelling-out of its desire for what is beneficial".

12 Ferrari (2007: 166) *contra* Kahn (1987: 82).

remain notoriously elusive; more importantly for the sake of the current paper, it is not even clear how the contemplation of this Form (if possible) has an impact on our good.[13] However, important as they are, these problems are not addressed by Cooper and Kahn. When it comes to the Good, both Cooper and Kahn quote the same text in support of their reading: The good is "that which every (*pasa*) soul pursues and for the sake of which it performs all its actions" (*Rep.* 505d11–e1).[14]

This sentence poses a challenge, as it conflicts with the assertion in Book 4 that the soul can also be dominated by spirit or appetite,[15] implying the existence of souls which are not inherently inclined towards the pursuit of the good.[16] One potential solution to this discrepancy could be to interpret the reference here as pertaining exclusively to the ordered soul, the soul governed by reason. Accordingly, *pasa* would then only apply to a specific configuration of the soul (see 438a). This interpretation might offer a plausible reconciliation. In any case, even leaving aside this problem, when we focus on the rational part (to which the aforementioned statement certainly pertains), it is crucial to emphasize that the concept of good (and soul's desire) does not inherently carry any moral, practical or political connotations. As David Sedley has rightly remarked, the good is always someone's good, and there is no indication in the dialogue that goodness will motivate anyone to act altruistically, against their own interest.[17] Not even the Form of the Good is a motivating factor transforming the philosophers (i.e., their reason) into pure altruists.

The rational soul pursues its own good and this is knowledge. Such a statement does not appear particularly controversial. In fact, it contains at least one controversial point, for it does not seem to indicate any practical or political commitment on the part of the soul, contrary to what Kahn claims (1987: 84 and 89). Indeed, such a possibility is never mentioned in the whole dialogue. The rational part has its own desire, to know and possess the Good. But this does not imply any practical commitment towards the others (and therefore also towards the other parts of the soul, as I'll come back to this point

13 See for instance the sound remarks by Annas (1981: 261–262 and 1999: 96–116); more in general on the Form of the Good see now Gerson (2020) and Broadie (2021). A good collection of articles, showing many different approaches to the Form of the Good is Reale – Scolnicov (2002).

14 See also *HiMa.* 297b, *Gorg.* 468b, 499e–500a, *Symp.* 205e–206a, *Phil.* 20d.

15 See the example of thirst with the emphasis on the fact that it is thirst of a drink without any consideration for what is good; on this see Kahn (1987: 85).

16 Vegetti (2003: 82 n. 116).

17 Sedley (2007a: 277–278) with further bibliography.

later) as part of its activities. The desire of the rational soul is to know, not to rule (Sedley 2007b: 103–104). Contemplation of the Forms is an end in itself.

Of course, this does not exclude that the rational part can rule, as it becomes even clearer in the more general case of the philosophers. On the contrary, since the rational part is the part which is in possession of the good, it is arguably the ideal ruler of the self; likewise the philosopher is the ideal ruler of the city for the very same reason that she is in possession of the good (see *Rep.* 517c4–6: "without having had a vision of this Form no one can act with wisdom, either in his own life or in matters of state", see also *Rep.* 484c–d and *Rep.* 540a–c).[18] But this is not the point. The desire of the rational soul is knowledge of the Good, not of how to best rule.[19] Like the philosopher outside the cave, the rational soul does not want (i.e., has no desire) to rule or to take care of the whole (because ruling is *work*: Ferrari 2007: 166); the rational soul desires to know. It might be the case that the rational soul by pursuing knowledge of the Good will possess knowledge of how to best rule – but this is secondary. Book 9 has taken over and reversed the account of the rational soul in Book 4.[20]

18 *Contra* see Annas (1999: 96–116). This has been also contested, from Aristotle to Hannah Arendt, because it is unclear how the contemplation of this metaphysical Good could serve as a practical guide in the human world.

19 This point is missing in Nightingale (2004: 127–131), who emphasizes the political and practical consequences of contemplation, but does not pay sufficient attention to the fact that there are two equally important problems at stake: not only whether the philosopher is the best ruler but also – and preliminary – whether the philosopher wants or desires to rule. A negative answer to the latter question has evident consequences on all the practical and political readings of Plato's *Republic*.

20 The parallel with the philosopher offers us a better understanding of what Kahn might mean by the verb "possess" in the above quoted passage (i.e. *Rep.* 517c4–6). Indeed, what is distinctive of the philosopher is the same desire – an erotic desire – to know truth and the real being that we will find in Book 9 in reference to reason: "So then, won't we make a sensible apology in saying that it is the nature of the real lover of learning to strive for what *is*; and he does not tarry by each of the many things opined to *be* but goes forward and does not lose the keenness of his passionate love nor cease from it before he grasps the nature itself of each thing which *is* with the part of the soul fit to grasp a thing of that sort; and it is the part akin to it that is fit. And once near it and coupled with what really is, having begotten intelligence and truth, he knows and lives truly, is nourished and so ceases from his labor pains, but not before" (*Rep.* 490a–b; see also *Rep.* 485d2–4: someone who loves learning must above all strive for every kind of truth from childhood on). In consequence of this drive the philosopher does not want or desire to rule: the philosopher wants or desires to know or contemplate the truth, as it is repeated in so many passages, and most notably at *Rep.* 517–521. In addition to the many passages when Socrates explicitly says that the philosopher does not want to rule/come back to the cave, see also the clear opposition at *Rep.* 476a–b: "I set apart and distinguish those of whom you were just speaking, the lovers of spectacles and the arts, and men of action

4 An Intellectualist Reading of the *Republic*

Another important contribution comes from Julia Annas.[21] As is well known, Annas has strongly opposed a political reading of the dialogue, insisting on the ethical dimension. At stake in the dialogue is not so much the construction of the ideal city (and the political commitment of the philosopher) as the possibility of making order in our soul and therefore living a good life. With regard to our topic, this would lead to the claim that the desire of the good is practically relevant at least at the level of the soul (a point shared also by Kahn and Cooper). Agreeing with Julia Annas that there is no political commitment, it could be argued that the rational soul, insofar as it is pursuing the good, is at least interested in ruling itself – meaning it seeks to rule over the other parts of the soul, as this aligns with its own interest: the rational ordering of the soul, which constitutes virtue. This is what matters most in the dialogue. Referring to the terms of Hellenistic debates, Annas has further argued that the main thesis of the dialogue is that virtue is sufficient for happiness. Reason, in other words, is not interested in ruling the others (thus there is no political task) and yet desires to rule over the whole soul, so as to promote order and harmony in it (and this is virtue), because this is the condition of a good life. Again, this is a vital claim in the *Republic*, because the possibility of achieving justice and therefore living a happy life depends on the possibility of having harmony within the soul. Annas is surely right when she underlines this point. And yet, there is a problem also with her interpretation, which leads us back to the previous discussion.

In her "Socratic" interpretation, reason is identified with practical knowledge.[22] But this leaves aside what is most importantly distinctive of the rational part in Books 8 and 9. In Annas' sufficiency-thesis there is no space for intellectual desire, the desire to know. But, as we know from Books 8–9 and 5–7 what really counts is precisely this desire. In other words, for Julia Annas it seems that the *Republic* ends at Book 4.

[πρακτικούς], and separate from them again those with whom our argument is concerned and who alone deserve the appellation of philosophers [φιλοσόφους]." On the difference practical-philosophic, see Irwin (1977: 236–237) and Annas (1981: 260–272).

21 See Annas (1999).

22 Annas (1999: 97): "[I]n the Socratic dialogues Socrates is searching for a special kind of wisdom which will give the person who has it overall knowledge of how to best live her life; this wisdom is identified with virtue, and it is thought of on the model of a skill or expertise, a practical ability to get things right which presupposed an overall understanding of the area and the way it is best handled".

To be sure, it could be objected in defense of Annas' reading that it remains true that virtue (that is order in the soul) is important – and the text should give us some indication of how to reach this order in the soul. Indeed, this is precisely what happens, but *thanks to intellectual desire and not to the practical commitment of the rational part.* How are these virtues and order in the soul achieved? By practicing moral virtues, as Annas argues?[23] This is not what the text says.

An eloquent testimony to this regard is *Rep.* 500b–c. The discussion refers to the philosopher as the human being in which the rational part is the strongest. Here there is a clear description of this desire of the rational part and the consequences for the whole soul:

> For presumably, Adeimantus, a man who has his understanding truly turned toward the things that *are* has no leisure to look down toward the affairs of human beings and to be filled with envy and ill will as a result of fighting with them. But, rather, because he sees and contemplates things that are set in a regular arrangement and are always in the same condition – things that neither do injustice to one another nor suffer it at one another's hands, but remain all in order according to reason – he imitates them and, as much as possible, makes himself like them. Or do you suppose there is any way of keeping someone from imitating that which he admires and therefore keeps company with?
>
> It is not possible, he said.
>
> Then it's the philosopher, keeping company with the divine and orderly who becomes orderly and divine, to the extent that it is possible for a human being.
>
> *Rep.* 500b8–d3; see also *Rep.* 613a

Julia Annas is one of the few scholars who have rightly emphasized the presence in Plato's dialogues of what was the most important thesis in the ancient Platonist tradition, the assimilation to God (Annas 1999: 52–71). This text is indeed one of the most relevant testimonies, together with other passages from the *Theaetetus*, the *Phaedrus* and the *Timaeus*. Again, as it appears from this text, harmony and justice are essential. As it will become explicit in Books 8 and 9, the philosopher is the happiest person because she has harmony in the

23 See Annas (1999: 105). Opposing an intellectualist reading, Annas argues that "theorizing about a virtue inhibits you from practicing it". But the philosopher "theorizes" not so much about virtues as about Forms.

soul with the rational part having the lead. But how is this achieved? Not by practicing moral virtues. The soul's ordering does not derive from the soul practicing virtues or wanting to rule, and not even because it was interested in the other two parts. Instead, it arises merely as a byproduct of the soul's intense intellectual desire. The philosopher's soul achieves order and harmony as a natural consequence of its desire to emulate the divine entities it contemplates. The ordering unfolds automatically and spontaneously, so to say. Imitation, in this context, emerges as a direct outcome of contemplation. How this is possible is further explained in another passage, the famous hydraulic analogy:

> [W]e surely are aware that when in a man the desires incline strongly to any one thing, they are weakened for other things. It is as if a stream had been diverted into another channel. [...] So, when a man's desires have been taught to flow in the channel of learning and all that sort of thing, they will be concerned, I presume, with the pleasures of the soul in itself, and will be indifferent to those of which the body is the instrument, if the man is a true and not a sham philosopher.
>
> *Rep.* 485d–e; see also *Rep.* 328d

The fundamental basis of reason's sway over the others is not its aptness for a supervisory role – though it has not lost this aptness – but the strength of its passion. As John Ferrari has rightly remarked, "this manifests itself not through internal conflict but through the atrophy of the other elements in the face of an all-consuming interest, as when water is channeled to a single irrigation ditch and the others run relatively dry (485d)".[24] As it has been remarked by Cornford and others,[25] from this passage it emerges an interesting view on human desire. It is as if there is only one desire, which can be directed in different directions bringing about different outcomes. The best outcome, and the most natural one according to Plato, is when this desire is channeled in the direction of the intelligible world – it is when it becomes rational, intellectual desire. The intensity of the desire to know, *to philosophon*, automatically dry all the rest – by consequence the philosopher is also morally virtuous – but this is secondary. In Books 8 and 9 we observe the contrary downward movement of those who are dominated by the other parts and therefore have their own desire channeled in another direction.

---

24 Ferrari (2007: 192); see also Sedley (2007b: 103). Interestingly, the metaphor of irrigation plays an important role also in Socrates' palinode in the *Phaedrus* (see for instace 251b–252a) and in the *Timaeus* (77c–78a); more in general, see Sassi (2007: 283–288).

25 See Cornford (1971); Sassi (2007); the same idea is also mentioned in the above quoted Kahn (1987: 98).

5 The *Republic* and Other Dialogues (and the *Republic* Again)

Surprising as it may appear, it is worth noticing that this reading of the tripartition and Plato's theory of motivation is confirmed by what we read in the *Phaedrus* and the *Timaeus*. In the famous metaphor of the winged chariot in the *Phaedrus*, it is the intensity of the charioteer's desire that wins over the black horse, without any external support (not to mention the white horse).[26] In the *Timaeus* the idea of assimilation is further developed, again along the same lines, insisting on the rational desire of the rational part.[27] And this leads us to a final proposal.

At the end of the journey, the question we started with – what are we? – appears to be even more difficult than it was at the beginning. Sure, we are our soul. But what does this mean exactly, given the tripartition? Given the tripartition, this means that we are our rational part, preeminently. As a matter of fact, it now turns out that this statement is more challenging than it appears at first sight. For the analysis of desire, and of the intensity of rational desire, seems to lead to an unexpected conclusion. Given the distance with all the rest, given the incompatibility between this rational, philosophical part and the *logistikon* or the other two parts, it could arguably be concluded that we are *just* this rational, contemplative, part.[28] Remarkably, this claim is explicitly established in the *Timaeus*, which is the dialogue which just follows and integrates the conclusions of the *Republic* in the dialogical fiction:

> Concerning the most authoritative kind of soul found in us, we must have the following thought. God has given it to each of us as a divinity (*daimon*) – this thing which we say dwells in the topmost part of our body and raises us up from the earth towards what is akin to us in the heaven, because we rightly call ourselves a heavenly plant, not an earthly one. For the divinity keeps our body upright by suspending our head and root from the place out of which our soul was first born. The consequences regarding mortality and immortality then follow: Hence if someone has devoted all his interest and energy to his appetites or to competition, all

26 See Ferrari (1985).

27 See now Ferrari (2022: lxxxi).

28 See also Annas (1999: 135–136). But she considers the two models as equally valid; besides, she gives her preference as the most Platonic to the composite model. To be sure, this is probably more appealing to our tastes. But it is not what emerges from Plato's dialogues.

> his beliefs must necessarily be mortal ones, and altogether, so far as it is possible to become par excellence mortal, he will not fall the least bit short of this, because it is the mortal part of himself that he has developed. But if someone has committed himself entirely to learning and to true thoughts, and it is these among the things at his disposal that he has most practiced, he must necessarily think immortal and divine thoughts, provided that he gets a grasp on truth. And so far as it is possible for human nature to have a share in immortality, he will not in any degree lack this. And because he always cultivates that which is divine, and has the divinity (*daimon*) that lives with him well ordered (*eu kekosmēmenon*), he will be preeminently happy (*eudaimon*).
>
> *Tim.* 90a–c

In this text the message is clear: "the highest achievement of a human life is to succeed in identifying the core of your being with the intellect, located in the head, setting aside the lower psychological drives, housed in the heart and gut. The implicit advice is: stop thinking of yourself as an entire amalgam of body and soul parts; identify yourself instead exclusively with the spherical thinking-hub perched on top of this rather complex vehicle. By doing so you can [...] hope to achieve a truly godlike existence" (Sedley 2007b: 92).

Remarkably, this is also what we find at the end of the long journey of the *Republic*, with a surprising image. Towards the end of Book 9 Socrates describes the soul as composed by a monster, a lion and a man:

> One of those natures such as the tales say used to come into being in older times – the Chymaera, Scylla, Cerberus, and certain others, a throng of them, which are said to have been many ideas grown together in one.
>
> Yes, he said, they do tell of such things.
>
> Well then, mold a single idea for a many-colored, many-headed beast that has a ring of heads of tame and savage beasts and can change them and make all of them grow from itself.
>
> That's a job for a clever molder, he said. But, nevertheless, since speech is more easily molded than wax and the like, consider it as molded.
>
> Now, then, mold another single idea for a lion, and a single one for a human being. Let the first be by far the greatest, and the second, second in size.
>
> That's easier, he said, and the molding is done.
>
> Well, then, join them – they are three – in one, so that in some way they grow naturally together with each other.
>
> They are joined, he said.

> Then mold about them on the outside an image of one – that of the human being – so that to the man who's not able to see what's inside, but sees only the outer shell, it looks like one animal, a human being.
>
> *Rep.* 588c–d

Scholars have been perplexed by the apparent contradiction of this description that seems to imply an infinite regress (every man containing inside another lion, another monster, and another man). But this risk is only real if we assume that the soul and the self are made of the three different parts. If we assume that we are, properly speaking, our rational part only (the man of the image), there is no contradiction anymore.[29] Interestingly, such an image also seems to rest on a similar model as the one delineated in the above-quoted passage of *Rep.* 485b: by feeding one part we starve and enfeeble (or dry, as in the hydraulic metaphor) the other parts.[30] It all comes to feed the correct part, that is reason – who we really are. It is a surprising conclusion, which does not of course explain everything. But it is the most consistent (and provisional) conclusion of the long journey in search of the soul.

6 Final Remarks

If this interpretation is correct, some intriguing implications emerge. Methodologically, it appears that Plato engages in the construction of different models that do not necessarily settle into a singular, rigid doctrine. Indeed, an effort to establish a static doctrine of the soul would contrast with the assertion that exhausting its complexity is nearly impossible.[31] Notably, this fluidity extends across different dialogues. The emphasis on the significance of the rational part does not appear to sharply contradict the soul theory found in other dialogues, particularly in the *Phaedo*. In other words, there are also important lines of continuity between the *Phaedo* and the *Republic* (or the *Phaedrus* and the

29 See also *Rep.* 611b–612a where Socrates discusses immortality by comparing the soul to the sea-god Glaucus (on which see Woolf 2012). As it is also stated in *Phdo.* 79c1–4, 80b2 what is composite can and will resolve into its parts: how can the soul be immortal then? Tripartition does not give us "the soul in its true nature" (*tei alethestatei physei*: *Rep.* 611a10–12). Again the alternative solution (that we are our intellect) prevents us from charging Plato with surprising inconsistencies (see Shields [2010: 148–149]).

30 See Cornford (1971: 123).

31 See *Rep.* 435c9–d4, 504b1–c2; *Phdr.* 246a4–5; on this point see the interesting remarks by Shields (2010: 168).

Timaeus). It may be wise to consider integrating the developmentalist interpretation with other readings of the dialogues, seeking parallels rather than solely focusing on progressions and evolutions. While this approach might not yield a perfectly closed and Platonic system, it can illuminate Plato's primary interests. Regarding the content, a comprehensive understanding of the importance of reason in the *Republic* underscores some radical innovations as it implies a diminished emphasis on the political dimension. However, one must question whether this constitutes a genuine innovation. Historically, the prevailing notion of a contemplative Plato dominated for centuries until the political turn in the 20th century. The goal of this paper was to reconsider that perspective on the basis of textual evidence – an attempt that does not certainly solve all the issues but warrants careful consideration.

Bibliography

Annas, J. 1981. *Introduction to Plato's Republic*. Oxford: Oxford University Press.

Annas, J. 1999. *Platonic Ethics Old and New*. Ithaca and London: Cornell University Press.

Broadie, S. 2021. *Plato's Sun-like Good: Dialectic in the Republic*. Cambridge: Cambridge University Press.

Cornford, FM. 1971. "The Doctrine of Eros in Plato's Symposium". In *Plato: a Collection of Critical Essays. II: Ethics, Politics, and Philosophy of Art and Religion*, edited by G. Vlastos, 119–131. New York: Anchor Books.

Cooper, J. 1984. "Plato's Theory of Human Motivation". *History of Philosophy Quarterly* 1: 3–21.

Cross, RC. and Woozley, AD. 1964. *Plato's* Republic. *A Philosophical Commentary*. London: MacMillan.

Ferrari, F. 2022. "Introduzione." In *Platone. Timeo*, edited by F. Ferrari and F. Petrucci. Milan: Rizzoli.

Ferrari, J. 1985. "The Struggle of the Soul: Plato, Phaedrus 253c7–255a1." *Ancient Philosophy* 5: 1–10.

Ferrari, J. 2007. "The Three-Part Soul." In *The Cambridge Companion to Plato's Republic*, edited by J. Ferrari, 165–201. Cambridge: Cambridge University Press.

Gerson, LP. 2020. *Platonism and Naturalism*. Ithaca: Cornell University Press.

Irwin, T. 1977. *Plato's Moral Theory: The Early and Middle Dialogues*. Oxford: Clarendon Press.

Lear, J. 1992. "Inside and Outside the Republic." Phronesis 37: 184–215.

Kahn, C. 1987. "Plato's Theory of Desire." *The Review of Metaphysics* 41: 77–103.

Lorenz, H. 2006. *The Brute Within: Appetitive Desire in Plato and Aristotle*. Oxford: Clarendon Press.

Nightingale, A. 2004. *Spectacles of Truth in Classical Greek Philosophy: Theoria in its Cultural Context*. Cambridge: Cambridge University Press.

Reale, G. and Scolnicov, S. (2002). *New Images of Plato. Dialogues on the Idea of the Good*. Sankt Augustin: Academia Verlag.

Renaut, O. 2014. *Platon: La mediation des émotions*. Paris: Vrin.

Robinson, TM. 1972. *Plato's Psychology*. Toronto: University of Toronto Press.

Roochnick, D. 2003. *Beautiful City: The Dialectical Character of Plato's* Republic. Ithaca, NY: Cornell University Press.

Santas, G. 2010. *Understanding Plato's Republic*. Oxford: Wiley-Blackwell.

Sassi, MM. 2007. "Eros Come Energia Psichica. Platone E I Flussi Dell'Anima." In *Interiorità e anima. La psychè di Platone*, edited by M. Migliori and L. Napolitano, 275–292. Milan: Vita e Pensiero.

Sedley, D. 2007a. "Philosophy, the Forms, and the Art of Ruling." In *The Cambridge Companion to Plato's Republic*, edited by J. Ferrari, 256–283. Cambridge: Cambridge University Press.

Sedley, D. 2007b. "Divinization." In *Plato's Symposium: A Critical Guide*, edited by P. Destrée and Z. Giannopoulou, 88–107. Cambridge: Cambridge University Press.

Shields, C. 2010. "Plato's Divided Soul." In *Plato's Republic: A Critical Guide*, edited by ML. McPherran, 147–169. Cambridge: Cambridge University Press.

Vegetti, M., ed. 2003. *Platone: Repubblica, vol. v: Libri vi–vii*. Naples: Bibliopolis.

Woolf, R. 2012. "How to See an Encrusted Soul." In *Plato and the Divided Self*, edited by R. Barney, T. Brennan, and C. Brittain, 150–173. Cambridge: Cambridge University Press.

CHAPTER 2

The Dark Side of *Theôria* as Cognition in Plato: Mathematics

Elisabetta Cattanei

1 First Picture: Plato's Mathematical Platonism – The Bright Side of Mathematics as a Complex of Theoretical Disciplines

"Arithmetic" – we can say mathematics in general – "must be discovered in just the same sense in which Columbus discovered the West Indians, and we no more create numbers than he created the Indians".[1] With this analogy Russell explains in an early essay what we still refer to as "mathematical Platonism" – a position in the fields of ontology and epistemology that first appears in Plato's dialogues. It is "a type of realism," as Julia Annas explains, "amounting to the belief that mathematical objects literally exist, independent of us and of our thoughts about them. Numbers are not, of course, in the world around us; they do not exist in space and time. But they exist none the less; they are there to be discovered before we think of the means of describing them."[2]

This kind of discovery – the intellectual discovery of the abstract entities in a new realm – can be seen as the essence of Plato's "reformed" view of the mathematical disciplines. These disciplines, which the Pythagoreans of the 5th century regarded as "sisters," were later categorized by Aristotle – not yet Plato – as "mathematical sciences," placing them among the "theoretical sciences."[3] In Book 7 of the *Republic*, Socrates and Glaucon discuss how arithmetic, *logistike* (the "science of calculation"), geometry (both plane and solid), astronomy, and harmonics – disciplines that were in a rather dismal state at the time – could actually be powerful forms of knowledge within the educational framework intended for future philosophers.

Two brief comedies are presented in the same book to illustrate this point: a Pythagorean trick against fractional unity and a satire of the pretentious jargon

1 Russell (1901: 312).
2 Annas (1987: 3–4). An updated picture of mathematical Platonism is to be found in Panza-Sereni (2013).
3 Cattanei (2003: 473–475).

of geometry. The absurd use of numbers and calculations – criticized by Socrates and Glaucon – is that of "amateurs," like "dealers and merchants, who use it for buying and selling." This stands in stark contrast to the serious use of mathematics, which "examines the nature of numbers by means of thought" and is pursued "in order to acquire knowledge," ultimately "facilitating the soul's conversion from the world of generation to essence and truth" (*Republic* 7.525c1–3). Socrates exposes the absurdity of using numbers and calculations instrumentally by bringing the Pythagoreans into the scene. He demonstrates that the qualities of unity/identity and indefinite multiplicity attributed to the one and to all other numbers should not be understood as physical properties but as objects of *dianoia*.[4]

For the *mathema peri tous logismous* (7.525d1) to be useful, one must not consider the simultaneous presence of unity and multiplicity in both the one and in all other numbers as a form of unity divided into multiple or infinite parts, as a physical body would be. Those who use mathematics for practical purposes do so empirically, accepting that the basic unity can be divided, leading to fractional numbers. They treat the basic unity as if it were a physical body that can be split into two, three, or four parts, and therefore assume that numbers, composed of units, can be divided in the same way. But now comes the ironic laughter of the "experts" (7.525d9–e1), likely the Pythagorean mathematicians. These experts mock those who try to divide the basic unity as if it were a physical object: the more one attempts to divide it, the more the experts multiply the unity, aiming to make the one truly one, not a collection of parts (7.525d6–e4). Those who conceive of unity empirically divide it into parts – halves, thirds, quarters, etc. – and thus produce fractional numbers (1/2, 1/3, 1/4, etc.), which are unacceptable in the Pythagorean framework, where unity must remain indivisible. To restore the unit, the experts simply multiply the fractions by the number of parts into which unity has been divided – 1/2 times 2, for instance, or 1/3 times 3 – thereby reestablishing the basic unity. "Suppose now, Glaucon," says Socrates, "someone asks them, 'My good friends,

4 See Pl. *Rep.* 525b11–526c4. In his paper in this volume (n. 6), Damien Storey suggests – though ultimately rejects – the idea that the behavior of unity and multiplicity in this passage alludes to a theory of calculation based on *antanairesis*, which is crucial for my understanding of *logistike techne* and the 'happily problematic' aspect of Plato's reformed mathematics (see below, § 2). I am also convinced that this passage has nothing to do with theoretical *logistike*, but rather with *arithmetike*, which is absurdly practiced by dividing numerical units as if they were physical bodies, when it should instead preserve the indivisibility of units grasped by *dianoia*. Theoretical *arithmetike*, and by extension theoretical *logistike*, require the absolute indivisibility of numerical units and do not admit fractional numbers (see Cattanei [2003: 477–485]).

what numbers are these you are talking about, in which the one is such as you postulate, each unity equal to every other without the slightest difference and admitting no division into parts?' What do you think would be their answer?" (7.526a1–5). Glaucon immediately responds with a crucial insight: "This, I think, that they are speaking of units which can only be conceived by thought, and which it is not possible to deal with in any other way" (7.526a6–7).

Socrates and Glaucon's satire spares few who practice plain or solid geometry. The "ridiculous" geometricians are those who, tied to the realm of becoming and to contingent concerns, oppose the approach of "reformed" scientists, who are driven by a completely different need: the pursuit of "contemplating the essence" (7.526e6) until they achieve the "vision of the Idea of the Good" (7.526e1). In the case of stereometry, the discipline's "ridiculous" state (7.528d9) is attributed to bad and arrogant astronomers who focus solely on the practical applications of their knowledge. This kind of knowledge stands in stark contrast to the new, "graceful," but misunderstood approach to practicing the discipline.

Ordinary geometry is depicted as an upside-down world, which Socrates and Glaucon find ridiculous, especially when considering the language used by its practitioners (7.527a9). Socrates notes that "this science" is in "direct contradiction with the language employed in it by its adepts" (7.527a1–4), because "their language is most ludicrous, though they cannot help it, for they speak as if they were doing something and as if all their words were directed towards action. For all their talk is of squaring and applying and adding and the like, whereas in fact the real object of the entire study is pure knowledge" (7.527a5–b1). Even if their drawings were exceptionally beautiful, such as those "made by Daedalus," a true expert "would think it absurd to examine them seriously in the expectation of finding in them the absolute truth with regard to equals or doubles or any other ratio" (7.529e1–530a1).[5]

From this grotesque situation, Socrates and Glaucon draw a crucial and serious conclusion. Socrates asks, "Must we not agree on a further point?" Glaucon responds, "What?" Socrates then clarifies, "That geometry is the knowledge of that which always is, and not of something which at some time comes into being and passes away." Glaucon agrees, "That is readily admitted, for geometry is the knowledge of the eternally existent" (7.527b3–8).

---

5 Plato likely refers to sophists like Hippias of Elis, who practiced a 'visual' geometry (see Cattanei [2003: 485–486] on Hippias; 511–512 on Hippocrates of Chios). The use of drawings in their approach persists in 'reformed' theoretical geometry as a 'limit,' which plays a crucial role in the transition from mathematics to dialectic, as demonstrated by Damien Storey in his paper in this volume.

This represents the positive view – the "official portrait," so to speak – of mathematical disciplines as presented by Plato: forms of intellectual knowledge of something intelligible, often called *dianoiai*, characterized by the rigorous nature of their *logoi*, and pursued to achieve theoretical goals (see 7.521c7–8; cf. 7.523a1–3 on numeracy; 7.524d3, 7.525d1–2, 7.525c1–6 on arithmetic and *logistike*; 7.527d9–10, 7.527a6–b1, 7.527b5 on geometry). This is the "bright side" of the mathematical realm discovered through "reformed" mathematics, as described by Socrates and Glaucon with intelligent lightness and ironic optimism, in contrast to the "laboratorial" state of mathematics in the 5th and 4th centuries BC.[6] However, reformed theoretical mathematics also has a darker side; it must possess another aspect to influence the human soul in our world. Mathematics are a discovery, but also entail treacherous navigation (Pl. *Symp.* 210e).

2 Second Picture: The "Paradoxical" Features of *Arithmetike, Logistike* and Geometry

What do theoretical arithmetic and *logistike* address, and what is their relationship? In 7.525a9–11, it is stated that "the science of calculation and arithmetic is wholly concerned with number." According to *Gorgias* 451b and *Charmides* 166a (as well as *Theaetetus* 198a and *Statesman* 259e), arithmetic examines numbers in themselves and their properties, whereas *logistike* studies the relationships (*ratios*) between numbers and their properties.[7]

2.1 Arithmetike

To grasp the meaning of "number," we can refer to the definitions in Book 7 of Euclid's *Elements*. The first and the second definitions are of unit and number: "a unit (*monas*) is that by virtue of which each of the things that exist is called one (*hen*)"; "a number is a multitude composed of units (*arithmos ... to ek*

6 The 'fluid' status of mathematical disciplines before Euclid was first described by Caveing (1990: 127–128) in terms of a 'laboratory.' This 'laboratorial' status has both a historical and onto-epistemological dimension (see Cattanei [2003: 488–493]) and is closely linked to the emergence of deductive demonstration as the scientific reasoning par excellence (see Mueller [1992]: 175–194; Netz [1999]).

7 Some scholars have argued that *logistike* is a form of 'applied mathematics,' based on a scholium to the *Charmides* (see Heath [1981: 14–15]). However, this interpretation is mistaken and has been corrected both mathematically since Klein (1968) and through textual evidence. Numerous passages in *Republic* 7 and other dialogues defend the theoretical nature of both *arithmetike* and *logistike*. See also Cattanei (2003: 493–495).

monadon sunkeimenon plethos)." Euclid's definitions expand on earlier formulations, likely from the Pythagoreans, which view the unit as indivisible and numbers as collections of indivisible units. Arithmetic, as Socrates describes in *Rep.* 7, is the "study of unity" (7.525a2: *mathesis peri to hen*), where the characteristics and properties of unity relate strictly to those of "any other number" (7.515a6–7). More broadly, the definitions of fundamental number properties and the subsequent classification of numbers for theorem demonstration rely on "Pythagorean" notions of number and unity. According to Socrates and Glaucon (6.510c4), the distinction between "even" and "odd," included in the definitions of *Elem.* 7, is seen as synonymous with "arithmetic" (see *Pol.* 262d–e, *Eutiphr.* 12d–e). Definitions 6 and 7 of Euclid's Book 7 state that "an even number is that which is divisible into two equal parts" and "an odd number is that which is not divisible into two equal parts, or that which differs by one unit from an even number," where "part" signifies a group of units within the number, a submultiple of it (see *Elem.* 7, def. 3). The *Elements* contain an extensive theory based on the definitions "even" and "odd", detailed in *Elements* 9, propositions 21–34. For example, propositions 21–30 establish that the sum of even numbers is even, the sum of an even and an odd number is even, and the sum of odd numbers is odd. These rules also apply to subtractions, multiplications, and divisions.[8]

Some aspects of the theory of "even-and-odd-numbers" suggest it underpins Socrates and Glaucon's reflections on mathematics. They evidently view the universe of numbers as coextensive with "evens-and-odds," establishing the rules for inclusion in this universe. However, what stands out in this theory is – as a historian and philosopher of mathematics noted – its "absolute superfluous character," because "its rules are not applicable to mathematical studies or everyday life."[9] The theory lacks practical applications, a point Plato saw as positive, though it might not have been so for the average person. Another striking aspect of the "arithmetic of evens and odds" is the behavior of the indivisible monad, which reveals a paradox: it is a full and perfect unity because it is indivisible, yet it can be infinitely multiplied, repeated infinitely in the series of natural numbers.

Socrates and Glaucon focus on the paradoxical nature of unity, which "appears at the same time as one and as infinitely multiple" (7.525a4–5), extending this paradox to "any other number" (7.525a6). However, they do not explain

---

8 According to Zhmud (1997: 158), the theory of 'even-and-odds' in Euclid's *Elements* 9 is among the oldest mathematical doctrines of the Pythagoreans and can be attributed to Pythagoras himself.

9 Toth (1998: 167–168).

how arithmetic and *logistike* handle units and numbers that are simultaneously unitary and multiple. They merely reject the idea that unity's multiplicity corresponds to a division into multiple parts (7.525d5–526a7), emphasizing that it is a challenging problem in the science of calculation (7.526c2). This suggests that a purely arithmetical explanation, where each unity is one but can be infinitely repeated, and each number is one in its species but can be infinitely repeated, is insufficient. Complex examples of theoretical *logistike* are needed to understand how unity and multiplicity coexist in the units and numbers mentioned by Socrates. The lack of precise considerations from Socrates may imply an implicit acknowledgment of a general attitude in arithmetic and the science of calculation: a tendency to contemplate paradoxical and contrasting conceptions of unity without being unsettled by its unusual aspects.

2.2 Logoi *and* Antanairesis

The primary theoretical difficulties arising from the indivisibility of unity and the concept of numbers as sets of indivisible units were likely addressed early on through a theory of *ratios* or *logoi*. This is what *logistike* represents: in the *Charmides*, it is defined as the science that "is concerned with the even and the odd in their numerical relations to themselves and *to one another*" (166a5–7). The terms *logismos*, *logizesthai*, and the activity of the *logistikon*, frequently used in the mathematical curriculum of *Republic* 7, pertain to *ratios* and the *relations* established between positive integers.

How can we understand which type of *logos* is examined by *logistike*? It is useful to revisit a stage preceding the theory of proportions in Euclid's *Elements* 5, traditionally attributed to Eudoxus of Cnidus. We find a definition of logos "closer to the concept of algorithm or a process of numerical computation," of which "fragmentary and uncertain testimonies" exist, although "Aristotle provides a clear and explicit formulation of it."[10] Book 8 of Aristotle's *Topica* offers the best definition of the *logos* likely examined by *logistike*, and denominated as *antanairesis* or *antiphairesis*: Aristotle explains that, for a parallelogram divided by a straight line parallel to one of its sides, both the bases and the surface are divided in "similar ways, since the surface and the bases have the same *antanairesis*, and this is actually the definition of *the same ratio*" (158b31–35). The term *antanairesis* suggests a form of "antagonism" (anti), "a mutual confrontation between magnitudes" and it also hints at "the resolution of this conflict". This reflects "the idea of finding a 'balance' between opposed parts" and

10 Zellini (1999: 131). Fowler (1999) demonstrates how the 'theory of ratios,' based on *antanairesis*/*antiphairesis*, can be reconstructed from the works of Plato, Aristotle, and Euclid, as well as its presence in Plato's dialogues.

outlines a process involving the "repetition of the same elementary operation of subtraction on different magnitudes related to each other by a recursive formula."[11]

This is generally referred to in mathematical texts as the "Euclidean algorithm." The basic example involves the ratio between two line-segments. Given two segments, *a* and *b*, with *a* being longer than *b*, the ratio can be determined by using *b* as the unit of measurement and performing "repeated subtractions": subtract *b* from *a* multiple times until *a* is completely measured, corresponding to a certain number of units. However, repeated subtractions may leave a remainder between the segment to be measured and a certain number of units. At this point, "reciprocal subtraction" is employed: the remainder is used as the new unit of measurement and is subtracted from the smaller segment. The process continues until the remainder measures the segment being subtracted from, without further remainders. The overall ratio between the measured segments is expressed as an ordered pair of positive integers from the two phases of "reciprocal subtraction," for example, [3, 2]. In this case, the initial segments are commensurable. For incommensurable magnitudes – such as the diagonal and side of a regular pentagon or square – it is impossible to find a remainder that measures both, making the measurement process infinite. Although the remainder becomes increasingly smaller and closer to the segment being measured, it never reaches an exact result. Geometric attempts to construct the "reciprocal subtraction" of incommensurable magnitudes result in a cyclical repetition of the same, increasingly smaller image. For example, the *antanairesis* of the side and diagonal of a regular pentagon leads to the construction of the "pentagram."[12] Even for incommensurable magnitudes, the number of subtractions performed can be represented by an infinite series of ordered pairs of positive integers, unlike the finite series representing ratios of commensurable magnitudes. In a theory of calculation based on *antanairesis*, the term *logos* denotes a ratio expressed as an ordered pair of integers, corresponding to one stage of a recursive measurement process of two magnitudes, represented in an ordered series (*taxis*) of *logoi*, either open or closed.[13]

It is evident how the development and study of *antanairetic logoi* can address some of the major limitations of Pythagorean arithmetic. The notions

---

11 Zellini (1999: 179).

12 Von Fritz (1965: 295–298).

13 This is what Proclus, commenting Plato's *Republic* (*In Pl. Remp.*, II, p. 27, l. 1–p. 28, l. 12 Kroll), called "Pythagoreans' elegant theorem"; see Fowler (1999: 100–104), Toth (1998 42–45, 76, 83–88).

of *logos* as an *ordered pair of integers* rather than a single number resolve the contradiction between a type of mathematics that rejects division of the unit and one that accepts it, such as "fractional mathematics." The assertion that "1," as a single *individual* object is indivisible, while the *logos* [1, 1], consisting of a *pair* of numbers, is divisible, does not present any contradictions or logical issues.

However, arithmetic does not seem to be the most promising field for advancing *antanairetic logistike:* the greatest achievements are found in the geometrical problem of incommensurable magnitudes. Among the various methods for discovering incommensurability, one particularly notable approach, reported by Aristotle (An. Pr.1.23, 41a23–30) and reconstructed through Alexander of Aphrodisias' commentary, involves the idea that the commensurability of a diagonal would imply a number that is neither even nor odd. Pre-Euclidean demonstrations of incommensurability, including a potential geometric variant of the Aristotelian method, "seem regularly interested in establishing unlimited sequences of similar geometrical shapes". These sequences were intended to be represented in the *taxeis* of *antanairetic logoi.*[14]

The *antanairetic logoi*, therefore, can represent the geometrical phenomenon previously labeled as "irrational" (*alogon*) and "inexpressible" (*arrheton*) in an open but ordered and "rational" way: the ratio between magnitudes without a common measure, which could not be expressed using positive integer numbers.[15]

Socrates and Glaucon do not provide explicit examples of the *logistike* they discuss, though they mention "symmetry" (7.530a1) and "irrational lines" (7.534d5). The *curriculum* of mathematical studies in the *Laws* extends that of the *Republic* by including the need for endless examination and resolution of "problems concerning the essential nature of the commensurable and the incommensurable" (*Law* 7.820b3–c9, with 7.817e5–820e7). The only concrete example of mathematical demonstration in the simile of the Line involves the square and its diagonal (*Rep.* 6.510d7–8). A relevant example of theoretical *logistike* familiar to Socrates and Glaucon might be identifying the numerical series related to the *antanairesis* of two incommensurable lines, such as the diagonal and the side of a square. In the context of discovering incommensurability, the mathematics of *antanairesis* offers "a way to put back together the separate pieces in which numbers and magnitudes had been divided," serving as a "type of reflection that entails a contraposition between 'same' and 'other'

14 Fowler (1999: 294–308).

15 Mathematical and philosophical implications of the "irrational" in Plato are presented and discussed by Toth (2011).

and ends up in a calculation (*logismos*) inspired to a principle of similarity and balance of opposites."[16] Thus, *logistike* – the link between arithmetic and geometry, as Archytas described[17] – would be characterized by a compresence of opposites, which Socrates associates with the *dynamis* of the *mathema* he seeks.

3 Third Picture: The "Confusions" of Mathematics and Their Power

3.1 *"Auroral" Power*

At the outset of the mathematical *curriculum*, Socrates and Glaucon describe the disciplines they seek using two light metaphors. Their *dynamis* is to draw human beings up (*anagein*) to the light, bringing them back (*anerchomai*) from Hades, the realm of shadows, to the Gods (7.521c2–3): Gods *are* lights, as they correspond to the stars in the analogy of the Sun (6.508a4). This power also turns the human soul around (*periagogein*) "from a day whose light is darkness to the veritable day" (7.521c6–7), in other words, from a starry night to full daylight. The sought *mathema* should effect a transformation from one state to its opposite. The initial state, represented by Hades and nocturnal light, parallels the condition where vision is obscured and only "shadows" are visible – a state referenced in the three major "analogies" of *Rep.* 6 and 7 (6.508c4–10, the Sun; 7.510a1, the Line; 7.515b, 7.516b, the Cave) and implies examining sensible objects to gain sensible knowledge. The transition from this condition to the opposite is described as an "ascent towards what really is" (7.521c7), a shift "from the world of becoming to that of being" (7.521d2–3), and thus "true philosophy" (7.521c8).

This transition is not simple: "so this, it seems, would not be the whirling (*peristrophe*) of the shell in the children's game" (7.521c4). It's not merely about spinning a shell or a shard in the game of *ostraka* – where black and white potshards or shells are thrown, moving from one side to the other by chance. Such a shift, occurring randomly, is akin to moving suddenly from night to day. In contrast, the transformation sought through the *mathema* must be neither immediate nor casual. It should be gradual and recursive, similar to the *periagoge* from a starry night to full daylight that occurs daily at dawn, as the stars set and the sun rises. The terms *anagein*, *anerchomai*, *epanodos*, and *periagoge* can also denote the periodic rising of the sun. Socra-

16 Zellini (1999: 161, 133).
17 See Archytas, 47 B 1–4 Diels-Kranz. The importance of "the generation of Archytas" in the early history of Greek mathematics was recently defended by Netz (2022: 53–106).

tes and Glaucon seek a *mathema* with "auroral" power to regularly bring together the opposite conditions of light and shadow, aiming to make them more similar and balanced. This power operates on human souls, merging the realms of the sensible and intelligible, of becoming and being, striving for a constant equilibrium.[18]

3.2 *A "Longitudinal Magnetism"*

Socrates and Glaucon also use the metaphor of a "drawing up lever" (*holkon*) to describe the power of arithmetic, *logistike*, and geometry. This lever lifts the soul "away from the world of becoming to the world of being" (7.521d3), "toward the essence" (7.524e1; cf. 7.523a2–3), and "toward the truth" (7.527b9: *holkon pros aletheian*). From "the world of becoming," the soul is elevated toward the opposite "world of being and *ousia*," "truth," and "pure thought (*noesis*)" (7.521d4, 7.523a3, 7.523d8, 7.524b4, 7.525d5, 7.524e11, 7.525a1, 7.525b5, 7.526b1–2, 7.526e4, 7.526e6, 7.527b5–6), and even towards the "Idea of the Good" (7.52 6e3). The tension between these opposing poles, from which the soul cannot escape, moves from the bottom to the top: *logistike* "directs the soul upward and compels it to discourse about pure numbers" (7.525d5), and geometry, being "knowledge of what always is" (7.527b7–8), rather than "what comes into being and passes away" (7.527b5–6), "draws the soul to truth," "directing upward the faculties that are now wrongly turned earthward" (7.527b10–11).

The rational part of the soul (*logistikon*), the "organ of the soul" more precious than "a thousand eyes" and capable of "seeing the truth," is influenced by the mathematical power that draws the *psyche* upward, healing it from the "blindness" caused by "other activities" (7.527d7–e3). The "study of the one" – only if the one is also presented as infinitely multiple – can lead the soul to focus on the opposite world, compelling it to "contemplate" (*thea*) the realm of being (7.525a1). Geometry, too, "whose entire purpose consists in acquiring knowledge" (7.527b1), despite its images, procedures, and language being closely tied to the realm of practice (7.527a6–9), has the power to enhance the soul's vision of the "Idea of the Good" (7.526e1; cf. 7.526e2–4, 7.526e6–7).

The image of the *holkon* defines the nature of the "auroral" power that mathematical disciplines must possess: the progressive and balanced alignment of sensible and intelligible dimensions, of becoming and being, involves a tension between two opposing poles. This creates an ascending attraction exerted by

---

18 For more details, see Cattanei (2003: 522–524) on the 'auroral' power and 524–527 on 'longitudinal magnetism'). Storey's paper in this volume demonstrates how the polarities and problematic contrasts that arise in mathematics can be 'preparatory' for dialectic, as 'mathematics fails to adequately reveal the true nature of the objects it studies, compelling students to turn to dialectic for answers.'

the second pole on the first – a kind of "longitudinal magnetism" that permeates the soul and enhances its intellectual capacity to "see" the truth.

3.3 *A "Vital" Disease*

The antagonism of opposites gives rise to the auroral *dynamis* that lifts the soul: Socrates and Glaucon explore this in detail, developing a new "analogy" (7.522e5–525d7, 7.524d9: *analogizou*). They argue that arithmetic and *logistike* induce a psychological condition characterized by a confused and simultaneous perception of opposites: one and infinitely multiple, small and big, sensible and intelligible. This "disease" is far from fatal; it "provokes the soul to reconsideration" (7.523b1, 7.523b9, 7.523c1, 7.523e1, 7.524b4, 7.524d3, 7.524d4; cf. Phaed. 89c7) and to "awaken *noesis*" (7.523e1, 7.524d5), the purely intellectual faculty, initiating a rigorous process of examination and discussion aimed at discerning the true nature of things.[19]

In mathematics, the power of "dawn" parallels the awakening process, where transitioning from Hades to daylight truly rouses the sleeper. Socrates draws a comparison with perceptions, focusing on those related to "seeing." Some objects of perception "do not provoke thought to reconsideration", while others "always invite the intellect to reflection" (7.523a10–b2). Objects that do not engage the intellect are those for which "judgment by sensation seems adequate" (7.523b1–2), whereas those that do provoke further examination are those whose "sensation yields nothing that can be trusted" (7.523b3–4). The illness that compels reflection creates a "confused state" where perceptions are blurred, but the lack of clarity is not general – as in distant views or shadow-painting (7.523b5–6). Instead, perceptions simultaneously shift between opposites, as "the perception no longer reveals one thing than its contrary" (7.523c2–3).

Socrates introduces here the famous example of the three fingers: "the little finger, the second, and the middle" (7.523c5–6).[20] After ruling out the possibility that the position, color, or shape of the fingers might obscure the percep-

---

19 The task of discerning the true nature of things is no longer the responsibility of mathematical disciplines but of dialectic, as Storey explains in his paper in this volume. Warren (2024) recently selected the passage 521c–525c to illustrate how concepts like 'one,' 'two,' 'a pair,' etc., represent an 'early stage' of knowledge, marked by a partial conceptual grasp of the intelligible in the ascent to pure intellectual knowledge. What I wish to emphasize is that this 'conceptual grasp' within mathematics is not necessarily, or solely, 'partial,' but rather problematic, as it involves pairs of contraries that mathematics cannot 'divide,' meaning it cannot define their nature with precision.

20 The example of the 'three fingers' has recently sparked significant interest among Plato scholars, as Marongiu (2025: chapter 5.2–3) demonstrates with extensive bibliographical references. My point, once again, is to emphasize how the mathematical disciplines in

tion of each as a finger, Socrates notes that "the faculty of sight never signifies to the soul at the same time that the finger is the opposite of a finger" (7.523d5–6). This is the standard, physiological position, unlikely to "provoke or awaken reflection and thought" (7.523d8–e1). Conversely, when the eye cannot precisely distinguish between small and large objects, the relative positions of each finger become irrelevant. The same issue affects touch and other senses, which can give the impression that the object of perception is at the same time both something and its opposite – big/small, light/heavy, etc. "In these circumstances," Socrates says, "the soul must necessarily be at a loss and wonder what significance this perception has for it" (7.524a6–7), grappling with the "strange communications" from the senses and "invite reconsideration" (7.524b1–2).

The soul is put to the test: it must awaken *logismon te kai noesin* (7.524b4) and determine whether what it perceives is "only one object or two" (7.524b4–5). If sight perceives two objects, each will appear distinct and individual (7.524b7–8). If the soul acknowledges two entities, it recognizes each as one and separate from the other. At this point, *noesis* distinguishes the opposites: it separates them and conceives each object independently, otherwise the two objects would not have been recognized as *two* but rather as *one* (7.524b10–c1).

The same occurs with the opposition between big and small: the eye perceives them simultaneously, but *noesis* can distinguish them as separate from each other. From here comes the need to examine what the quality of being "big" and what that of being "small" consists in; and this is the origin of the designation "intelligible" for the one, and "visible" for the other (7.524c13). Socrates concludes that even the distinction between intelligible and sensible (cf. 6.509d4) arises from a separation achieved by *noesis* between opposing states that appear simultaneously, whether in the "ill perception" of two objects or in that of a big and a small object. This distinction also parallels what happens in the "preparatory" and "dianoetic" disciplines of mathematics and *logistike* (7.525a9–10).

3.4 *One and Infinitely Many*

Socrates draws an analogy between "the number and the one" and the two cases of different sensations described earlier (7.525d2–525b10). The "healthy condition" which does not stimulate the mind was the one in which each finger was satisfactorily perceived simply as a finger. Similarly, if the one is

Plato's *Republic* 7 are portrayed as forms of rational knowledge (*dianoiai*) that confront 'antagonisms' and 'confusions of contraries' (such as one-infinite, small-big, etc.). These challenges 'awaken' the human soul to the highest form of intellectual knowledge (*noesis*), where these polarities and confusions will ultimately be resolved.

perceived merely in itself, whether through sight or another sense, it lacks the force to "draw us toward the essence" (7.524e1). For the one to possess this power, it must be perceived in conjunction with "some contradiction, so that it no more appears to be one than the opposite" (7.524e2–3). In such a situation, the soul begins to question and examine the problem, distinguishing between opposites. This process leads the soul to investigate what is "one" in itself. Thus, *mathesis peri to hen* proves to be a discipline capable of guiding the soul and transforming it to contemplate what always is.

At this point, Socrates could have applied the same examples from above to arithmetic and *logistike*, such as counting two objects or assessing whether something is big or small. Instead, he shifts the focus to a different opposition: the contrast between unity and infinite multiplicity. He transitions from Glaucon's observation that "we see the same thing at once as one and as an indefinite plurality" (7.525a6–8) to a broader consideration of the one and the number (7.525a6–8), and ultimately to their theoretical exploration in arithmetic and *logistike* (7.525a9–11). Thus, disciplines that examine numbers, having demonstrated their ability to "lead the soul toward the truth," will form at least part of the sought *mathema* (7.525a6–b3).

Socrates and Glaucon do not specify exactly *how* mathematics manages to embrace both the balanced presence and the reciprocal overlapping of unity and multiplicity, which are crucial for establishing its power. While they take this capability for granted, they emphasize that it is far from trivial; rather, it is considered a very refined task (7.525d1) that involves, as previously noted, significant efforts to master (7.525c1–3). This acknowledgment reflects the depth and complexity of the mathematical discipline they are exploring, suggesting that the integration of these concepts into a coherent framework requires substantial intellectual rigor and precision.

But did not theoretical arithmetic indeed demonstrate a notable inclination toward examining various characteristics of the basic unity, including its most paradoxical aspects? Was not *logistike*, conceived as the theory of *antanairetic* calculations, dedicated to finding a balance between unity and multiplicity, small and large, "same" and "other," in order to establish a progressive and reciprocal fading of the opposites into each other? Did it not similarly address unity and indefinite multiplicity, the infinitely large and the infinitely small, the identical and the different, as well as the rational and irrational, thereby offering an "arithmetic" translation of the geometrical phenomenon of incommensurability?

Perhaps, if we were part of the culture represented by Socrates and Glaucon, we would be so well-acquainted with these aspects of mathematics, which they describe as "sick," that the need for them to be explicitly stated would seem unnecessary. In such a context, as members of the Academy, we would likely accept that even within mathematics, there is no "pure theory," no "discovery of the intelligible," without undergoing what Socrates and Glaucon refer to as "absurd efforts of interpretation for the soul" – *atopoi tei psyche hai hermeneiai* (7.524b1) – to use once again Socrates' and Glaucon's words.

Bibliography

Annas, Julia,. 1987. *Aristotle's Metaphysics Books M and N*, with introduction and notes. Oxford: Oxford University Press.

Cattanei, Elisabetta. 2003. "Le Matematiche al tempo Di Platone e la loro Riforma." In *Platone. La Repubblica Libri VI–VII*, edited by Mario Vegetti, 473–540. Roma-Napoli: Bibliopolis.

Caveing, Maurice. 1990. "Platon, Aristote Et Les Hypothèses Des Mathématiciens." In *La Naissance de la raison en Grèce*, edited by Jean-François Mattéi, 119–128. Paris: PUF.

Fowler, David H. 1999. *The Mathematics of Plato's Academy: A New Reconstruction*. 2nd ed. Oxford: Clarendon Press.

Heath, Thomas. 1981. *A History of Greek Mathematics* I–II. 2nd ed. New York: Dover Publications.

Klein, Jacob. 1968. *Greek Mathematical Thought and the Origin of Algebra*. Cambridge, MA: MIT Press.

Mueller, Ian. 1992. "Mathematical Method and Philosophical Truth." In *The Cambridge Companion to Plato*, edited by Richard Kraut, 170–195. Cambridge: Cambridge University Press.

Marongiu, Laura. 2025. *Un piccolo esempio. La psicagogia matematica in Platone*. Leiden: Brill.

Netz, Reviel. 1999. *The Shaping of Deduction in Greek Mathematics: A Study in Cognitive History*. Cambridge: Cambridge University Press.

Netz, Reviel. 2022. *A New History of Greek Mathematics*. Cambridge: Cambridge University Press.

Panza, Marco, and Andrea Sereni. 2013. *Plato's Problem: An Introduction to Mathematical Platonism*. London: Palgrave Macmillan.

Russell, Bertrand. 1901. "Is Position in Space and Time Absolute or Relative?" *Mind* 10: 293–317.

Toth, Imre. 1998. *Lo Schiavo di Menone*. Milano: Vita e Pensiero.

Toth, Imre. 2011. *Platon et l'irrationnel mathématique*. Paris: éditions de l'èclat.

Von Fritz, Kurt. 1965. "Die Entdeckung der Inkommensurabilität Durch Hippasos Von Metapont." In *Zur Geschichte der Griechischen Mathematik*, edited by O. Becker, 271–307. Darmstadt: Wissenschaftliche Buchgesellschaft.

Warren, James. 2024. "Early Learning in Plato's Republic 7." In *Conceptualizing Concepts in Greek Philosophy*, edited by Gábor Betégh and Vasilis Tsouna, 56–73. Cambridge: Cambridge University Press.

Zellini, Paolo. 1999. *Gnomon: Un'indagine sul numero*. Milano: Adelphi.

Zhmud, Leonid. 1997. *Wissenschaft, Philosophie und Religion im frühen Pythagoreismus*. Berlin: De Gruyter.

PART 2

From Sense Perception to Theôria

CHAPTER 3

From Mathematics to Dialectic in Plato's *Republic*

Damien Storey

The guardians' preparation for becoming rulers includes ten years devoted, apparently exclusively, to the study of mathematics, and Socrates spends much of Book 7 explaining why this is necessary and what subjects it will include. But despite these explanations, mathematics' contribution remains puzzling. Socrates says it is a 'necessary preparatory education [*propaideia*] for dialectic' (536d4–6), and he gives us a basic account of how it prepares students: it 'turns the soul towards the study of what is' (525a1–2).[1] That is, mathematics turns students' attention away from the sensible world towards intelligibles, where the subject-matter of dialectic lies. But many have felt that to justify ten whole years, mathematics must do more than just get students to look in the right general direction. Thus, commentators have sought other benefits of mathematics, less explicit in the text. It has been suggested, for example, that a thorough mathematical training is needed to protect students from the dangers of premature dialectic,[2] or that Plato's Good is itself, in whole or part, a mathematical concept, so must be studied by those already well versed in mathematics.[3]

In this paper, I'm going to focus on a more immediate textual question: what exactly does Socrates tell us about students' transition from mathematics to dialectic? My aim is to show that, while he does not say a great deal, the little he does say already suggests that mathematics' contribution is more than its ability to make students familiar with intelligibles. Specifically, I will explore the proposal that Socrates' careful description of the transition from perception to mathematics provides us with a model for understanding the less fully described transition from mathematics to dialectic.[4] I will argue that, in line with the problem–solution structure of the earlier transition, dialectic overcomes the two limitations of mathematics – its use of hypotheses and its reli-

1 Translations in this paper are adapted from Reeve's (2004) translation of the *Republic*.
2 Sarah Broadie (2021: 185–95). See also n. 6 below.
3 The most influential defence of this view is Myles Burnyeat (2000). See n. 24 below.
4 This is a proposal I encountered from Elisabetta Cattanei, to whom I was a respondent at the First Middle East Regional Meeting of the International Plato Society, '*Theôria* as Cognition in Plato'. I think her proposal is right, although I reach this conclusion via a very different route. See n. 8 below.

HTTPS://DOI.ORG/10.1163/9789004747128-005

ance on sensible images – and students of mathematics come to realise this through the 'synoptic vision' with which their studies culminate.

1 From perception to mathematics

Mathematics is introduced in Book 7 as the solution to a pedagogical problem. The goal of the rulers' education is knowledge of the Good, which will be the subject matter of dialectic, but getting students to this goal turns out to be a momentous task, requiring years of careful moral and intellectual preparation. Mathematics' contribution, or at least that part of it that is explicitly highlighted in the text, is to take students who have grown up relying only on perception, and who are unaware of anything non-perceptibles, and reorient them towards intelligibles, so that they come to appreciate these as a more 'real' and worthy object of study. As Socrates puts it, mathematics 'turns the soul towards the study of what is' (525a1–2) or 'draws the soul from what is becoming to what is' (521d4–5).

How does mathematics do this? In the so-called 'finger' passage, Socrates explains that there are two kinds of properties: those that do not 'summon the understanding to investigate them' (523b1) and those that do. Properties that do not summon understanding are those that are adequately grasped by perception. For example, by perception alone, we can grasp that something is a finger, so a person is not 'compelled to ask the understanding what a finger is' (523d2–4). Properties that summon understanding are those for which perception is inadequate. Socrates gives the example of perceiving a ring finger as both large and small: larger than my little finger, smaller than my middle finger. What's inadequate about this? Perception is evidently able to grasp *that* the finger is large and small, since that is precisely what it reports. There is also nothing paradoxical about a finger that is large and small: that's how relational properties are supposed to work, and a moments reflection makes it obvious that it is only the alternative – i.e. one of my fingers being *unqualifiedly* large or small – that would be absurd. The problem is rather that perception cannot give an adequate account of what properties like largeness or smallness *are* (524a5–9).[5] So while my ring finger is a decent (if imperfect) guide to what a finger is, it is not a helpful guide to what largeness and smallness are. Socrates puts the lesson in a simple way: we already know that largeness and smallness are different, so when we perceive them 'not as separate, but as mixed up together' in one and the same sensible, we are compelled to separate them in

5 See Terence Irwin (1995: 159), and Gonzalez (1996: 255).

thought, as two things that are distinct from any one perceptible large-and-small thing (524b3–c9). In this way, we are taking the first steps towards recognising that properties like largeness and smallness can only be grasped by thought alone and that they are, therefore, intelligible rather than sensible.

The point, of course, is that the properties studied by mathematics are among those that summon understanding. For example, any perceptible is both one and many things at the same time – the Parthenon is made of many chunks of marble – and this can be easily extended to all numbers. What makes summoning properties like number special, and different from those like big and small or light and heavy, is that they have subjects devoted not only to studying them, but studying them *as* intelligible entities.[6] Calculation, for example, 'compels [the soul] to discuss the numbers themselves, never permitting anyone to propose for discussion numbers attached to visible or tangible bodies' (525d5–8). Socrates goes on to describe four further branches of mathematics – geometry, stereometry, astronomy, and harmonics – that each in their own way compel students to see the properties they study as intelligible and separate from their instantiations in the sensible world. So mathematics is a systematic study of properties that we are compelled to grasp in thought rather than with our senses, and as such it has the power to turn students' intellect from sensibles to intelligibles.[7]

This side of mathematics' contribution is well studied. But what we've seen so far is how mathematics turns us *from* sensibles, which tells us very little about the *to* part of Socrates' various from–to phrases: 'from a kind of night ... to the true day' (521c5–6) or 'from what is becoming to what is' (521d4–5; see also 524c5–6 and 529a2). In other words, there is a gap to fill between the end of mathematics and the beginning of dialectic: how do students of mathematics discover that they need to stop doing mathematics and start doing dialectic? My proposal is that the transition from mathematics to dialectic has the same problem–solution or question–answer structure that characterises the transition from perceptibles to mathematics: mathematics fails to adequately show

6 Socrates later, at 533a10–c3, is explicit that mathematics and dialectic are the only subjects that study intelligibles; all others study sensibles.

7 Why not go directly to dialectic, since it also studies intelligibles? One reason is that dialectic has the danger of corrupting practitioners that are young or unsuitable (537e1–539d1). But also (and arguably relatedly) it is not clear that dialectic deals with the right kind of 'summoning' properties. That is, a person can see Achilles as *brave* or a pile of money as *good* without simultaneously seeing them as cowardly or bad, so those properties are dangerously easy to identify with their sensible instances. On this difference between mathematical and ethical properties, see Storey (2022: 303–5).

students what the objects it studies are, and, thus, they realise that they must turn to dialectic for the answer.[8]

2 The limitations of mathematics

As we would expect if this reading is right, mathematics does not just precede dialectic, but is an inferior or flawed subject in comparison. The idea is not that it is less useful for their current purposes or that it studies truths that are less important or profound. Rather, Socrates describes it as an essentially limited theoretical discipline: it employs a method that is deficient in certain respects and, as a result, is unable to fully grasp the nature of things, which means that it ends not in 'knowledge' (*epistēmē*) but in the inferior, less clear cognitive state that Socrates calls *dianoia*.[9] Put another way, mathematics does not even reach truths about mathematics as successfully as dialectic reaches truths about the Good. So the move from mathematics to dialectic is not (or not only) a sidestep, from one subject matter to another, but an advance from a worse to a better kind of study.

Socrates identifies two features of the mathematical method that hold it back: its use of images and its use of hypotheses. These leave mathematicians unable to get a direct apprehension of intelligibles, and, thus, they bar it from knowledge. He says:

---

8 Elisabetta Cattanei, who sees the same problem–solution structure, proposes that just as a puzzling compresence of properties in sensibles is what turns us from sensibles to the mathematical study of intelligibles, there are analogous puzzles arising within mathematics that turn us deeper into mathematics and ultimately beyond it to dialectic and the study of the good. For example, she identifies a case of compresence of opposites in the simultaneous unity and multiplicity arising in the *antanairesis* of incommensurable magnitudes, which is an application of 'calculation' (*logismos*). However, since Socrates gives few descriptions of specific mathematical proofs or procedures, this is hard to confirm. Perhaps his description of how calculation maintains unity – 'if you divide it, they multiply it, taking care that the number one never appears to be, not one, but many parts' (525c8–526a7) – is a reference to *antanairesis*, but here, rather than a persistent tension between multiplicity and unity, calculation successfully preserves unity. Moreover, the 'power' mentioned in this passage is still the earlier 'from' part of the turn, which gets students to see numbers as separate from their sensible instances (from 'numbers attached to visible or tangible bodies'; 525d7–8).

9 For reasons to think that *epistēmē* here might be better understood as one species of knowledge, explanatory understanding, see Storey (2022: 271–2). For the purposes of this paper, I set this issue aside.

> While [mathematicians] do dream about what is, they cannot see it while wide awake as long as they make use of hypotheses that they leave undisturbed, and for which they are unable to give an account.
>
> 533c1–3

Dreaming was earlier described as apprehending, unawares, mere images of something, rather than the thing itself (476c3–5), so mathematicians appear to be stuck with a relationship to intelligibles that, like dreaming, involves indirectly grasping them through images, and which is, at the same time, related to their reliance on hypotheses. So what are images and hypotheses? Hypotheses are the more straightforward of the two.[10] Mathematics begins from a set of hypotheses, which the mathematicians take for granted as if they are something known (510c2–d1), even though they cannot give an account of them. But as long as they start from such hypotheses, Socrates says, they cannot achieve knowledge, because knowledge needs to be fully explanatory, right down to its starting points:

> When the starting point is unknown, and the conclusion and the steps in between are put together out of what is unknown, what mechanism could possibly turn any agreement reached in such cases into knowledge (*epistēmē*)?
>
> 533c3–6

As a result, mathematicians are only able to achieve a lesser kind of cognition, *dianoia*.

Mathematics' dependence on images gets less attention, and is sometimes overlooked. But this is certainly a mistake. Socrates consistently mentions images alongside hypotheses, as a closely related and equally important limitation. Consider, for example, the following two descriptions of *dianoia* during Socrates' exposition of his Line analogy:

> This, then, is the kind of thing that I said was intelligible, though the soul is forced to (a) use hypotheses in the investigation of it, not travelling up to a starting point, since it cannot escape or get above its hypoth-

10 Disagreements usually concern what Socrates means when he says, for example, they 'hypothesise the odd and the even, the various figures' and so on (510c3–5): do these hypotheses assert their existence, provide definitions of them, or propose a set of propositions that hold true of them? For my purposes here, all that is important is that hypotheses are not true by definition: if they were, Socrates would hardly claim that they are in need of an account.

> eses, and it (b) uses as images those very things of which images were made in the section below.
>
> 511a4–8[11]

> (b) Using as images the things that were imitated before [sc. ordinary sensibles], the soul is forced to (a) investigate from hypotheses, proceeding not to a starting point (*archē*) but to a conclusion.
>
> (510b4–6)

In these passages, both (a) and (b) appear to be essential features of Socrates' characterisation of *dianoia*. If we take our cue from the second passage, it seems that the relationship is '(a) because (b)': the soul is forced to investigate from hypotheses because it investigates using sensibles as images of intelligibles.[12] Methodologically, hypotheses are mathematics' starting points, but it is important to recognise that hypotheses are just methodological tools, which exist in the minds of mathematicians. Since hypotheses presumably do not arise in their minds spontaneously, the question of where mathematics gets its starting point can be pushed back one step further: from where do mathematicians get their hypotheses? The most promising proposal is that (a) the hypotheses are derived from (b) sensibles used as images of intelligibles.

One way to reach this conclusion is by considering the distinctive cognition of mathematicians, *dianoia*. *Dianoia* is one of four cognitions described in the Line analogy, each of which is partly characterised by what kind of objects they make direct contact with, be it Forms, sensibles, or mere images of sensibles. Furthermore, as many have noticed (with Socrates' encouragement: 534a3–5), the relationship between the two species of knowledge, *noēsis* and *dianoia*, parallels the relationship between the two species of belief, *pistis* and *eikasia*. *Pistis* directly apprehends sensible things, but *eikasia* can only apprehend them indirectly through images, like shadows or reflections. Similarly, while *noēsis* directly apprehends intelligibles, *dianoia* can only apprehend them indirectly through images, specifically by using *sensibles* as images of intelligibles ('using as images the things that were imitated before'). So *dia-*

11 Here I change Reeve's (2004) translation slightly to take the μέν at a4 to be concessive.

12 This reading of their relationship is most clearly seen in Richard Robinson (1941: 160-61), R. C. Cross and A. D. Woozley (1964: 224–46), and Julia Annas (1981: 278–9), David Ross (1951: 52) and Ian Mueller (1992: 185–9), defend much the same view, but understand 'images' narrowly as diagrams. Moon-Heum Yang (2005: 285–312) discusses a variety of views of the relationship between images and hypotheses; contrary to his objection (2005: 296–8), I think this reading is neutral on, first, whether the images are of Forms or intermediates and, second, whether or not its use of images is a defect.

noia's starting point, or where it first makes contact with the world, is the sensible world, which it uses as images of intelligibles, and given this, we should expect this also to be where it finds its formal or methodological starting points, hypotheses.

Put more simply, hypotheses must come from somewhere, and since mathematicians have no direct contact with intelligibles, the only other source is sensibles. The proposal, then, is that mathematical hypotheses arise from observation. Mathematicians observe certain invariable characteristics of the messy but not altogether irregular sensible world and then formalise these into a set of hypotheses from which they can deduce conclusions. These will be basic observations 'obvious to everyone' (510d1) like the fact that any set of observables we count either is or is not divisible into two equal parts, from which we can hypothesize that numbers are either even or odd.[13] So hypotheses are mathematical judgements made on the basis of experience. This helps explain why mathematicians are 'unable to give an account (*logon*)' (533c3; see also 510c2–d1) of their hypotheses, since it is a familiar Platonic assumption that while perception can tell us that something is the case, it cannot give us explanatory accounts.[14]

This reading allows us to understand why diagrams are important in mathematics, as one kind of image, without attributing to Plato the implausible view that it is *simply* mathematicians' use of diagrams that precludes knowledge.[15] Diagrams are simplified, manipulable examples of the mathematical regularities found in the sensible world.[16] A geometrical diagram, after all, will follow the same rules as any observed sensible – parallel lines will never meet and triangles will have three sides – because it is, when drawn, itself a sensible. For the same reason, when drawn, it will follow them imperfectly: drawn par-

13 Compare Robinson (1941: 162): 'what made the mathematicians so convinced of their "hypotheses" was that they seemed to be directly given in sensible intuition. They were "plain to all" or παντὶ φανερά in the physical sense of being their to see in the geometer's sand.'

14 See Annas (1981: 280–82), who makes a similar connection between mathematics' reliance on experience and its failure to provide explanatory accounts.

15 Commentators who closely identify the relevant images with diagrams include Mueller (1992: 182), Burnyeat (2000: 66 ff.), Richard Patterson (2007), Andrew Payne (2017), and Tamsin de Waal (2022). For reasons why this is problematic, see Hugh Benson (2010: 196–7).

16 Socrates' describes the movements of the celestial bodies as themselves models of intelligibles, comparable to Daedalian diagrams (529d7–e3), with the implication that at least some of the regularities they exhibit will be recreated in diagrams of mathematical astronomy. See also *Euthydemus* 290b10–c6: 'none of these <mathematicians> make their diagrams; they simply discover those that already exist'.

allel lines will appear to converge and will never be perfectly straight. Diagrams have the advantage, however, of a degree of abstraction that any collection of observations lack. In a proof, we should understand the diagram to be the conceptual construction, not the drawn diagrams that might illustrate it: i.e. the conceptual object described by 'a triangle ABC with a square on AB', not the many possible drawings of this, using different lengths and angles. A diagram of a triangle functions both as an abstraction from observed triangles and, at the same time, a representation of an intelligible triangle, since observed triangles *are* images of intelligible ones. The problem with what Socrates calls 'images', then, is not specific to diagrams; rather, as abstractions from sensibles, diagrams simply inherit the deficiencies that Plato commonly attributes to sensibles, chief among which is their inability to provide explanatory accounts, which is exactly the problem Socrates attributes to mathematics.[17]

In summary, mathematics' two limitations are part of a single account of why it lacks explanatory starting points. Methodologically, this is seen in its use of hypotheses that are unknown because they lack accompanying explanatory accounts. But the origin of the problem is the fact that mathematics lacks any direct access to intelligibles and must, instead, study them indirectly, starting from those features of intelligibles that are reflected in the sensible world. Since sensibles do not provide explanatory accounts, mathematics will never achieve knowledge, since knowledge, for Plato, is fully explanatory.

3 From mathematics to dialectic

The suggestion this paper is exploring is that the limitations of mathematics that I've just described are what forces students to rise higher and examine intelligibles through dialectic, just as earlier it was the imperfections in the sensible world that forced students to study mathematics as a theoretical,

17 Patterson (2007) offers the most detailed exploration of the view I am rejecting here, namely, that Socrates is referring to problems specific to diagrams. While I cannot do justice to his many careful arguments here, it strikes me that the problems he identifies are the wrong *kind* of problems. He proposes ways in which pre-dialectic mathematicians would be liable to be misled by diagrams into having mistaken beliefs about the epistemological and metaphysical foundations of mathematics (26–32). But Socrates seems to offer images and hypotheses as limitations on mathematics itself, not mathematicians. In other words, even if some pre-dialectic mathematicians were not misled (having been corrected, say, by dialecticians), they would *still* not be able to use mathematics to achieve knowledge: that, I take it, is the problem Socrates is explaining.

rather than empirical, discipline. If this is right, we should expect dialectic to overcome the limitations of mathematics, just as mathematics overcame the limitations of perception, and that is exactly what we find. Socrates describes dialectic by comparing it with mathematics and explaining how it manages to overcome its limitations. First, dialectic treats hypotheses as *mere* hypotheses and uses them instead as a 'stepping-stone' to an *un*hypothetical starting point (511b2–c2; 533c8–d1). In this way, it 'destroys' them: that is, the propositions that were merely hypothetical in mathematics, become not hypotheses in dialectic, but propositions explained in relation to an unhypothetical starting point.[18] Second, it replaces images with Forms: it proceeds 'without using the images used in the previous section, using forms themselves and making its investigation through them' (510b7–8; see also 511c1–2). Crucially, in all of the relevant comparisons, the parallel is between *images*, not hypotheses, and Forms: where mathematics 'uses' sensible images, dialectic 'uses' Forms. Socrates' point cannot be that they are, respectively, *about* images and Forms, since he is explicit that although mathematicians 'use visible forms and make arguments about them, their thought isn't directed to them', but to intelligibles like 'the square itself' (510d5–7). The point is rather about their starting points: mathematics argues *from* sensible images, since it derives its hypotheses from them, and *about* Forms, whereas dialectic argues *from* Forms and *about* Forms – it makes 'no use of anything visible at all, but only of forms themselves, moving on through forms to forms, and ending in forms' (511c1–2).

So dialectic and mathematics have a kind of problem–solution relationship, where dialectic in some sense 'fixes' the methodology of mathematics. This brings the two subjects surprisingly close together. At first, we might have thought they are entirely unconnected subjects, with different roles in the guardians' education and different subject matters: after all, the future rulers train in dialectic primarily because it gives them the *ethical* knowledge needed to rule, and that seems very different from mathematics. Yet even if it is not entirely explicit, Socrates very much appears to be saying that dialectic grounds the hypotheses *of mathematics*. This makes dialectic look like a more fundamental investigation of the same subject matter as mathematics.

I think the best way to make sense of this is to understand Socrates to be taking for granted that dialectic is, unlike mathematics, a methodology that is neutral with respect to subject matter.[19] So although he describes dialectic in

18 This understanding of 'destroying' the hypotheses is now standard. For a good discussion, see Cross and Woozley (1964: 247–8).

19 An alternative reading from Francis C. Cornford (1932: 37–52, 173–190) proposes two methodologically distinct types of dialectic: mathematical dialectic (1932: 176–184), which takes the One as the unhypothetical starting point, and moral dialectic (1932:

relation to mathematics, he expects his listeners to understand him to be describing a method of achieving the kind of knowledge of Forms that is the basis of any knowledge whatsoever. In other words, at least part of his aim is to make an entirely general epistemological point about how knowledge is achieved, and for this purpose mathematics is a useful foil. The future guardians, then, will use dialectic in the way he has described, but they will eventually use it to address the ethical topics that are relevant to ruling. When understood in this more general way – as, in short, how Forms are known – we see that even when dialectic concerns mathematics, it is not a way of *doing* mathematics. Rather it is a way of grounding mathematics in relation to the relevant Forms, and ultimately the Good, so that its conclusions are explained in relations to an unhypothetical starting point. Equally, mathematicians do not need dialectic to do mathematics. After all, Socrates recognises many mathematicians among his contemporaries, none of whom are trained in dialectic. Mathematicians make the different mistake of assuming that they have knowledge (510c2–d1), when in fact they only have the inferior grasp of their conclusions that Socrates calls *dianoia*.[20] So mathematics is not 'fixed' by dialectic in the sense of offering an alternative method of doing mathematics itself, but it is nonetheless true that only a dialectician doing mathematics will have knowledge of the conclusions it reaches.

Nonetheless, it is highly significant that Socrates chooses to describe dialectic via a comparison with the mathematical method. It suggests that there is indeed a natural pedagogical progression from mathematics to dialectic. Students of mathematics are not, it seems, simply told by the existing guardians to stop doing mathematics and start doing dialectic, as two unlinked subjects. Rather, they will understand what dialectic is and why it is needed through a comparison with mathematics, just as Socrates explains it to us. Socrates speaks at times as if mathematics has within itself the resources to direct students to dialectic, and even to give them momentum towards their

181–4), which takes the Good as the unhypothetical starting point. But Socrates' initial description of dialectic during his exposition of the Divided Line gives the clear impression that it is just one method that reasons from a single starting point: 'what is unhypothetical and the starting point of everything' (511b5–6; see also 532a1–b2).

20 Burnyeat (2000: 37–42) denies that Socrates is claiming that mathematicians mistakenly assume they have knowledge. Part of the problem is the ὡς phrases in 510c2–d1. E.g. they treat their hypotheses *hōs eidotes*: either '*as* known' or '*as if* known'. Burnyeat favours the latter, which allows him to say that they are not in fact *claiming* to have knowledge, but the former strikes me as more likely. It makes better sense, for example, of Socrates' description of mathematicians as 'dreaming' (533c1–3), discussed above. See Burnyeat (1987: 151 n. 18).

goal of understanding the Good.[21] If we continue the comparison with the transition from perception to mathematics, we will expect mathematics to do this by making its own limitations apparent, so that students are motivated to find something that overcomes these limits. In other words, for it to have a pedagogical benefit, it is not enough for mathematics to have problems that dialectic solves – this fact must also be revealed to students. This revelation is, I propose, one of the effects of the enigmatic final stage of the students' mathematical studies.

4 The 'synoptic view'

In two passages, Socrates describes how the students' mathematical studies will culminate in a 'synoptic view' (*synopsis*) of the five mathematical subjects:

> I take it that if the investigation of all the subjects we have mentioned arrives at what they share in common with one another and what their affinities are, and draws conclusions about their kinship, it does contribute something to our goal [sc. of understanding the Good] and is not labour in vain; but that otherwise it is in vain.
>
> 531c9–d3

> The subjects they learned in no particular order in their education as children, they must now bring together into a synoptic view of their kinship with one another and with the nature of what is. … [This] is also the greatest test of which nature is dialectical and which is not. For the person who can achieve a synoptic view is dialectical, and the one who cannot isn't.
>
> 537b8–c7

Up to this point, mathematics' contribution was said to be its ability to encourage students to see mathematical properties as intelligible and separate from their sensible instantiations. The synoptic vision is something different. Rather than separating intelligibles and sensibles, it appears to have more to do with the relationships between intelligibles, and Socrates clearly sees this as

21 The mathematical crafts 'make it easier to see the form of the good' (526d7–e2) and 'lead the best part of the soul up towards the contemplation of the best among the things that are' (532b6–d1); dialectic doesn't just come after mathematics, but proceeds 'using the crafts we discussed to help it and cooperate with it in turning the soul around' (533d1–4).

a unique contribution to the goal of knowing the Good. Consonantly, the synoptic view doesn't appear to be itself a conclusion of mathematical reasoning in any normal sense. There is no evidence, for example, that ancient mathematics included a kind of meta-subject that compares different areas of mathematics and 'draws conclusions about their kinship'. Note that the vision is not just of the subjects' kinship with each other but also of their kinship 'with the nature of what is', which sounds more metaphysical than mathematical. Mathematicians were said to only 'dream about what is', where, as we saw, this means that they fail to realise that their reliance on sensible images prevents them from fully and directly grasping intelligibles. Insofar as it reveals mathematics' relationship to what is, the synoptic view is perhaps a first step towards waking up and recognising that one has been studying real things only indirectly.[22]

The second clue is Socrates' description of the synoptic view's contribution. While he says that it 'contributes something to the goal' of understanding the good, one could insist on reading this rather weakly: strictly speaking, even the childhood musical and gymnastic training might be said to 'contribute' to this goal, in the sense that it is a necessary step in that direction. A more telling claim, however, is that 'a person who can achieve a synoptic view is dialectical' (*ho men gar synoptikos dialektikos*). That they are 'dialectical' does not necessarily mean they are already doing dialectic, but certainly it means they are ready and able to do so. Socrates earlier said that being 'clever' in the mathematical subjects does *not* make one dialectical (531e3–5). So the synoptic view seems again to be different from the five mathematical subjects and to uniquely enable a student to go beyond mathematics and start dialectic. In some sense, then, the synoptic view is the bridge between mathematics and dialectic.

But what exactly is it? The little evidence we have makes it difficult to say. Like so much in relation to dialectic and the Good in Books 6 and 7, Socrates does not seem to be even trying to give us a complete or literal understanding. We are left, then, with speculations. The few that have been proposed focus on the fifth and final subject, harmonics, which immediately precedes the synoptic view. Robins, for example, argues that it is the realisation that the properties studied by harmonics, such as ratio and proportion, are pervasive and

---

22 Burnyeat (2000: 70–72) has a very different reading of the phrase 'with the nature of what is'. He sees a connection with the idea that a student's soul is itself assimilated to the ordered system of mathematics, and so becomes like it. But I find it hard to see what the connection is supposed to be. The passage concerns the mathematical subjects' relationship with each other and what is; it makes no mention of their effect on students, except that they come to recognise these relationships.

fundamental to mathematics and the basis of the 'nature and excellence of each mathematical object'. Once their attention is turned to the fundamentality of ratio and proportion, he argues, students realise that these properties themselves are in need of an account, which requires them to turn to dialectic.[23] Others have argued that it involves a recognition that properties like harmony, proportion, or unity are constitutive of, or at least deeply related to, goodness, and so they naturally lead to the main subject matter of dialectic.[24]

But leaving aside the specific interpretations of the synoptic view, we can say some general things about the realisations to which it is likely to lead. Taking a 'synoptic view' of the five subjects suggests taking a step back from the nitty-gritty of solving problems and instead looking at mathematics from a distance, as it were. From this perspective, seeing 'what they share in common with one another' will presumably include an explicit recognition that all five subjects share the same method – an awareness, that is, of what unifies them as branches of mathematics. That in itself is already a progression from mathematics to the philosophy of mathematics. Specifically, it requires addressing the question 'what is mathematics?', and an adequate answer to this question, for Socrates, entails an understanding of mathematics as a hypothetical method and a method with significant limitations. So a student who progresses from unquestioningly using hypotheses, to investigating what hypotheses are and how they have been using them, will inevitably discover that they do not really know what they previously thought they knew, since their hypotheses are derived from sensible images rather than Forms and are, thus, a step removed from what is. With that discovery, students will recognise that if they are truly going to understand what they have been studying, they need to ground their hypotheses in first principles, leading them to see the need for dialectic. This general understanding of the synoptic view's pedagogical benefit can, I take it, be common across a wide range of accounts of the conclusions it leads students to draw about, say, harmony or unity.[25]

---

23 Ian Robins (1995: 387–90). See also Mitchell Miller (1999: 84–87), who adds the suggestion that Forms, including the Good, are expressible in ratios.

24 The most thorough defence of this position is Burnyeat (2000), though suggestions of it go back a long way; see, e.g., Cornford (1932: 181–2). Christopher Gill (2007) offers excellent descriptions of a range of related views; see p. 270 for discussion of the role of the synoptic vision. For recent objections, see Broadie (2021: 176–85), and Nicholas Smith (2019: 167–73).

25 See Robins (1995: 359): 'this vision is essential because it helps mathematicians become aware of the limitations of the mathematics they have been engaged in while showing how they can begin to overcome those limitations'.

We can conclude, then, that although Socrates' initial rationale for the mathematical subjects is that they turn students in the general direction of intelligibles, by the end of Book 7 he has given mathematics the more specific role of helping students come to recognise what dialectic is and why it is needed. It does this, first, because there is a methodological continuity between mathematics and dialectic, in the sense that the dialectic addresses the specific limitations of the mathematical method, and, second, because the curriculum of five mathematical subjects that Socrates describes are intended to culminate – albeit in a way that is difficult to pin down – in a 'synoptic view' that reveals the need to progress to dialectic.

5 Mathematics after *dianoia*

I want to conclude by briefly considering mathematics' role after students move to dialectic and, eventually, become mature philosophers. I have so far more or less identified mathematics with the cognitive stage of *dianoia*, but presumably the mature philosopher is also able do mathematics, with the added benefit of having full explanatory understanding of their conclusions. But will they in fact do any mathematics? In other words, will mathematics be included in philosophical *theôria*, which I understand to be the philosophical contemplation of the mature philosopher and an essential ingredient of the philosophers' distinctive happiness.[26]

Let me begin by sketching why someone might think that mathematics ends with *dianoia*. Socrates recommends purifying mathematics of its practical or empirical side and instead practising it for the sake of knowledge. But his reason is not that he thinks mathematics is an intellectual activity that is valuable for its own sake. He says that calculation 'is useful for our purposes, provided you practice it for the sake of knowledge rather than trade' (525d1–3; 527a9–10), and what he means by 'useful for our purposes' is useful for achieving their current pedagogical goal, which is to turn the students towards being,

26 While the pleasures of philosophical *theôria* often appear to concern the contemplation of truths already known (e.g. 486a8, 500c4, and 517d5), this is tied closely to the pleasure of learning. In Book 9, for example, the philosophers' pleasures are those of 'knowing where the truth lies and always enjoying some variety of it while he is learning' (581e1–2). There does not, then, appear to be a need to sharply distinguish the discovery and the contemplation of mathematical truths. However, it might be argued that the director who oversees the development of stereometry (528b6–c8), who is presumably a mature philosopher, does his work for the sake of the city's educational curriculum and not the pleasures of *theôria*.

so that they can progress to dialectic. Rather than being intrinsically valuable, then, the point is that if mathematics is practised as a purely theoretical discipline, it is *instrumentally* valuable for the education of the future philosopher rulers. Socrates leaves no doubts about this in his closing remarks on mathematics. Harmonics, he says, 'is useful in the search for the beautiful and the good. But pursued for any other purpose it is useless' (531c6–7) and a few lines later, as we saw in the previous section, he says that if it results in a synoptic view, mathematics 'does contribute something to our goal and is not labour in vain; but that *otherwise it is in vain*' (531d2–3; my emphasis).

Nonetheless, I think there is still room for mathematics after *dianoia*. Consider again the 'purpose' or 'goal' guiding the discussion in Book 7. Socrates' goal is not to nurture the philosophical life or create the ideal conditions for philosophical *theôria*, even if these overlap with the actual goal. Sometimes, the goal is said to be knowledge of the Good, but even that is not quite right. Rather, Socrates' and his interlocutors' goal, which has been their preoccupation since Book 6 (502c10–d3), is to discover how they can educate rulers suitable for a just and happy city. While the saviours turn out to be philosophers, it is not because philosophy is valued for its own sake. Consider, for example, the following description of their task:

> SOC: It is our task as founders, then, to compel the best natures to learn what was said before to be the most important thing: namely, to see the good, completing the ascent. ... [But we will not allow them] to stay there and refuse to go down again to the prisoners in the cave to share their labours and honours, whether the inferior ones or the more excellent ones.
>
> GLA: You mean we are to treat them unjustly, making them live a worse life when they could live a better one?
>
> SOC: You have forgotten again, my friend, that the law is not concerned with making any one class in the city do outstandingly well, but is contriving to produce this condition in the city as a whole.
>
> 519c9–e3

The crucial point here is that while laying down the rules of the *kallipolis*, Socrates is not speaking *in propria persona*, but in the restricted role of a 'founder' with a duty to bring about the best city. After all, Socrates himself, who famously avoided politics, would likely be among those reluctant to re-enter the cave. Whatever Socrates and Glaucon think as individuals, in performing their 'task as founders', they have to value philosophy only for its contribution to the city. What extracurricular philosophy the rulers do in their spare time, simply for knowledge's sake, is not their concern, except insofar as it contributes something to the well-being of the whole city. Philosophical *theôria* should instead be understood as the guardians' personal goal, for the sake of their own happiness. This personal goal is not only different from the political goal of being good ruler, but the two goals compete for their time, so they must be compelled to rule, viewing it as an unwanted distraction from philosophising. In short, it turns out that Socrates, *qua* founder, must treat not only mathematics as mere means to the good governance of a just city, but philosophy too.

There will still be a great deal of philosophy to be done in the city, and mathematics too. But not all of it will be done for the sake of the city. Philosophers value truth for its own sake and find their highest pleasures in learning, so they have their own reasons to tackle intellectual problems, including mathematical ones. This is a sufficient reason to think that mathematics will continue after its role in the educational curriculum, even if it has ceased to contribute to the goal of the city's founders. So we do not need to doubt that philosophical *theôria* will include the contemplation of mathematical truths, which for the mature philosopher, unlike the student at the cognitive level of *dianoia*, can be appreciated as truths fully known.[27]

Bibliography

Annas, Julia. 1981. *An Introduction to Plato's Republic*. Oxford: Oxford University Press.

Benson, Hugh. 2010. "Plato's Philosophical Method in the Republic: The Divided Line (510b–511d)." In *Plato's Republic: A Critical Guide*, edited by M. L. McPherran. Cambridge: Cambridge University Press.

Broadie, Sarah. 2021. *Plato's Sun-Like Good*. Cambridge: Cambridge University Press.

Burnyeat, Myles. 1987. "Platonism and Mathematics: A Prelude to Discussion." In *Mathematics and Metaphysics in Aristotle*, edited by A. Graeser, 1–81. Bern: Paul Hapt.

Burnyeat, Myles. 2000. "Plato on Why Mathematics is Good for the Soul." In *Mathematics and Necessity in the History of Philosophy*, edited by T. Smiley, 145–172. Oxford: Oxford University Press.

27 I am grateful to Elisabetta Cattanei for stimulating discussion on the topic of this paper and to Rod Gow and Irem Cessur for comments on earlier drafts.

Cornford, Francis C. 1932. "Mathematics and Dialectic in the Republic VI–VII (I & II)." *Mind* 41 (161): 37–52, 173–190.

Cross, R. C., and A. D. Woozley. 1964. *Plato's Republic: A Philosophical Commentary.* London: Macmillan Press.

de Waal, Tamsin. 2022. "The Mathematicians' Use of Diagrams in Plato." In *Knowledge, Number and Reality: Encounters with the Work of Keith Hossack*, edited by N. Kürbis, B. Assadian, and J. Nassim. London: Bloomsbury Publishing.

Gill, Christopher. 2007. "The Good and Mathematics." In *Pursuing the Good: Ethics and Metaphysics in Plato's Republic*, edited by D. Cairns and F. G. Hermann. Edinburgh: Edinburgh University Press.

Gonzalez, Francisco J. 1996. "Propositions or Objects? A Critique of Gail Fine on Knowledge and Belief in Republic V." *Phronesis* 41 (3): 245–275.(*Added initials; Bib 1 gives full names where available.*)

Irwin, Terence. 1995. *Plato's Ethics*. Oxford: Oxford University Press.

Miller, Mitchell. 1999. "Figure, Ratio, Form: Plato's Five Mathematical Studies." *Apeiron* 32 (4): 73–88.

Mueller, Ian. 1992. "Mathematical Method and Philosophical Truth." In *The Cambridge Companion to Plato*, edited by Richard Kraut. Cambridge: Cambridge University Press.

Patterson, Richard. 2007. "Diagram, Dialectic, and Mathematical Foundations in Plato." *Apeiron* 40 (1): 1–34.

Payne, Andrew. 2017. *The Teleology of Action in Plato's Republic*. Oxford: Oxford University Press.

Plato. 2004. *Republic*, trans. by C. D. C. Reeve. Cambridge: Hackett Publishing.

Robins, Ian. 1995. "Mathematics and the Conversion of the Mind: Republic VII 522c1–531e3." *Ancient Philosophy* 15 (2): 359–391.

Robinson, Richard. 1941. *Plato's Earlier Dialectic*. Ithaca: Cornell University Press.

Ross, David. 1951. *Plato's Theory of Ideas*. Oxford: Oxford University Press.

Smith, Nicholas. 2019. *Summoning Knowledge in Plato's Republic*. Oxford: Oxford University Press.

Storey, Damien. 2022. "Dianoia & Plato's Divided Line." *Phronesis* 67 (3): 303–305.

Yang, Moon-Heum. 2005. "The Relationship between Hypotheses and Images in the Mathematical Subsection of the Divided Line of Plato's Republic." *Dialogue* 44 (2): 285–312.

CHAPTER 4

Plato on Noticing

Andy German

1 Introduction

We begin with two quotes from the Platonic *Phaedo*:

1. We've seen sticks or stones or some other things that are equal, and from these we've noticed (ἐνενοήσαμεν) it [the Equal Itself], although it is other than these.[1]
2. Whenever somebody who's seen something notes (ἐννοήσῃ), "What I'm now seeing wants to be of the same sort as something else among the things that are; yet it falls short (ἐνδεῖ) and isn't able to be that sort of thing but is inferior," I suppose it is necessary that the one who notes (ἐννοοῦντα) this had occasion to see beforehand that thing he says it's like but falls short of.[2]

Both these passages come from Socrates' "second" argument for the immortality of the soul, which purports to demonstrate it by means of the experience of recollection. It is a demonstration that has been thoroughly worked over by a massive scholarly literature. Readers have investigated the cogency of the proof, the role of Form in it and the gist of its overall identification of learning with recollection. For example, is ἀνάμνησις involved in the ordinary learning of which everyone is capable, more or less? Or is it the special privilege of the philosopher?

This is well-plowed ground and will not concern me here. Instead, I am interested in two questions that arise from the passages above, and from the whole recollection argument. First, what exactly are the mechanics of ἀνάμνησις? It is all very well to say that perceiving the purported equality of two sticks, or the beauty of our beloved's face, "triggers" the recollective act. But when all is said and done this is metaphorical language, and potentially misleading at that. For if I tell you that a certain movement of my finger triggered

1 *Phdo.* 74b3–6. I have worked with the Greek text of the *Phaedo* from the Oxford edition by Duke et al. (1995). The edition of the *Republic* used here is that of Slings (2003) and all other texts come from the standard Burnet (1900–1907) OCT edition of the *Platonis Opera.* Translations are my own, but in translating the *Phaedo* I have largely followed the choices of Brann et al. (1998).

2 *Phdo.* 74d9–e4.

HTTPS://DOI.ORG/10.1163/9789004747128-006

the firing of the bullet, you can at least hope to receive an account of the interrelation of trigger, action, firing pin, primer, gunpowder and so forth which takes you from the cause to the effect. But, when one tries to build an equivalent account of the movement leading from sense perception as the "trigger" to the recollective grasp of Form, matters are by no means as straightforward. And this leads directly to our second question.

No minimally attentive reader of the *Phaedo* account of recollection can miss the outsized role played in it by the verb ἐννοεῖν: to consider, have in mind, form a notion of, or simply – to notice.[3] Indeed, between 73b6 and 77a5 one could be forgiven for wondering if this was the all-purpose Greek verb for thinking. In various forms, the word appears fully thirteen times in these four pages. Nowhere else in Plato's corpus is it used with such frequency. Why should this be so? And why, moreover, should the same word be used to denote two very different acts which are somehow involved in recollection but whose relation is quite obscure? On the one hand, as we can see in the first quote, ἐννοεῖν refers to the act of coming to have some *particular thing* "in" mind: the Equal itself, for example (ἐκεῖνο ἐνενοήσαμεν). On the other hand, in the second quote, it describes noticing a *state of affairs* (ἐννοήσῃ ὅτι) – more specifically, it requires noticing a *lack*, an absence.

We will focus on these questions in the hopes of shedding some light on the broader topic of what thinking is for Plato and more specifically, what kind of thinking θεωρία is.

There is, of course, no shortage of evidence that Plato assigns θεωρία to a faculty which is somehow separate or discontinuous from all others. Νοῦς is variously described as the "divine in us", as immortal, instantaneous and intuitive rather than discursive in its operation, as "akin" to the deathless, unmoved perfection of its objects. But A can *be separate* from B and nevertheless share certain important continuities with B as regards its *mode of being* – its τρόπος τοῦ εἶναι. I suspect that something like this is true of noetic contemplation. Close attention to Socrates' account of recollection, and the role of noticing

3 This verb points to a variety of closely related meanings and can be translated in a variety of ways as evidenced by the different choices of Hackforth (1972) (who goes in for "conceive"), Rowe (1993) (who prefers "have in mind" in his commentary to the dialogue) and Cooper (1997) ("realize"). I agree with Brann and Kalkavage (1998) that "notice" is the better translation because of the way it preserves the basic duality we are about to discuss: the activity of having a notion (which Aristotle might describe as a kind of coming to rest – *De An.* 407a33–4) and "noticing", or paying attention, which is an active, inquisitive, attending to things. "Consider" and "consideration", or "reflect" and "reflection" are not as successful in conveying this.

in it, reveals a basic continuity underlying the various ways that ψυχή is open to world – a continuity shared in by νόησις.

The continuity runs still deeper, however. Noticing, in all its forms, is a kind of *striving*. This, I argue, explains why Socrates emphasizes this quite ordinary cognitive act in this particular passage. But if to notice is to *strive* to get something clearly in view, then it is the psychic manifestation of Plato's core metaphysical insight into the nature of the whole: all things (and not only souls) can be said, in a sense, to strive – they strive for the formal perfection expressed by their εἶδος and ultimately for the Good as the source of perfection. That is why contemplation can be both good for *us* and yet an act by which we transcend ourselves, and why it can be both an activity unique to ensouled beings and yet the highest expression of the kinship of such beings with everything else.

2 Sensation as a Mode of 'Having the World in View'

Before turning to a detailed examination of the steps involved in recollection, we ought to spend some time on sense perception as such, since it is so central to how Socrates begins his demonstration for Simmias at 73c5–11:

> So then, do we also agree to this, that whenever ἐπιστήμη comes to be present in the following way, there is recollection? I mean something like this: Whenever someone having either seen or heard something or grasped it by some other sense, not only recognizes (γνῷ) that thing, but also takes note of another (ἐννοήσῃ), of which the knowledge (ἐπιστήμη) is not the same but different – don't we justly say that he has recollected that of which he grasped the notion (τὴν ἔννοιαν ἔλαβεν)?

The picture here is a complex one:

Sense perception is described as a kind of grasp which enables recognition of something and it is accompanied by a different "knowledge" which is made possible by an act of noticing standing in some unspecified relation to the original sense perception. Later in the dialogue, the link between sense perception and knowledge is tightened considerably since Socrates will say that the knowledge (of the Equal Itself) comes "from nowhere else" (μὴ ἄλλοθεν) but the perception of sensible equal sticks or stones (74c7).[4] This is ample

4 And cf. *Phdo.* 75a5–8.

reason, surely, to study Socrates' treatment of αἴσθησις. However, there are other reasons, arising from the fact that what Socrates has to say to Simmias from 73c onward is unexpected (to grossly understate matters) in light of what he had to say about αἴσθησις to this same Simmias only a few pages earlier.

The Socratic charge sheet against the senses (and the body more generally) at 64c–67b is justly famous. It is the seemingly inescapable conclusion to be drawn from his description of philosophy as the practice of nothing else but dying and being dead – that is, "dead to" the body in the sense of being separated from it (64a4–9, c4–8). But in assessing this passage one must carefully distinguish the problematic state of *being embodied* from the specific problem with sensation as an *embodied mode of grasping the world*. These are not the same problem.

As Socrates describes it, *being embodied* is a high-maintenance affair yielding very dubious returns. The servicing of the body (τὰς περὶ τὸ σῶμα θεραπείας) requires an inordinate amount of work, of drudgery really (64d8–e1), while this same body that we spend our time servicing is also "an impediment in the quest to attain" what we really want: φρονήσεως – thoughtfulness (65a9). The reason given is that bodily seeing and hearing are not precise or clear (μὴ ἀκριβεῖς…μηδὲ σαφεῖς) to say nothing of the other senses which are almost wholly worthless in this regard. Therefore, when the soul searches or "strives" for what is (ὀρέγηται τοῦ ὄντος) it must avoid all communion with the body (65a9–c9).

And Socrates continues piling it on at 65eff. We do not behold (θεωρεῖται) the true *being* of things "through" the senses of the body (διὰ τοῦ σώματος). With its diseases, its loves and desires and fears, its "images and nonsense" the body deprives us of leisure. It produces wars, factions and battles, and constantly interrupts our psychic peace, making a great racket, driving us crazy (66b–d). Knowledge requires getting free of τῆς τοῦ σώματος ἀφροσύνης (67a7) – in this life as far as possible and after death, definitively.

This is a litany of horrors delivered with such heat that one is apt forget that it is actually rather ambiguous regarding sensation's specific shortcomings. Though we are told that the senses lack precision or clarity, most of the passage is concerned with something quite different – not with the senses, but with the interference of the many bodily needs and desires which are evidently unrelated to the ὄρεξις τοῦ ὄντος altogether. But what is the relation of the senses to τοῦ ὄντος? Are they, too, unrelated? Not exactly. Sight and hearing are imprecise and unclear, and they are incapable of beholding what is "truest" (ἀληθέστατον) about, say, "Bigness Itself" or "Health Itself". But note that this does not mean that they behold nothing at all true about such things. On the contrary, Socrates' criticism makes sense only if perception *is* in some way directed toward ἁπάντων τῆς οὐσίας. The body may very well be a prison, and the prisoner may curse the little window in his cell for being so dirty that he cannot see through it, or for being so small that it lets in very little light and shows only a narrow, unsatisfying slice of the visual field. But he would make no such critique of a stretch of wall, since the wall is neither open to the world nor was it meant to be. Sensation, faults and all, remains a mode of "having the world in view".

What, then, does its lack of precision or clarity consist in? Here too, Socrates is uninformative. For example, at no point does he deny what we all know – viz., that eyes, for example, can distinguish fine shades of color, that hearing is attuned to the slightest distinctions of melody and harmony, and that even smell can discern differences among scents with amazing precision.[5] His focus throughout is on "body" as a whole, with the senses taking the hit mostly as collateral damage. We need to keep this in mind as we turn to the recollection argument that begins at 73b5, since otherwise Socrates' language there about knowledge of Form coming *from* the senses will make it seem like he has flatly contradicted himself in the space of a few minutes' conversation. But only in the recollection argument, and not in the attack on sense perception we have just studied, do we get a fuller account of what αἴσθησις can and cannot do, and of the metaphysical status (and deficiency) of the αἰσθητόν.

What, then, does Socrates think sensation *can* do within its own proper purview? What are its limitations and what explains them? How does sensation instigate recollection? Why should its role in such instigation express itself in two different acts denoted by the same word? We turn now to these questions.

3 From Sensation to *Anamnêsis*

Throughout his conversation with Simmias, Socrates takes it for granted that we can, for example, "see" a lyre, or a cloak, or even a set of sticks or stones which are equal to one another. In other words, although Socrates is in the process of demonstrating that the soul must have at one point existed and learned without a body, he never denies that when we sense now we are

5 See Rowe (1993: 140) on this point: "…once again…the argument is singularly lacking rigour; we should need more than poetic hyperbole (b2–4) to convince us that the evidence of the senses is *always* deceptive as S. here implies."

sensing some particular *thing*.[6] This is not a trivial point, and Socrates' example of the "three fingers" in *Republic* VII can help elucidate why.

At *Republic* 523aff, Socrates is attempting to explain to Glaucon how the "lowly business" of distinguishing and counting numbers can "by nature lead to intellection" and is "in every way apt to draw men toward being" – that is, toward the consideration of pure intelligible structure.[7] Some perceptible objects, says Socrates, are adequately "judged" or "discerned" by the senses (ἱκανῶς ὑπὸ τῆς αἰσθήσεως κρινόμενα).[8] When one looks at the ring, middle and index fingers, for example, Glaucon agrees that the sense of sight adequately grasps that each of these colored shapes "looks like a finger". Nothing here compels the soul to ask "What is a Finger?" I pause to note the oddity in this statement. It is unclear whether Socrates means to say that sensation itself is somehow capable of judging something *as* a finger (which would seem to import a conceptual element into the act of sensation) or whether he means that, in this case, sensation instigates something like a recollective apprehension of what it is to be a finger in one fell swoop without the soul being compelled to ask and investigate. But for our purposes we can leave this aside. The more interesting point is that the senses are indeed capable of "revealing" (δηλοῦσιν) something to the soul as *some one thing*. Whatever else we may say about it, sense perception taken just as such is capable of presenting to us unities within experience. This entails that it is directed toward a kind of discerning and judging (κρινεῖν) at a considerable level of complexity and sophistication. Socrates is even willing to go so far as to say that the senses "indicate" (σημαίνει). Sight can indicate "finger" adequately and touch (even lowly touch) indicates distinct qualia such as "hard" and "soft".[9]

Sensation goes off the rails as a form of discerning, however, when it is compelled, against its grain, to face the following problem:

> Now what about this? Does sight see their bigness and smallness [i.e. – of the fingers] adequately, and does it make no difference to it whether a finger lies in the middle or on the extremes? And the same goes for touch, for thickness and thinness or softness and hardness? And do the other senses reveal such things without insufficiency? Or doesn't each of them do the following: first, the sense set over the hard is also

6 *Phdo*. 73c6: ἐάν τίς τι ἕτερον ἢ ἰδὼν ἢ ἀκούσας.

7 *Rep*. 523a1–3.

8 *Rep*. 523b1–2.

9 Glaucon here refers to the deliverances of sensibility as "interpretations" (ἑρμενεῖαι) which it sends to the soul (*Rep*. 524b1). Not Socrates' language to be sure, but Socrates doesn't correct him either, as he is wont to do elsewhere. See *Rep*. 509a9 and 523b7 as examples.

> compelled to be set over the soft; and it reports to the soul that the same thing is sensed by it as both hard and soft.....Isn't it necessary, I said, that in such cases the soul be at a loss as to what this sensation indicates by the hard, if it says that the same thing is also soft....?[10]

The difficulty, we note, is the compresence of opposite properties in the same sensed unity; here sensation is of no help. Sensation "indicates", the soul thereby is cast into *aporia,* and this is the occasion for summoning intellection. A conative impulse to discern is present in sensation but carries it only so far; sensation discerns, but does not distinguish and relate. Senses are narrow because they are *inarticulate* in the original Latin meaning of *articulare*. In this lies their insufficient clarity and precision.

With this in hand, we can return to the text of the *Phaedo* to study three crucial Socratic assertions: First, sense perception is somehow able, alongside presenting the sensible unity, to cause us to note or have in mind something else. We can "see" a lyre and then "grasp in thought" (ἐν τῇ διανοίᾳ) the form of our beloved. We can see equal sticks and from these we notice (ἐνενοήσαμεν) the Equal Itself. Second, however, we are able to do this despite the fact that the Equal Itself is not quite what we see in seeing two sticks as equal.[11] There is a "stick equal to stick" or "stone to stone" but the Equal itself is "something other, beyond all these" (ἀλλὰ παρὰ ταῦτα πάντα ἕτερόν τι, 74a11), other than any sensible manifestation of equality. This is repeatedly emphasized. Third, we have a further ability, either coterminous with sensation, or instigated by it, or in some other way related to it: we can "note" a certain *lack* and *striving* within sensible objects in their relation to that which is other than or beyond them but is nonetheless that which they lack – their Form (75a1–3).

These assertions and their interrelation must now be explained. But in order to do so we must be on guard against making life too easy for ourselves, specifically with regard to the second of them. Sensation presents us with a spatio-temporal particular or a congeries of particulars. To say of any such particular that it is "equal", however, is to say that it stands in a relation of equality *to* something. But by virtue of this it is also in a relation of being unequal to some other thing. The spatio-temporal particular has the property of being equal only under a certain aspect or in certain relations. The Equal Itself however, cannot be like this. As Lee Franklin writes:

10 *Rep.* 523e1–524a3.

11 Cf. *Phdo.* 74b6–7, c4, d4–7.

> Both equal things and Equality can be said to be equal, but not at all in the same sense......Once we grasp this difference, we can see why the perception of equal particulars is insufficient to bring us to have Equality in mind. A stick manifests one way of being equal, but it does not manifest Equality itself, the property or relation common to absolutely all equal things. Coming to behold the form *F-ness* is coming to grasp *what-it-is-to-be-F*. No single act of perception could ever communicate this essence to us....[12]

So, it can't quite be that we "note" the Equal Itself because we "see" it in the particular, just as we can never note the Circle Itself by somehow seeing it "within" the drawn circle. To know what-it-is-to-be-equal is to grasp an articulated formal structure (a collection, or community, of Forms or γενῆ, as they are called in the *Sophist*) the *very being* of which is entirely indifferent to accidental properties like being a wooden stick, or an igneous rock, or having a certain color, or a certain spatio-temporal location in relation to something else, etc. Clearly, we do not, in the ordinary course, pass directly from an act of sense perception to this kind of knowledge. Instead, ἐννοεῖν τι – "taking notice of" – must be a kind of inferential transition from perception to the inchoate presence-to-mind of the Form as that which is somehow responsible for what was perceived.[13] This is not yet the completely articulate grasp of the Form, but it can be the occasion for a "recovery" of that grasp – or *knowledge* – of the Form's whatness, which was "in us" but then lost before our first ever use of the senses (75b5–6).[14]

But if taking notice of the Equal Itself is not some kind of "seeing" of a magnificently, paradigmatically equal stick lurking "in", or "behind", or "above" the sensible sticks, so that we can't help stumbling on the former when looking at the latter, what exactly is it about the sense experience that triggers the recollective process? On this point, Socrates is unclear because alongside "taking note of" or "coming to have in mind" the Equal Itself he also mentions, five times, the ability to note how the sensible object falls short of something, or strives (ὀρέγεται, προθυμεῖται) to be like something else but is incapable (οὐ

---

12 Franklin (2005: 305–306).

13 Or perhaps we might say that what is involved here is an initial noticing *that* there must be something responsible for the stable determinacy which makes any sense perception possible in the first place, without any articulate grasp of *what* this is.

14 See also *Phdo.* 74c7–9: "it's nevertheless from these equals [i.e. the sticks], although they are other than that Equal, that you noted *and* grasped the knowledge of it (ἀυτοῦ τὴν ἐπιστήμην ἐννενόηκάς τε καὶ εἴληφας)". The *te kai* construction here binds the noticing to the grasping of the knowledge, but does not eliminate the fact these are two mental acts.

δύναται) or inferior (φαυλότερον).[15] Is this second kind of ἐννοεῖν, then, the trigger we've been seeking for the recollective movement? Does it merely happen simultaneously with ἐννοεῖν as "coming to have F in mind" or is it some altogether different act, distant in time?

A fine analysis of this point is found in Franklin's article mentioned above. For the sake of brevity, I will summarize his main argument as it relates to our present study. Since Franklin sees (correctly, in my view) that one cannot simply "read off" the Equal Itself from equal sticks, some other resource must be at play when one comes to have the Equal Itself in mind upon seeing equal sticks; one *must* be recollecting. But, according to Franklin, "noting" the shortfall or lack which the equal sticks suffer from is neither the act which causes this recollection nor simultaneous with that act. It is, rather, part of Socrates' *explanation* to Simmias of why given that (i) we do have knowledge of Forms (something Simmias accepts without further ado at 74b1), and (ii) this knowledge certainly could not come from ἄισθητα (because of their metaphysical deficiency), therefore (iii) it must be the case that we're drawing on knowledge which we previously had. Socrates' talk of "noting of shortfall" is not a reference to something that happens *in* or simultaneously alongside the act of coming to have the Form in mind (i.e. in the everyday act of recollecting).[16] It is an explanation of why that act has the character it does.

This reading is ingenious but, as far as I can tell, it faces two difficulties. First, it must be confessed that the preponderance of textual evidence is against it. Socrates seems to be describing here an act which is quite common, not some move in a rarified discussion between two philosophers. For example, at 74d4–7 Socrates asks Simmias:

> Do we undergo something like this concerning equals among sticks and the other equals we were talking about just now: Do they appear to us to be equals in just the same way as the Equal Itself, the equal that *is*? Or do they fall somewhat short of being the sort of thing the Equal is?.

15 *Phdo.* 74a, 74e, 75a, b1 and b7.

16 Franklin (2005: 303) and especially Franklin's interpretation of *Phdo.* 75b4–8 (2005: 312–313), where he translates the Greek as follows: "Then, before we began to see and hear and have other perceptions, somehow we took up the knowledge of it, the Equal, what it is, if we were going to refer the equals from perception to it, since all such things desire to be such as it is but are deficient compared to it." This reading goes against Vahlen's addition (noted in the apparatus of Duke, Hicken, Nicoll, Robinson, Strachan, 1995) of καὶ ἐννοήσειν at line b7 to try to make sense of the ὅτι clause. Franklin wishes to argue that the MS reading can stand as is and without any interpolation and that when read this way we do not need to posit that recollection and comparison of sensible particulars to Forms are two thought-moments that must occur simultaneously.

Now it is true that we have here a conversation between two people who already agree that there is a Form of equality and that we know it.[17] And it is also true that Socrates is referring to something he and Simmias been discussing "just now" (νυνδὴ). But what *have* they been discussing just now? Nothing but the perfectly ordinary act of seeing equal looking sticks and stones. Surely, when Socrates says, at 74e9–75a3: "Then it's necessary that we saw the Equal before that time when we first saw equal things and noted: 'All these things are striving to be like the Equal'", he is referring to something our common humanity is capable of, not to some unique accomplishment of Simmias or Socrates.[18] This becomes even clearer in a slightly later passage at 76a1–7:

> For surely this appeared possible: A person senses something either by seeing or by hearing it or by grasping it with some other sense and – starting from this thing – notes (ἐννοῆσαι) some other thing he'd forgotten and which the sensed thing approaches, whether it's dissimilar or similar.

Part of what underlies Franklin's argument is the intuition that *without* something like his reading, passages like this, along with 75b4–8, ought to strike us as phenomenologically false, since they seem to imply that every act of perception which initiates recollection must also be accompanied by a *propositional* belief about the perceptible object as "straining" bravely toward its Form, only to fall flat on its face. But I don't think we need to take it this way.[19] Socrates and Simmias are perfectly aware, as we all are, that thinking would be impossible if we had to talk to ourselves like this all the time. Socrates seems, rather to be verbalizing, and therefore making concrete for Simmias, something that does occur together with the act of sense perception, but need not be discursive. It is rather an unquiet sense that something is missing, an experience or impetus which accompanies our various perceptive and cognitive relations to the world.[20] So, the "mechanics" should be under-

---

17 As David Sedley points out, however, even this need not necessarily imply a specialized knowledge available only to philosophers. See Sedley (2006: 318).

18 In this sense, I side with what Franklin (2005: 290–291) calls the "ordinary", rather than the "sophisticated" interpretation of recollection. *Anamnēsis* is part of our common human patrimony, and not an accomplishment unique to philosophers.

19 On this, see Kelsey, (2000: 95–97). And cf. also the thorough analysis in Morgan (1984: esp. 247–251).

20 Shorey (1967: 124–125) seems to argue along the same lines when he writes, "There are no perfect circles or equalities in nature. Yet we do conceive them, and we *feel* [my emphasis, A.G.] how far concrete circles and equalities fall short of the ideal toward

stood as follows: (i) sense perception, accompanied by (ii) noticing some lack which characterizes the perceptible object, then (iii) coming to notice, or have in mind, the Form in some inchoate way, and finally (iv) the possibility of fully articulate, recollective knowledge of Form.

I think that this also will help us to address the further problem arising from Franklin's reading. Given that Form is metaphysically dissimilar to and therefore not "present" in the sensuous particular (in any simple sense) one might wonder *why we are driven to recollect at all*? It is not enough to say that just as a painting of Simmias is an image meant to be a likeness of the original, flesh-and-blood Simmias, so are sensed particular somehow images of their original, the Form.[21] Why are we constitutionally incapable of just making do with the images as if they were the originals? Indeed, in the Cave Allegory which is, after all, a description of "our nature" (τὴν ἡμετέραν φύσιν), this is precisely the condition of that part of the soul represented by the bound prisoners who can neither see themselves, nor each other, nor the fact that the shadows on the wall facing them are images of something else. Life is literally happening behind their backs and they have not the slightest inkling. The "of-something-else" relation is lost on them.[22] Evidently, the allegory is trying to make Glaucon see that there is something in us which is like this, and that makes it all the more striking when Socrates later asks Glaucon what the release of these prisoners from their bonds would be like if "some such thing were to happen to them φύσει – by nature".[23] We are not told what it could be that would make chains fall away "naturally" and are thus faced with a version of the same question asked above: What impels us past images and toward originals?

It would seem obvious that the answer must be ἔρως. Ἔρως is the great connector, the psychic force which awakens the soul to the distinction between the visible realm of Becoming and the intelligible realm of Being, granting the soul no rest until it is able to complete the transition from our ordinary comportment in the first realm to contemplation of the second:

> So, then, will we not make a sensible apology in saying that it is the nature of the real lover of learning to strive for what *is*; and he does not tarry by each of the many things opined to be but goes forward and does not lose the keenness of his love nor cease from it before he

---

which they strive…". In this regard, note the uses of προσπάσχειν and πάσχομέν at *Phdo.* 74a6 and 74d4 respectively.

21 As argued by Kelsey (2000: 99 and 111–112).

22 *Rep.* 515c1–2.

23 *Rep.* 515c5.

> grasps the nature of each thing which is with the part of the soul fit to grasp something of that sort....And once near it, and coupled with what really *is*, having begotten intelligence and truth, he knows and lives truly, is nourished and ceases from his labor pains, but not before.[24]

As it stands, however, this answer is insufficient. Ἔρως is a highly ambiguous psychic force, a "magician, druggist, and sophist" Diotima calls it, which can just as easily lead us away from the intelligible realm as towards it. Eros is love, but not exclusively or necessarily love of the intelligible. What we require is some understanding of how ἔρως is manifest in the different forms of human awareness, such that it can be present *both* in our relation to the sensible realm and to the intelligible realm thus binding the two together.[25] An examination of Plato's uses of ἐννοεῖν throughout the corpus can help in this regard.

4 What Is Noticing?

In the dialogues, ἐννοεῖν carries a variety of meanings: to have a person or thing in mind, to understand the meaning of something, to notice something about a situation, to take a second look, etc. In fact, it often has no technical valence whatsoever. Socrates, for example, asks Gorgias whether he is the kind of person who is willing to refute and be refuted in turn out of concern for the truth. If he is, they could have a conversation; if not, they cannot. Gorgias responds:

> I do, indeed, claim that I myself am the sort of person you indicate, Socrates; but maybe we ought to give some thought (ἐννοεῖν) to the preferences of those present. Because, I tell you, even before you two got here, I gave those already present a display of many things for a long time and now we'll be drawing things out too far....[26]

The sense here is almost equivalent to "being considerate" of others, or "reading the room", as we might say. Elsewhere (including, of course, at *Phdo.* 74b4–8 mentioned above), one notes an intimate relation between ἐννοεῖν and verbs of "looking", such as σκοπεῖν or βλεπεῖν. Thus, for example, in order to know what the soul is, Glaucon must look (βλεπεῖν) towards its love of

24 *Rep.* 490a8–b8. On the connective role of ἔρως see *Symp.* 202e3–203a8.

25 *Symp.* 202e6–7: ὥστε τὸ πᾶν αὐτὸ αὑτῷ συνδεδέσθαι.

26 *Gorg.* 458b5–6.

wisdom, and note (ἐννοεῖν) with what the soul longs to keep company.[27] Or, elsewhere in the *Republic,* Glaucon is exhorted: "Take a look (Σκόπει δή) then, and see whether I'm saying anything. In my opinion, we must consider it (δεῖν ἐννοῆσαι) from this point of view."[28] First we look, then we try to notice, or focus in more closely on what there is in the direction in which we are looking.[29] In general, outside of the *Phaedo,* ἐννοεῖν is not used in any context related to contemplation of Forms, with the singular exception, it seems, of *Symposium* 210a8–210b4, where κατανοεῖν and ἐννοεῖν are essentially synonymous and refer to noticing, or "realizing", that the beauty in any body is akin to that in any other body. Nor, except for the *Phaedo,* is it common to find the word in contexts having to do with the doctrine of recollection. It does not appear in the *Phaedrus* or the *Meno* passages, for example.[30]

In fact, in some dialogues its use is quite distant from that of the *Phaedo.* In the *Laws,* for example, the Athenian Stranger speaks of a device (μηχανή) which the legislator of Magnesia must "think up" (ἐννοεῖν), that is, invent,[31] while in the *Cratylus* Socrates once uses the verb for his "suspicion" or "hunch" that many Greek words have a barbarian root.[32]

Careful study does, however, reveal a striking frequency of *negative* constructions of the word, i.e. ones where Socrates asks his interlocutor: "Haven't you ever noticed ....?" (οὐκ ἐννοεῖς;). And this, I think, is significant. For noticing really does have something secondary about it. Socrates is often trying to get someone who is looking in a certain direction to look again and better, since there is something there which he is, as it were, looking at but not seeing.[33] Glaucon, for example, knows a great deal about the courage of guard dogs, but "hasn't he ever noticed" how irresistible and unbeatable θύμος is?[34] Or we can take Thrasymachus as an example. Evidently, he spends a lot of time thinking about ruling and rulers, and yet "hasn't he ever noticed" that no one willingly wishes to rule, but rather demands wages for it?[35]

---

27 *Rep.* 611d6–e2.

28 *Rep.* 578d1–2.

29 See also *Rep.* 584b–c, *Symp.* 208c1–5, *Law.* 823b5.

30 It is, however, occasionally used together with verbs of "forgetting". Cf. *Tht.* 201c9, *Rep.* 493e2 and appears in a parallel discussion of a likeness falling short of its original – namely at *Crat.* 432d1–3.

31 *Law.* 798b5.

32 *Crat.* 409d9–e1.

33 Compare with Isaiah 6:9–10 and Matthew 13:13.

34 *Rep.* 375b1.

35 *Rep.* 345e6. This usage is very much present in the *Republic* (*Rep.* 507d7, 519a1, 522d2–3, 537e1, to name a few) but by no means is it unique to that dialogue. Cf. *Euth.* 15c5, *Tht.*

Every stand-up comedian knows how to make use of this tactic in provoking a laugh from an audience. But what is crucial here is that we do not always *need* to be nudged or provoked by someone else into trying to get a better look. On the contrary, noticing is something we are driven to do by ourselves, from an innate desire to get a better grasp of everything relevant to the object field we are concerned with. Noticing, in this general sense, is simply the effort to be clear and precise, and it can indeed be found in sensation taken by itself, as when eyes adjust to a change in light, or sharpen focus in the move from viewing things near at hand to those at a distance.

It is this common usage that helps explain why, in addition to being the word for "coming to have" the Form in mind, Socrates uses ἐννοεῖν for noticing something else about sensible objects – namely, that they are ἐνδεής; they lack something (in the particular case discussed in the *Phaedo*, they lack the mode of being of the Equal Itself) and are striving toward it. One thing, though, should be clear: lack here means a lack, by the sensible, for something which is *properly* its own, not for any old property. What it is to be Equal is what the equal sticks try to instantiate, but cannot, since the sticks instantiate equality only in relation to something, but not in relation to some other thing, or only from one point of view but not from another, to say nothing of the fact that they do so only temporarily and accidentally.[36] The Equal Itself is perfectly determinate *being-what-it-is-to-be-equal*; but qua perceptible objects our equal sticks necessarily lack this perfect determinacy. They are always changing, always exhibiting other properties which are unrelated and accidental, as far as being equal is concerned. They are more indistinct and less determinately what they are. In all of these senses, the Equal Itself is other than the equal sticks but is nevertheless, at the same time, the *true being* of their equality – that which they are *supposed to be qua equal*, what they "desire" to be but are incapable of.[37]

But put this way, it becomes clear why we cannot help noticing this characteristic of the sensible realm, and why this noticing should be the trigger for the kind of thinking called ἀνάμνησις: because this lacking and striving is

161b1–2 and 196d8. Though this construction is more expressive of Socrates' interrogative rhythm, it is also used by the "Strangers" in *Soph.* 238d5 and *Law.* 788d4.

36 *Phdo.* 74b4–9. See Nehamas (1975: 109). See also p. 113, where he writes: "....Plato is not maintaining the implausible idea that....if two things are similar to each other, and one is a copy of the other, the copy can never reproduce exactly the very characteristic which makes it a copy in the first place. Plato believes that if one thing is a copy of another then it must lack some *characteristic* of its model, and that one must realize this if one is to realize that he is confronted with a copy in the first place."

37 *Phdo.* 65d12–e1: τῆς οὐσίας ὅ τυγχάνει ἕκαστον ὄν.

analogous to what, in countless ways, characterizes what it is to be the ensouled beings we are. We, too, only imperfectly and occasionally exhibit our nature at its fullest, though the presence of ἔρως is proof of our striving toward what would genuinely complete us (the good and the beautiful).[38] Our ordinary relation to the world, and this includes all our cognitive comportments, is a ceaseless effort to gain clarity and distinctness and we have seen that even sensation as a power of discernment is such an effort. The use of a single verb to describe two different moments in the act of recollection is therefore neither a mistake nor an impediment to understanding the *Phaedo* passage.[39] It is Plato's way of pointing to a continuity that runs through our psychic faculties and that links our mode of being with that of our objects.[40]

5 *Theôria* as Purified Noticing

What of our highest psychic faculty and its objects? After all, ἀνάμνησις is not θεωρία; it is an activity undertaken as part of (indeed, because of) our embodied

38 Lindenmuth (1988: 15): "...for the philosophic lover of truth, the sensible world engenders awareness of that world's deficiency in comparison to the truth."

39 Nor can I agree with Rowe that it is just a dead metaphor. See Rowe (1993: 171–172).

40 If this interpretation sounds suspiciously like what David Sedley (2006: 324–325) calls a Platonic anticipation of the Aristotelian understanding of form as *energeia*, it is because I happen to think this is the correct understanding of the Form-particular relation, or at the very least the most philosophically promising one, though I cannot argue for it fully here. One comment, however, in passing: Sedley points out that the language of the particular "striving to be like" the Form should be understood as making sense primarily in the context of geometrical examples of particular-Form relations, where it is a useful, and indeed intuitively comprehensible, metaphor that explains the deficiency of the sensible triangle in relation to the formal ideal. However, after the *Phaedo* this language of sensible particulars "striving to be like" their paradigmatic Forms disappears. It is not to be found, Sedley notes, in the *Timaeus* or the *Republic* despite the heavy emphasis in both dialogues on the "Form qua paradigm" relation. Therefore, he sees this language as primarily pedagogical, and restricted to a very particular, introductory, context. It should be abandoned as a general metaphysical framework for the interpretation of Plato. I fail to understand this. While it is true that the language of striving does not appear elsewhere, there is another massively important theme in Plato which makes the same point in a different register and I have already mentioned it: *eros*. For an insightful analysis of just this point see Kosman (1976) and Lindenmuth (1988). Indeed, I would say that the evidence brought by Sedley himself for the absence of "striving" language in the *Timaeus* or the *Republic* might best be explained by the way *eros* is suppressed as a theme in both these dialogues, in one case by Timaeus' mathematized conception of reason and in the other by the political necessities of the *kallipolis*. But the fact that it needs to be suppressed points rather to its presence, not its absence.

existence. Θεωρία, by contrast, is often described as an activity we enjoyed either before or after our embodied existence, or at the very least as a release from it, however temporary. And Forms, too, are not indistinct and deficient like sensible objects. Forms are "always in the same condition", and set in a regular arrangement, according to reason.[41] It would seem that in θεωρία we are dealing with a kind of non-conative absorption in the noetic object. There may be a continuity running throughout our psychic faculties, but θεωρία, the divine activity in us, would seem to be χώρις.

I do not think this is the case however. θεωρία is the activity of νοῦς, and this holds true even for the divine θεωρία described in the *Phaedrus*.[42] While I am well aware of the prominence of metaphors of vision (and, occasionally, of instantaneous "grasp") in Plato's attempts to describe what happens during νόησις – both in the dialogues and in the Seventh Letter[43] – and while θεωρία is described, in the myth of Er or in the description of the ὑπερουράνιος τόπος in the *Phaedrus*,[44] as a kind of beholding that is free of all (or almost all) of the usual constraints, noetic intellection is generally not described as mere passive staring. Νοῦς, rather, is an active discerning that *both* "sees" the ὅ ἐστιν ἑκάστου τῆς φύσεως,[45] but also thinks through the complete interrelation among intelligible objects. This is what Socrates tells us at the top of the Divided Line when he describes the "other segment of the intelligible" (later called νόησις at 511d8) as "logos itself" which *by means of the power of dialectic* (τοῦ διαλέγεσθαι δυνάμει) works its way to and from the first principle, through forms. Whatever else it is, then, νόησις involves some conceptual ability: the ability of "logos itself" to relate the intelligible elements it has distinguished to intelligibility as a whole, thus rendering their being completely perspicuous. The evidence for νοῦς as a kind of purified discursivity, rather than passive vision, was marshalled with impressive force by Monique Dixsaut (1997), and she concludes her analysis by noting the following passage from *Rep.* VII 533a5–10:

> But that there is some such thing [viz., the truth itself as it looks to Socrates] to see (ἰδεῖν) must be insisted on. Isn't it so?.... And also that the power of dialectic alone (ἡ τοῦ διαλέγεσθαι δύναμις μόνη) [and, hence, not a silent gazing – A.G.] could reveal it to a man experienced in the things we just went through, while it is in no other way possible?

41 *Rep.* 500c3–5.

42 See especially *Phdr.* 247d1 – where the θεοῦ διάνοια is nourished by νοῦς.

43 A document which I accept as authentically Platonic, *pace* the strictures of Burnyeat and Frede (2015).

44 Or in the *Timaeus*, for that matter. Cf., *Tim.* 41e1.

45 *Rep.* 490b3.

Intellection may be likened to vision, but it is a kind of vision which not only *can* give an account of itself, but will not rest until it does so completely.[46] And if so, contemplation is the same kind of striving for clear, distinct (but in this case genuinely comprehensive) articulation which begins in sensation, is found in practical circumspection and διάνοια, and triggers recollection. We might say that in θεωρεῖν the same intentional, seeking impetus which is present throughout the whole soul and uniquely concentrated in the dialectical capacity of νοῦς, is released from the various constraints on its full operation and finally set free to roam, as it were, on the pastureland most suited to it, what the *Phaedrus* calls the "Plains of Truth".[47] Θεώρια then, is not mere "gazing", but a *purified* activity of noticing.[48]

What about the objects of noetic contemplation – the Forms? Surely, they cannot in any sense be said to desire or even to strive? This is true. But Forms do *implicate* other Forms and ultimately "depend", as the Sun Image and Divided Line tell us, on the Form of the Good, the source of their being and without which they themselves cannot be known adequately.[49] In the noetic realm, too, there is a certain sense – not identical with what we find among sensibles in relation to form, or in the activity of ψυχή in relation to its objects, but not wholly unconnected with it either – in which the very being of the "things known" (τοῖς γιγνωσκομένοις) already points beyond them.

Striving should be understood as a *pollachôs legomenon*, then. It is said in many ways, and one of those, central to Plato's doctrine of recollection in particular but also to thinking in general, is to say that all things point beyond themselves, and the soul cannot fail to notice and understand this fact because it is all too familiar with it, from within the whole experience of its own life.

Bibliography

Brann, Eva, Peter Kalkavage, and Eric Salem, eds. 1998. *Plato's Phaedo*. Newburyport, MA: Focus Press.

Burnet, John, ed. 1900–1907. *Platonis Opera*. 5 vols. Oxford: Clarendon Press.

46 *Rep.* 534b–d.

47 *Phdr.* 248b6: Τὸ ἀληθείας…πεδίον.

48 Or, alternatively, had I chosen to translate ἐννοεῖν as conceiving, contemplation could be called a purified *conceiving*, a purified "taking together" (*concipere*) of all the elements, and only those elements, comprising the intelligibility of the *noēton eidos*. I thank Prof. Mauro Bonazzi for this comment.

49 *Rep.* 506a4 and 509b5–9.

Burnyeat, Myles and Frede, Michael. 2015. *The Pseudo-Platonic Seventh Letter.* Edited by Dominic Scott. Oxford: Oxford University Press.

Cooper, John, ed. 1997. *Plato: Complete Works.* Indianapolis: Hackett.

Duke, EA., WF. Hicken, WSM. Nicoll, DB. Robinson, and JCG. Strachan, eds. 1995. *Platonis Opera: Tomus I.* Oxford: The Clarendon Press.

Dixsaut, Monique. 1997. "What is it Plato Calls Thinking?". *Proceedings of the Boston Area Colloquium in Ancient Philosophy* 13: 1–27.

Franklin, Lee. 2005. "Recollection and Philosophical Reflection in Plato's Phaedo." *Phronesis* 50 (4): 289–314.

Hackforth, R. 1972. *Plato's Phaedo.* Translated with an Introduction and Commentary. Cambridge: Cambridge University Press.

Kelsey, Sean. 2000. "Recollection in the Phaedo". *Proceedings of the Boston Area Colloquium of Ancient Philosophy* 16 (1): 91–121.

Kosman, Aryeh L. 1976. "Platonic Love." In *Facets of Plato's Philosophy,* edited by WH. Werkmeister, 53–69. Assen: Van Gorcum.

Lindenmuth, Donald C. 1988. "Love and Recollection in Plato's Phaedo." *Ancient Philosophy* 8 (1): 11–18.

Morgan, Michael L. 1984. "Sense Perception and Recollection in the Phaedo." *Phronesis* 29 (3): 237–251.

Nehamas, Alexander. 1975. "Plato on the Imperfection of the Sensible World." *American Philosophical Quarterly* 12 (2): 105–117.

Rowe, Christopher J. 1993. *Plato. Phaedo.* Cambridge: Cambridge University Press.

Sedley, David. 2006. "Form-Particular Resemblance in Plato's Phaedo." *Proceedings of the Aristotelian Society* 106 (1): 311–327.

Shorey, Paul. 1967. *What Plato Said (Abridged Edition).* Chicago: Chicago University Press.

Slings, SR., ed. 2003. *Rempublicam.* Oxford: Clarendon Press.

CHAPTER 5

Plato on the Ascent to Knowledge

Béatrice Lienemann

1 Introduction

In the last chapter of the second book of the *Posterior Analytics*, Aristotle asks two different questions: first, how do we come to know first principles? And, second, of all the epistemic states by which we grasp the truth, to which does knowledge of first principles belong? Many commentators interpret the first question as asking for "the process or method by which we gain knowledge of the first principles" (see, for example, Barnes 1993, 268).[1] In answering this question, it is supposed that the ascent to first principles of knowledge proceeds, on Aristotle's view, via induction (*epagôgê*). The state that is identified with knowing first principles is subsequently *nous*, "comprehension" (in Barnes's translation).

My aim in this text is to consider how *Plato* would answer Aristotle's first questions from the end of the *Posterior Analytics*. How does the ascent to the highest of form of cognition – cognition of the "first principle(s)" – proceed, on Plato's view? In adopting this approach, I hope to come closer to understanding whether *theôria* has a specific method according to Plato and what role the senses, sense perception and memory play in the pursuit of attaining knowledge (*epistêmê*).

I will examine Plato's account of knowledge from a specific perspective by investigating it within the framework of Aristotle's questions at the end of the *Posterior Analytics*. Unlike many scholars discussing the distinction between knowledge (*epistêmê*) and belief (*doxa*) in Plato, I will be less concerned with what we might call a "static" conception of knowledge (and belief). According to the classical approach, Plato distinguishes knowledge from belief in the middle books of the *Republic* on the basis of their distinct objects and their respective capacities (*dynameis*), namely knowing and believing. While know-

1 This interpretation has been challenged by David Bronstein who argues that the first question concerns "the original prior knowledge from which first principles become known" (2016: 227). According to Bronstein, the answer to this question is perception; perception is the kind of knowledge from which the principles first become known, although perception is not all that is required for getting to know them.

HTTPS://DOI.ORG/10.1163/9789004747128-007

ledge refers exclusively to Forms as unchangeable objects that are only accessible by thought, beliefs refer to particular and changeable objects that can only be perceived through the senses and only yield belief, but never knowledge. I call this picture "static", because it strictly separates knowledge from belief based on their two distinct and mutually exclusive classes of objects, making the transition from belief to knowledge (and *vice versa*) seem impossible, given that objects of thought and objects of perception differ categorically.[2]

My approach, by contrast, examines the possibility of a transition from belief to knowledge – and back (?) – in Platonic terms. In the Allegory of the Cave in Book 7 of the *Republic*, Plato himself seems to adopt a "dynamic" (or processual) perspective on the distinction between belief and knowledge. Rather than focusing exclusively on the description in the famous allegory, I wish to discuss other textual evidence that supports the idea of a less categorical and more flexible relationship between *doxa* and *epistêmê* – one that allows for a gradual transition from one to the other (and back). I will argue, drawing on the *Phaedo*, the *Republic* and the *Theaetetus*, that sense perception plays an important role in the process of striving for knowledge, including knowledge of Forms, for Plato.

2 Plato and Aristotle on the Knowledge of First Principles

In a paper on the *Posterior Analytics* 2.19, Peter Adamson explores the Platonic background of the last chapter of Aristotle's treatise.[3] While there is widespread agreement that the chapter contains at least two allusions to Plato, Adamson argues that one should add at least a third reference to Plato. The two uncontroversial allusions are, first, the so-called "*Meno* Paradox" regarding how learning or inquiry is possible. The second concerns Aristotle's description of our ability to reach universals based on sense perception. The third reference, proposed by Adamson, is Aristotle's appeal to *nous* as the highest form of cognition, a kind of knowledge that is superior to *epistêmê*.

2.1 *How Do We Grasp First Principles? Perception, Memory, and/or Recollection*

In this paper, I will focus on the first two of these suggested allusions. I will begin by looking at the main passage in which Aristotle describes original

2 A "two-world" view has been recently defended by Jessica Moss (2021) again.

3 Peter Adamson (2011).

prior knowledge and the successive steps by which first principles become known.[4] In my reading of the last chapter I try to stay neutral with respect to whether Aristotle's project in *APo.* 2.19 is an inquiry into original prior knowledge or into the method by which we gain knowledge of first principles. In my eyes, the chapter deals with both issues, although it starts with the question of the nature of original prior knowledge, namely perception.[5]

> And so from perception there arises memory, as we say, and from memory (when it occurs often in connection with the same thing) experience; for many memories form a single experience. And from experience, or [rather] from the entire universal that has come to rest in the soul (the one apart from the many, whatever is one and the same in all those things), [there arises] a principle of craft and scientific knowledge – of craft if it concerns what comes to be, of scientific knowledge if it concerns what is.[6]
>
> *APo.* 100a3–9; Bronstein translation in BRONSTEIN [2016]

Aristotle here identifies four stages prior to the highest form of knowing, *nous*, and he calls these preceding steps a "principle (*archê*) of craft (*technê*) and scientific knowledge (*epistêmê*)". The first stage is perception, from which, when stored in the mind, memories arise. When many memories are repeated, a single experience (*empeiria*) ensues. In the fourth stage, an entire universal (*pantos katholou*), which "has come to rest in the soul", is grasped. This entire universal is not itself a first principle; it is rather a preliminary account, a proposition of the form "All *A*s are *B*s", which is required for scientific inquiry in the form of demonstrative syllogisms.

---

4 We find approximately the same description in *Met.* A.1., 980a27–981a3: "By nature animals are born with the faculty of sensation, and from sensation memory is not produced in some of them, though it is produced in others. [...] The animals other than man live by appearances and memories, and have but little of connected experience; but the human race lives also by art and reasonings. And from memory experience is produced in men; for many memories of the same thing produces finally the capacity of a single experience. Experience seems to be very similar to science and art, but really science and art come to men through experience." (trans. WD. Ross).

5 Cf. for discussion Bronstein (2016) who criticizes, for instance, Barnes (1993).

6 Ἐκ μὲν οὖν **αἰσθήσεως** γίνεται **μνήμη**, ὥσπερ λέγομεν, ἐκ δὲ μνήμης πολλάκις τοῦ αὐτοῦ γινομένης **ἐμπειρία**· αἱ γὰρ πολλαὶ μνῆμαι τῷ ἀριθμῷ **ἐμπειρία** μία ἐστίν. ἐκ δ' ἐμπειρίας ἢ ἐκ παντὸς ἠρεμήσαντος **τοῦ καθόλου** ἐν τῇ ψυχῇ, **τοῦ ἑνὸς παρὰ τὰ πολλά**, ὃ ἂν ἐν ἅπασιν ἓν ἐνῇ ἐκείνοις τὸ αὐτό, **τέχνης ἀρχὴ καὶ ἐπιστήμης**, ἐὰν μὲν περὶ γένεσιν, **τέχνης**, ἐὰν περὶ τὸ ὄν, **ἐπιστήμης**; (For the Greek I have used the OCT of the *Analytics*, W.D. Ross and L. Minio-Paluello [1964] edition).

This passage bears a striking resemblance to Socrates' report concerning his early engagement with natural philosophy in the *Phaedo*:[7]

> Do we think with our blood, or air, or fire, or none of these, and does the brain provide our senses of hearing and seeing and smell, from which come memory and belief/opinion, and from memory and belief/opinion which has come to rest, arises understanding/knowledge.[8]
>
> *Phaedo* 96b3–9, Grube translation, in COOPER [1997]

Comparing the passages from the *Phaedo* and the *Posterior Analytics*, there are striking similarities between their descriptions of the stages prior to the highest form of knowing – *nous* for Aristotle, and *epistêmê* for Plato. In both passages, perception (*aisthêsis*) is recognized as a starting point in the acquisition of further cognition. Plato has Socrates further specify hearing, seeing and smelling as three kinds of perception. Aristotle, in turn, states just before the quoted passage that perception is an "innate discriminatory capacity" that belongs to *all* animals, i.e. that is not specific to rational animals.[9] Furthermore, memory (*mnêmê*) is not unique to humans either, as some animals also have the ability to store perceptions, which, according to Aristotle, allows us to ascribe *gnôsis* to them.[10] Significantly, Aristotle remarks that the faculty that gives us know-

7 Cf. Barnes (1993: 259): The source of *APo*. 2.19 is to be found in Plato (i.e. *Phaedo* 96b and *Phaedrus* 249b).

8 Καὶ πότερον τὸ αἷμά ἐστιν ᾧ φρονοῦμεν, ἢ ὁ ἀὴρ ἢ τὸ πῦρ; ἢ τούτων μὲν οὐδέν, ὁ δ' ἐγκέφαλός ἐστιν ὁ τὰς αἰσθήσεις παρέχων τοῦ ἀκούειν καὶ ὁρᾶν καὶ ὀσφραίνεσθαι, ἐκ τούτων δὲ γίγνοιτο μνήμη καὶ δόξα, ἐκ δὲ μνήμης καὶ δόξης λαβούσης τὸ ἠρεμεῖν, κατὰ ταῦτα γίγνεσθαι ἐπιστήμην; For the Greek I have used either the text of John Burnet, in *Platonis Opera*, OCT, five volumes (Oxford: 1900–1907), or more recent Oxford Classical Texts, as for the *Phaedo* the text of Duke et al. (1995) and for the *Republic* Slings (2003).

9 *APo*. 2.19, 99b32–100a3: "We must therefore possess some sort of capacity, but not one that will be more valuable that these states in respect of exactness. And this is clearly true of all animals: they have an innate discriminatory capacity, which is called perception. Given that the capacity to perceive is present in [them], some animals retain what they have perceived, and others do not. Those [animals] that do not [retain it] have no knowledge (*gnôsis*) when they are not perceiving (either in general or about the things they do not [retain]). But those [animals] that do [retain what they have perceived] still have [it] in their soul even after perceiving. When many such things are [retained] there is then a further difference: some animals come to have reason (*logos*) from retention of such things, and others not."

10 Note that the term "*gnôsis*" covers a wide range of cognitive states in Aristotle, from "our perceptual awareness of sensible particulars to our intellectual grasp (i.e. *nous*) of necessary and eternal first principles" (Bronstein 2016: 17). Yet, here, I take it, Aristotle uses "*gnôsis*" not in this wide sense as a term for the genus of cognition in general (with *epistêmê* [scientific knowledge] as a species), but in a more restricted sense as a term

ledge of first principles (i.e. "these states", referring to "instances of knowing more exact than demonstration", 99b27: ἀκριβεστέρας ἔχοντας γνώσεις ἀποδείξεως) is a "less valuable" (*mê timiôtera*) faculty that belongs even to the lowliest animals.

A further apparent similarity is that both passages mention memory (*mnêmê*), which arises from perception, as the second stage in the ascent to knowledge. Furthermore, both Aristotle and Plato invoke the image of something's "com[ing] to rest" (*hêremein*) in the soul. Plato mentions memory and belief coming to rest. According to Aristotle, it is the entire universal that comes to rest in the soul (*êremêsantos*). Moreover, Aristotle seems to refer implicitly to Plato when he uses the phrase "the one apart from the many" (*henos para ta polla*), which clearly hints at Plato's description of a Form as a "one over many" (*hen epi polla*).

Before evaluating the genuine similarities between Plato and Aristotle on this issue, let us examine their points of divergence. Only Plato mentions belief (*doxa*), alongside memory. By contrast, Aristotle leaves out belief but mentions experience (*empeiria*) – apparently equivalent to the "entire universal"[11] – which arises from repeated memories. Furthermore, Plato mentions as a final stage only knowledge (*epistêmê*), whereas Aristotle refers to craft *and* knowledge – *technê* and *epistêmê* – as two kinds of cognition that are based on the preceding stages. This difference – the absence of *doxa* as a lower form of cognition in Aristotle – may not be significant and could be contextual,[12] especially since Aristotle's explanation of the difference between craft and knowledge sounds thoroughly Platonic: craft has to do with "what comes to be" (*peri genesin*), whereas knowledge refers to "what is" (*to on*).[13] Thus, one might suspect that Aristotle at first omits *doxa*, only to replace it – or bring it on board – shortly afterwards with *technê*.

With this brief sketch of the similarities in hand, we can now investigate whether Plato actually subscribed to Aristotle's empiricist picture of the acquisition of knowledge. When this question is viewed through the lens of APo.

referring to a cognitive state different from and at a lower level than *epistêmê* (*epistasthai*).

11 Reading the "ἢ" at 100a6 epexegetically.

12 *Pace* Adamson (2011). Adamson points out that Aristotle might consciously eliminate *doxa* in order to prevent the suggestion that *doxa* is a necessary step on the way towards knowledge of principles as we find it, for instance, in the *Meno* (97e–98a) when Plato has Socrates distinguish knowledge from true belief by the requirement of "giving an account" which is able to tie down the true *doxa*.

13 Cf. the Allegory of the Divided Line in distinguishing between *doxa* and *noêsis* (*Republic* 7. 5333–534a) or see *Tim.* 27d–28a.

2.19, one gets the impression that Aristotle at least meant to offer an alternative to Plato in giving his specific empiricist explanation of how it is possible to gain knowledge of first principles. Adamson even goes so far as to speak of an "*ad hominem* polemic" against Plato in the passage. Can this reading be supported? In what way, if at all, does Aristotle criticize Plato here? And to what extent, if any, does his criticism weakens Plato's position?

2.2 *Assessment of the Genuine Consensus between Aristotle and Plato*

I will begin by briefly looking at the broader context of 2.19. Aristotle starts by setting out three problems (*aporiai*), the third of which asks whether the states of knowing the principles "are not in us and come to be or are in us without our being aware of them" (*APo.* 2.19, 99b25–26: […] καὶ πότερον οὐκ ἐνοῦσαι αἱ ἕξεις ἐγγίνονται ἢ ἐνοῦσαι λελήθασιν). When it comes to how we attain knowledge of first principles, Aristotle presents two options,[14] thus going beyond the previous discussion of the same problem in *APo.* 1.3.[15]

APo. 1.3 discusses two views that assume that all knowledge, including knowledge of principles, results from demonstration (i.e. all knowledge is demonstrative). The first view assumes that there must be first principles in order to avoid an explanatory infinite regress, but since these first principles are non-demonstrable by nature, they cannot be understood and nothing can be based upon them. Thus, the first view amounts to the radically sceptical conclusion that knowledge is impossible.[16] The second view rejected by Aristotle in *APo.* 1.3 accepts circular demonstration, which allows us to say that everything is understood but thereby makes first principles unnecessary. Chapter 1.3. thus shows that views that hold that all knowledge results from demonstration do not permit knowledge (*epistêmê*) of first principles.

In contrast to 1.3, Aristotle uses "knowledge" (*epistêmê*) in 2.19 in a more restricted sense and reserves the expression exclusively for demonstrative knowledge, i.e. knowledge that results from demonstration. This allows for the possibility of another kind of knowledge, besides *epistêmê*, that is non-demonstrable and relates to first principles. In other words, *APo.* 2.19 refers back to the discussion in *APo.* 1.3 (by asking for knowledge of first principles) but also goes beyond it in assuming that *there* actually *are* "states in which we know first principles" that are different from *epistêmê*.

14 *APo.* 2.19, 99b20–32.

15 *APo.* 1.3, 72b18–25.

16 I say "radically" here because I think that Plato also holds a sceptical stance on knowledge of first principles, but that his scepticism is less radical since it amounts to a kind of practical scepticism that does not completely rule out the mere possibility of knowledge of first principles, as the radical form of scepticism does.

Aristotle presents two explanations of how we attain knowledge of first principles which are still not his own but are closer to his own view. The two options are introduced as if they constitute a dilemma: either we acquire first principles from nothing – "they are not in us and come to be" – or we already had them without being aware of them. The subsequent text shows that Aristotle rejects both options, leading us to assume that the dilemma is in fact a false dichotomy. Against the first option, he maintains that the acquisition of knowledge always presupposes pre-existing knowledge. In the first chapter of the *Posterior Analytics*, which explicitly mentions the learning paradox from the *Meno*, he had already argued "in the case of demonstration", i.e. *epistêmê*, that prior knowledge is required in order to gain new and further knowledge. There, he justified this position by distinguishing between knowing something *absolutely* (*haplôs*) and knowing it *universally* (*katholou*). It is possible, for example, for me to know universally that every triangle has internal angles equal to two right angles without knowing it of *this* triangle absolutely because I do not yet know that this triangle exists.

Thus, in *APo.* 1.1, Aristotle avoids a similar dilemmatic situation by assuming a possibility that allows us both to know something and not to know it at the same time. In *APo.* 2.19, by contrast, he uses a different strategy to reject the dilemma between acquiring knowledge either from nothing or from already possessed knowledge. He now replies by saying that we are born with a power (*dynamis*), namely perception, that eventually enables us to reach knowledge of principles. Being born with a power that potentially permits the acquisition of principles represents another sense in which we both are and are not in a state of knowing first principles.

What about the other horn of the dilemma – the claim that we can acquire knowledge of first principles because we already possess such knowledge without being aware of it? Aristotle abruptly dismisses this solution by saying it "is strange (*atopon*)". It is tempting to associate the rejected solution with Plato's theory of recollection, according to which we had actual knowledge of the Forms before birth – knowledge which must be recovered by recollection in this life because it has been forgotten.[17] To be sure, Plato also rejects the claim that we *actually* have knowledge of the Forms from birth on, throughout our life. Instead, he has Socrates argue in the *Phaedo* that we had actual knowledge of the Forms only *before* birth, but have lost it and must recollect it.[18]

17 Cf. *Phaedo* 72e–73a; *Meno* 85c.

18 Cf. *Phaedo* 76a4–6 (translation Grube; in COOPER [1997]): "[...] either we were born with the knowledge of it, and all of us know it throughout life, or those who later, we say, are learning, are only recollecting, and learning would be recollection."

One might say, then, that Aristotle thinks that although both he and Plato provide an answer to how we gain knowledge of first principles, only his own answer is successful. Both answers are on a par in proposing a way of knowing principles; whereas Plato views recollection as the answer, Aristotle argues for perception as the crucial prerequisite.

I have mentioned that Adamson suggests that there may be an *ad hominem* polemic against Plato in 2.19. Why is this? As he argues, Aristotle "is deliberately exploiting a Platonic resource to answer a Platonic question, even though it was not so exploited by Plato himself".[19] Adamson is referring to Socrates' description of how he acquired knowledge, starting from perception, attaining memory and opinion, and finally reaching knowledge. Aristotle seems to be exploiting this passage for his own purposes by ignoring the fact that it is part of an "autobiography" of Socrates' in the *Phaedo*, framed in the form of a question, and therefore does not represent what Plato has Socrates argue for positively in other passages of the *Phaedo*, starting with the theory of recollection. But what exactly do Plato and Aristotle disagree on, and on what do they agree in fact? The crucial difference, of course, is that Plato assumes prenatal knowledge and the possibility of recollecting it after birth. But what about perception? Does Plato really disagree with Aristotle on whether perception plays a necessary role in the process of gaining knowledge?

Plato also ascribes an indispensable role to perception in how we arrive at knowledge of the Forms by recollection. To show this, I will discuss three examples. These cases demonstrate that sense perception does not only play the negative role of offering deceptive impressions but also takes a constructive part in motivating us to improve and in pushing the process of cognition forward.

2.3 *Intermediate Summary of the Parallels between Plato and Aristotle*

I will briefly summarize the parallels that can be drawn between Plato and Aristotle, as shown thus far. Aristotle draws a connection between the first question in *APo.* 2.19 regarding how we can gain knowledge of first principles and the *Meno* paradox mentioned in *APo.* 1.1 While the paradox from the *Meno* applies to the much more general question of how learning is possible at all, the acquisition of first principles can be considered a special case of this more general issue.

At first glance, it seems that the answers that Plato and Aristotle offer to the question of how learning is possible turn out to be very different: for Plato it is recollection, whereas for Aristotle it is by means of induction and eventually

---

19 Adamson (2011: 10).

nous. Yet this first impression is misleading. It is crucial to keep the different questions and corresponding answers apart and to bear in mind which reply Plato himself endorses and which reply Aristotle ascribes to him.

On a first level, we can distinguish between the three following questions: firstly, how learning or inquiry is possible at all, secondly, how the acquisition of demonstrative knowledge (*epistêmê*) is possible, and, thirdly, how learning or inquiry into first principles is possible, taken as a special case of learning in general.

First Aristotle. He deals with the second question in *APo* 1.1 and explains the possibility of acquiring demonstrative knowledge (*epistêmê*) by introducing the distinction between knowing something *absolutely* (*haplôs*) and knowing it *universally* (*katholou*), which allows for inquiry into what one only knows universally. This clearly does not answer the first question of how learning as such is possible, because the process of gaining demonstrative knowledge already presupposes prior knowledge. It therefore requires an already experienced knower. By contrast, the scope in 2.19 is much broader, because Aristotle includes *how* the experienced knower reaches his necessary prior knowledge and *which* state he is in when he has access to the highest knowledge, which is not accessible by demonstration. As we have seen, he describes the way to the required prior knowledge in terms of the stages of perception, memory, experience and the *katholou* (universal), where the universal seems to be a necessary presupposition for getting demonstrative inquiry started. The state that knows the highest principles is *nous*.

Now Plato. Aristotle seems to view Plato as having answered all three questions in the same way, namely by appealing to recollection. Learning as such is possible because it is a recollection of knowledge one already possessed but has forgotten, and the same explanation equally applies to demonstrative inquiry and the acquisition of first principles.

However, this description overlooks certain distinctions that Plato clearly makes. On the one hand, he also describes the way to the highest form of knowledge in terms of various successive phases, in which perception also plays a crucial initial role at the beginning. On the other hand, although it is correct that for Plato recollection plays an indispensable role in the process of attaining knowledge, he sets himself apart from the assumption that one possesses this prior knowledge in this life, since we are unable, according to Plato, to give an account of it. Recollection is the Platonic explanation of how it is possible to both have and not have knowledge of the same thing at the same time: it is formerly possessed knowledge that needs to be recovered in this life in order to become actual. I will argue that for Plato sense perception plays an important role in recollecting prior knowledge. Yet for Plato, of course, in addition to perception, *elenchus*, the "method of hypothesis" and dialectic also have essential functions on the way to knowledge that I will have to set aside here. What I would like to look at more closely, however, is the extent to which perception is a crucial and necessary prerequisite for these advanced stages on the way to knowledge in Plato.

3 Plato on the Role of Sense Perception for Attaining Knowledge

3.1 *The* Phaedo*: Sense Perception as a Necessary Trigger for Knowledge*

The first example occurs in the passage containing the recollection argument in the *Phaedo*, which is supposed to show the prenatal existence of the soul. Cebes introduces the argument by referring to the discussion in the *Meno*, giving a brief summary of its results: In the *Meno*, the argument addresses a paradox concerning the possibility of learning and inquiry (if one can learn neither what one already knows – because one already knows it – nor what one doesn't know – because one does not know what one should learn or inquire into – then how is learning possible?). It provides an answer to the paradox by saying that learning is nothing other than recollection, namely recollection of latent knowledge that one previously possessed but has forgotten. Because the possibility of acquiring such knowledge during this life is ruled out, it turns out that the soul must have existed before birth and acquired knowledge that it would later forget.

In what follows, at the request of Simmias, Socrates gives a new explanation of the doctrine of recollection. He discusses the example of knowledge of the Equal and how it is evoked by the perception of certain instances, i.e. equal sticks or stones. Here Socrates asks: "Wherefrom have we acquired the knowledge of the Equal itself?" (*Phd* 74b4). In response, Socrates points out four times that it is initially the sight of equal and unequal things that causes one to reflect on the Equal and Equality itself and to wonder whether it is the same or different from the equal things. Socrates repeats many times that in the search for knowledge, perception (here always sight) or experience (*paschomen* in 74d4 and *peponthamen* in 74e6) always takes place first, and then thinking is engaged, prompted by the puzzling information gained by perception and experience.[20]

20 *Phaedo* 74c7–e4: Socrates: "But it is definitely from equal things, though they are different from the Equal, that you have derived and grasped the knowledge of equality?" – "Very true, Socrates." [...] "As long as the sight of one thing makes you think of another, whether it be similar or dissimilar, this must of necessity be recollection?" – "Quite so". [...] Socrates: "Whenever someone, on seeing something, realizes that that which he now

Thus, the passage containing the recollection argument supports an interpretation of the connection between perception and thinking (*ennoein*) according to which perception functions as a necessary catalyst for thinking by triggering recollection, which can lead to knowledge. The analysis of the necessary link between perception and reason culminates in the following passage:

> SOC: We must then possess knowledge of the Equal before that time when we first saw the equal objects and realized that all these objects strive to be like the Equal but are deficient in this.
>
> SIM: That is so.
>
> SOC: Then surely we also agree that this conception of ours derives from seeing or touching or some other sense perception, and cannot come into our mind in any other way, for all these senses, I say, are the same.[21]
>
> *Phaedo* 74e9–75a8, Grube translation

The argument departs from what Socrates had previously said at the outset of the *Phaedo* – namely that philosophers should make themselves completely independent of their sense impressions. Here, Socrates wholeheartedly appreciates the role that the senses play in attaining knowledge. He even states that one "cannot" reach knowledge of the Equal "in any other way" than by sense perception. Thus, according to the argument, sense perception – seeing equal or unequal objects – seems to have a necessary function in acquiring knowledge of the Equal and the Unequal. Commentators often focus on the difference between perception and thought when discussing the recollection argument: perception, which is directed towards sensible objects, is not a reliable source of knowledge because sensible objects provide changing and puzzling information; by contrast, only Forms provide reliable and stable evidence. However, the argument can also be read as acknowledging the productive connection between perception and reason: without the countervailing infor-

---

sees wants to be like some other reality but falls short and cannot be like that other since it is inferior, do we agree that the one who thinks this must have prior knowledge of that to which he says it is like, but deficiently so?" – "Necessarily."

21 Ἀναγκαῖον ἄρα ἡμᾶς προειδέναι τὸ ἴσον πρὸ ἐκείνου τοῦ χρόνου ὅτε τὸ πρῶτον ἰδόντες τὰ ἴσα ἐνενοήσαμεν ὅτι ὀρέγεται μὲν πάντα ταῦτα εἶναι οἷον τὸ ἴσον, ἔχει δὲ ἐνδεεστέρως. – Ἔστι ταῦτα. – Ἀλλὰ μὴν καὶ τόδε ὁμολογοῦμεν, μὴ ἄλλοθεν αὐτὸ ἐννενοηκέναι μηδὲ δυνατὸν εἶναι ἐννοῆσαι ἀλλ' ἢ ἐκ τοῦ ἰδεῖν ἢ ἅψασθαι ἢ ἔκ τινος ἄλλης τῶν αἰσθήσεων· ταὐτὸν δὲ πάντα ταῦτα λέγω.

mation that perception supplies, reason would lack the motivation to contribute to clarification and eventually to knowledge.

Nevertheless, we must also bear in mind the limitations of the argument. First, even if perception is necessary for acquiring knowledge of the Equal, this does not mean that perception is necessary for knowledge of *every* kind of object, i.e. for every Form. Second, the argument does not show that sense perception is *sufficient* for knowledge of the Equal. Third, although his appreciation of sense perception brings Plato closer to Aristotle in that perception constitutes a first stage on the way to knowledge, it is crucial to note that that what sense perception enables is different in each account: for Aristotle, perception results in memory (*mnêmê*) and, finally, experience (*empeiria*), whereas for Plato it allows for recollection (*anamnêsis*) of the Forms that one once knew but has forgotten.[22]

3.2 Republic *7: Objects of Perception that Do and Do Not Summon the Understanding* (Noein)

My second example is in Book 7 of the *Republic*. The context of the passage is a description of the type of education that is suitable for the prospective guardians and rulers of the polis. At the outset of Book 7, the Allegory of the Cave introduces the topic of education and the lack thereof (*paideia kai apaideia*) in the form of a simile. Socrates then derives the conclusion that education is not "putting knowledge into souls that lack it, like putting sight into blind eyes" (*Rep.* 518b9–c2), but rather akin to "turning the whole soul until it is able to study that which is and the brightest thing that is, namely, the one we call the good" (*Rep.* 518c9–d2). Socrates thus proceeds to examine how this turning of the soul can be performed through education so that the good polis will be led by well-educated guardians. The aim of the conversion of the soul is

22 Plato as well as Aristotle are completely aware of this difference and both distinguish terminologically between memory and recollection. Plato distinguishes memory from recollection in that memory can take the form of a persisting memory of a previous experience, whereas recollection is characterized as a recovery of some experience or knowledge after a period of absence. Aristotle discusses the difference between memory and recollection in the treatise *On Memory* at the beginning of the second chapter: "recollection is neither the recovery nor the acquisition of memory. For when someone first learns or experiences something, he does not recover any memory, since none has preceded." (451a20–23, Sorabji translation). The idea is that recollection is recovery of some experiences that were absent before, whereas memory can directly derive from an experience without any absence. Another difference for Aristotle is that memory belongs essentially to *phantasia* and is therefore present also in non-human animals (*On Memory* 449b28–30). Recollection in contrast occurs only in humans because Aristotle characterizes it as a sort of reasoning (453a7–10).

characterized as "turning the soul from a day that is a kind of night to the true day – *the ascent to what is, which is true philosophy* [my emphasis]".[23] In other words, if we succeeded in attaining a proper education, we will also have an idea of what the ascent to knowledge consists in for Plato.

What are the appropriate educational means of turning the soul? First, the methods of physical training, music and poetry, previously discussed in Books 2 and 3, are dismissed as insufficient. The mechanical crafts are likewise expelled, presumably because they are not sufficiently measurable or systematic. What Socrates is looking for instead is "the common thing that every craft (*technai*), every type of thought (*dianoiai*), and every science (*epistêmai*) uses and that is among the first compulsory subjects for everyone".[24] The relevant discipline that is identified as being crucial for all crafts, thought and sciences is mathematics, described as "distinguishing the one, the two, and the three; in short [...] number and calculation" (*arithmos kai logismon*). Mathematics thus belongs to the subjects (*mathêmata*) "that naturally lead to understanding (*tên noêsin*), [but] no one uses it correctly, namely, as something that is really fitted in every way to draw one towards being."[25] Because Glaucon does not understand what this means, Socrates uses a comparison to explain the difference between things that do lead and things that do not lead to understanding. The distinction he introduces is that between sense perceptions which do not summon understanding and sense perceptions which encourage us to reflect further and therefore lead to understanding.[26] Let us look at the first part of the distinction, namely things for which the judgment of sense perception alone is adequate (*Rep.* 7. 523b1–2: *hôs hikonôs hypo tês aisthêseôs krinomena*):

> SOC: The ones [i.e. sense perceptions] that don't summon the understanding are all those that don't go off into opposite perceptions at the same time. But the ones that do go off in that way I call *summoners* – whenever sense perception doesn't declare one thing any more than its

23 *Rep.* 7. 521c5–8, all translations from the *Republic* are Grube's, revised by Reeve, in COOPER (1997).

24 *Rep.* 7. 522c1–3.

25 *Rep.* 7. 523a1–3.

26 *Rep.* 7. 523: Socrates: "I will point out, then, if you can grasp it, that some sense perceptions don't summon the understanding to look into them, because the judgment of sense perception is itself adequate, while others encourage it in every way to look into them, because sense perception seems to produce no sound result. – "You are obviously referring to things appearing in the distance and to *trompe l'oeil*. – "You are not quite getting my meaning. – Then what do you mean?"

> opposite, no matter whether the object striking the senses is near at hand or far away. You will understand my meaning better if I put it this way: These, we say, are three fingers – the smallest, the second, and the middle finger. [...] It's apparent that each of them is equally a finger, and it makes no difference in this regard whether the finger is seen to be in the middle or at either end, whether it is dark or pale, thick or thin, or anything else of that sort, for in all cases, an ordinary soul isn't compelled to ask the understanding what a finger is (*anankazetai tôn pollôn hê psychê tên noêsin eperesthai ti pot'esti daktylos*), since sight doesn't suggest to it that a finger is at the same time the opposite of a finger.
>
> GLAU: No, it doesn't.
>
> SOC: Therefore, it isn't likely that anything of that sort would *summon or awaken* (*paraklêtikon oud' egertikon*) the understanding.
> *republic 7*. 523b9–d10; emphasis added

Objects that, once perceived, do not compel the soul to ask what they are classified by Socrates as those that do not summon understanding (*parakalounta tên noêsin*). They do not force the soul to examine things further because they offer clear evidence of what they are. These are things that do not have an opposite and therefore do not give rise to opposite impressions at the same time. A finger, for example, has no opposite, and in seeking to gain an adequate understanding of what a finger is, it is irrelevant which finger – the middle or the little finger – we perceive. In the case of things without opposite, it is unlikely, according to Socrates, that perception "would summon or awaken understanding", and this is also unnecessary, as we grasp what a finger is solely based on our perception of it.

Socrates then goes on to introduce objects of perception that *do* summon understanding because they give rise to opposite impressions:

> SOC: But what about the bigness and smallness of fingers? Does sight perceive them adequately? Does it make no difference to it whether the finger is in the middle or at the end? And is it the same with the sense of touch, as regards the thick and the thin, the hard and the soft? And do the other senses reveal such things clearly and adequately? Doesn't each of them rather do the following: The sense set over the hard is, in the first place, of necessity also set over the soft, and it reports to the soul that the same thing is perceived by it to be both hard and soft?

> GLAU: That's right.
>
> SOC: And isn't it necessary that in such cases the soul is puzzled as to what this sense means by the hard, if it indicates that the same thing is also soft, or what it means by the light and the heavy, if it indicates that the heavy is light, or the light, heavy?
>
> GLAU: Yes, indeed, these are strange reports for the soul to receive, and they do demand to be looked into.
>
> SOC: Then it's likely that in such cases the soul, summoning calculation and understanding, first tries to determine whether each of the things announced to it is one or two. [...] This, then, is what I was trying to express before, when I said that some things summon thought, while others don't. Those that strike the relevant sense at the same time as their opposites I call summoners, those that don't do this do not awaken understanding. *republic* 7. 523e1–d4

Objects that, once being perceived, do summon understanding are characterized as things, or rather qualities, that have opposites. When we look at a finger, what we see allows us to say that it is large (in comparison to the little finger, for instance), but also that it is small (in comparison to the arm). Here, the senses give rise to contradictory impressions, and it turns out that we cannot adequately grasp what largeness or smallness is on the basis of our sense perception. The same applies to all senses equally: a piece of wood that we touch can give the impression of being hard (in comparison to grass) and of being soft (in comparison to a rock) at the same time. As long as the soul relies only on the results delivered by the senses, it cannot resolve the puzzle of contradictions in perception; it is therefore compelled to reflect on these "strange reports" in order to dissolve its contradictory findings and to determine what the real nature of largeness and smallness (and the other pairs of opposites) is. Thus, the soul gets calculation and understanding going so as to eventually gain an adequate grasp of such things, i.e. knowledge.

What is important about this passage is that here, too, the description of perception can be interpreted in different ways. In one way, the juxtaposition of the two types of objects of perception hints at the unreliability of sense perception. If we were to confine ourselves to the senses, we would be unable to dissolve the puzzling perceptions and to determine whether the finger is large or small, or, more generally, what the large and the small are. Thus, in order to gain definite knowledge, we need a reliable and invariable foundation, such as the Forms. Yet one can also turn the other side of the coin and view the issue from that perspective, for one can equally ascribe to sense perception a constructive role on the way to knowledge: if we were not confused by the ambiguous or even contradictory findings of perception, we would have no reason or motivation to engage in thinking or to examine further the nature of largeness and smallness. Rather than disregarding sense perception as a source of deception, we might view it as an enabling condition for inquiry and understanding. Without these contradictory sensory reports, we would not be prompted to awaken from our state of ignorance and to activate our thinking.

I leave aside the question of whether we should deduce from the distinction between two kinds of objects of perception an argument as to what the range of the Theory of Forms is. To be sure, the argument may suggest that Forms are only necessary for things or properties that have opposites, but one still should not, and need not, draw the conclusion that Forms are limited to the range of pairs of opposites. Rather, one might suspect, as Julia Annas does, that terms or qualities that have opposites "might only be the most *obvious* examples, and therefore the examples with which we begin".[27] But one can also ask, as I have, whether in some cases perception is an impetus to further thinking and understanding, while in other cases – e.g. in the case of the *megista gênê* such as Being, Difference, and Identity – it is rather the thinking itself that raises puzzles and is thereby a catalyst for further investigation.

3.3 *The* Theaetetus*: The Interdependence of Perception and Reasoning*

The third and last example I would like to consider is a text where precisely those qualities that are not specific to a single sense (or to no sense at all) are at stake (as for example beauty and ugliness or identity and difference). This occurs in the *Theaetetus*, at the end of a long discussion on the first definition of knowledge, according to which "knowledge is perception". The passage I am interested in starts at 184b4.[28] In a first step, Socrates distinguishes between what we perceive *by means of* the senses (expressed by the instrumental dative, e.g. "ᾧ"), and what we perceive *with* the soul (expressed by the preposition "διά", in the genitive). Initially, it is suggested that each sense has its own proper object, which is inaccessible to all other senses: colours for sight, sounds for hearing. The different sense perceptions thus seem to correspond to different senses and sense organs. Nevertheless, it would be misleading to say that it is our *eyes* that see or our *ears* that hear, because this would suggest

27 Julia Annas (1988: 222).

28 *Tht.* 184b4–185c3.

that there are several separate senses in us – just as there are many warriors in the Trojan (Wooden) Horse – each performing its own perceiving without coordinating with the others. To prevent this, Socrates proposes that we understand the soul as the unitary centre in which the separate sense perceptions converge. Thus, the phrase "by means of which" depicts the function of the senses as instruments through which we perceive uniformly "with" the soul.

In the next step, Socrates considers features to which no sense or sense organ gives us access, namely the so-called "commons", or common features (*koina*), such as being and not-being, unity, likeness and unlikeness, oddness and evenness, identity and difference, as well as beautiful and ugly, good and bad. He asks Theaetetus to explain how these common features are revealed to us if not by any "special instrument" (*organon idion*), i.e. by any sense:

> SOC: Good. Now through what does the power (*dynamis*) function which reveals to you what is common (*koinon*) in the case both of all things and of these two – I mean that which you express by the words 'is' and 'is not' and the other things mentioned in our questions about them just now? What kind of instruments (*organa*) will you assign for all these? Through what does that which is percipient in us perceive (*aisthanetai*) all of them? – Theaet.: You mean being and not-being, likeness and unlikeness, same and different; also one, and any other number applied to them. And obviously too your question is about odd and even, and all that is involved with these attributes; and you want to know through what bodily instruments we perceive all these with the soul. – Soc.: You follow me exceedingly well, Theaetetus. – Theaet.: But I couldn't possibly say. All I can tell you is that it doesn't seem to me that for these things there is any special instrument at all (*organon idion*), as there is for the others. It seems to me that in investigating the common features (*ta koina*) of everything the soul functions through itself (*autê di' hautês hê psychê*). – Soc.: […] *while the soul considers some things through the bodily powers, there are others which it considers alone and through itself.* This was what I thought myself, but I wanted you to think it too.
>
> *Theaetetus* 185c4–e9, Lewett translation in COOPER [1997]

Socrates gradually gets Theaetetus to agree that some qualities (such as colours and sounds) are such that the bodily instruments, i.e. the sense organs, convey them to us. The common qualities, by contrast, are such that the soul reveals them alone and through itself, since there is no bodily power through which the soul could consider them. Socrates goes on to contrast cases in which perception and experience suffice to reach a decision about the qualities of an entity and cases where they are unable to settle the matter:

> SOC: […] Wouldn't you say that it is through touch that the soul perceives the hardness of what is hard, and similarly the softness of what is soft? – Theaet.: Yes.
>
> SOC: But as regards their being – the fact that they are – their opposition to one another, and the being, again, of this opposition, the matter is different. *Here the soul itself attempts to reach a decision for us by rising to compare them with one another.* – Theaet.: Yes, undoubtedly. – Soc.: And thus there are some things which all creatures, men and animals alike, are naturally able to perceive as soon as they are born; I mean, *the experiences which reach the soul through the body* (*hosa dia tou sômatos pathêmata epi tên psychên tenei*). But *calculations* (*analogismata*) *regarding their being and their advantageousness come*, when they do, only as the result of a long and arduous development, involving a good deal of trouble and education. – Theaet.: Yes, that certainly is so.
>
> *Tht.* 186b2–c5

When the bodily powers of the sense organs perceive their own proper objects, such as colours, sounds or hardness and softness, the soul does not need to investigate further. Where this is not the case, however, the soul has to reflect and to "attempt to reach a decision […] by rising to compare" the various common features with one another. This requires "calculation", a power that one achieves as the "result of a long and arduous development, involving a good deal of trouble and education". The argumentation ends with the conclusion that perception cannot be identical to knowledge, because perception and experience are unable to reveal the truths of all things. There are at least some things, such as being, of which we can attain knowledge only by reasoning (*syllogimo*[*i*]).

> SOC: Now is it possible for someone who does not even get at being to get at truth?– Theaet.: No; it's impossible. – Soc.: And if a man fails to get at the truth of a thing, will he ever be a person who knows that thing? – Theaetetus: I don't see how, Socrates. – Soc.: Then knowledge is to be found not in the experiences (*pathêmasin*) but in the process of reasoning (*syllogismo*[*i*]) about them; it is here, seemingly, not in the experiences, that it is possible to grasp being (*ousia*) and truth (*alêtheia*).
>
> *Tht.* 186c7–d5

This passage from the *Theaetetus* adds a further element to Plato's conception of the ascent to knowledge, namely the idea of a kind of interdependence and cooperation between perception and the rational power of the soul. Although the soul as a cognitive unifying centre can trust the results of the senses for some perceptions, when it comes to other experiences it will be challenged to reason by itself, to make comparisons, and to reach a decision.

4 Conclusion

In this paper, I have dealt with one of the stages of the ascent to knowledge in Plato. I have begun the discussion with Aristotle's counter-proposal to the question of how to grasp the first principles in *APo.* 2.19. It turned out that, despite the mainly critical allusions to Plato, there seemed to be more similarities between the Aristotelian position of the grasp of first principles and the Platonic position than one might expect. Contrary to what Aristotle might suggest, Plato does not see in recollection a universal remedy that will explain the possibility for every kind of inquiry and the possibility of acquiring every type of knowledge. In contrast, I have argued that sense perception in particular also plays an important role in the acquisition of knowledge for Plato. Drawing on three pertinent examples, I have tried to make it clear that sense perception can play a crucial role in the acquisition of some sorts of knowledge. First, the countervailing information that perception supplies can motivate to further reflect and to press thought to attain clarification. Second, looking at the finger passage from the *Republic*, one might even speculate whether Plato has Socrates accept cases where sense perception alone is sufficient to reach understanding because it is not exposed to contradictory findings. Third, it is worth considering whether sense perception might be sufficient for other forms of knowledge, for instance knowledge of scientific facts in the material world (which might be called "scientific knowledge"), which is different from knowledge of the corresponding Form, but nevertheless more than a true opinion due to its stability and reliability. My aim in this text was to trigger such further discussions on the significance and function of sense perception in Plato' epistemology by bringing his position closer to Aristotle's by focusing on empiricist elements in Plato.

Bibliography

Adamson, Peter. 2011. "Posterior Analytics 2.19: A Dialogue with Plato?". *Bulletin of the Institute of Classical Studies* 54: 1–19.

Annas, Julia. 1988. *An Introduction to Plato's Republic*. Oxford: Clarendon Press.

Barnes, Jonathan. 1993. *Aristotle. Posterior Analytics*. Translated with a Commentary. 2nd ed. Oxford: Oxford University Press.

Bronstein, David. 2016. *Aristotle on Knowledge and Learning. The Posterior Analytics*. Oxford: Oxford University Press.

Burnet, John, ed. 1903. *Platonis Opera, Tetralogiae V–VII*. Oxford: Oxford University Press.

Cooper, John, ed. 1997. *Plato. Complete Works*. Indianapolis: Hackett Publishing.

Duke, EA., WF. Hicken, WSM. Nicoll, DB. Robinson, and JCG. Strachan, eds. 1995. *Platonis Opera, Tetralogiae I–II*. Oxford: Oxford University Press.

Moss, Jessica. 2021. *Plato's Epistemology. Being and Seeming*. New York: Oxford University Press.

Ross, WD. and L. Minio-Paluello, eds. 1964. *Aristotelis. Analytica Priora and Posteriora*. Oxford: Clarendon Press.

Slings, SR. 2003. *Platonis Rempublicam*. Oxford: Clarendon.

Sorabji, Richard 2006: *Aristotle. On Memory*. Chicago: University of Chicago Press.

CHAPTER 6

Plato and Aristotle on Knowledge

Hikmet Unlu

1 Introduction[1]

Although Aristotle disagrees with Plato on several important issues, their philosophies are more similar than is believed. This chapter will focus on their theories of knowledge. I will first provide an analysis of passages that we encounter at the end of Aristotle's *Posterior Analytics*, where Aristotle discusses the acquisition of first principles. I will then turn to Plato and show that in several dialogues Plato adopts a similar framework. Next, I will examine what is often referred to as Plato's theory of recollection, comparing it with Aristotle's statement that the soul is in a way all things. Lastly, I will say a few words about the reception of Aristotle's texts in the early Middle Ages, which could partly explain why commentators in that era and ever since have overstated the extent to which the two philosophers disagree with one another.

2 From Perception to Thinking

In *Posterior Analytics* II.19, Aristotle briefly describes the journey undertaken by the human soul, the goal of which is the acquisition of first principles, i.e., of essences. According to Aristotle, humans are animals, and like all animals we partake in the activity of perception (αἴσθησις). Then comes memory (μνήμη), which allows us to store what comes from perception; memory (which according to Aristotle belongs under the imagination) enables the retention of what has come from the sensible object even in the absence of this object. Next, when sufficiently many instances of the perception of some kind of thing have been stored in memory, the soul groups them into one collective unit, which Aristotle calls experience (ἐμπειρία). None of these faculties are unique to human beings. All animals – human or non-human – can perceive, but *some*

1 Throughout the chapter, translations of Plato are from Cooper (ed.), Plato: Complete Works, translations of Aristotle from Barnes (ed.), The Complete Works of Aristotle: The Revised Oxford Translation.

HTTPS://DOI.ORG/10.1163/9789004747128-008

non-human animals *also* have memory and experience. In a word, neither the faculty of perception nor that of memory nor that of experience is exclusive to the human animal, which is not surprising: perception, memory, and experience are all directed toward *sensible particulars*,[2] which in turn do not lie outside the reach of non-human animals. What lies outside their reach are *essences*, so even the most complex non-human animals must content themselves with experience; none of them can attain the kind of knowledge that reveals the very essences of things.[3]

Also important in this context is the fact that in both perception and thinking, how the soul relates to its object is by some communion, wherein the object of the soul – be it a sensible or an intelligible object – serves as the agent, while the soul serves as the patient. Let us first discuss perception, as it is the easier of the two cases. Insofar as perception is concerned, we have on the one hand a sensible object, or more precisely a sensible form that inheres in that object, which serves as the agent. What we can ascribe to the agent (i.e., the sensible form) is a capacity for *acting* (δύναμις τοῦ ποιεῖν), in which case a sensible form is at bottom a power to produce some effect. Conversely, the animal soul is first and foremost a capacity for suffering (δύναμις τοῦ πάσχειν); it is a power to *undergo* the effect produced by the agent. Let us make this more concrete. Say that a green plant lies before me. This "green," which is a quality, inheres in this object as a capacity for affecting, but the animal soul (or the animalistic part of the human soul) is a capacity for being affected by such forms. In a way, colors are already in the mind, albeit potentially. The presence of the sensible object before me serves to activate this potentiality by forging a communion between the agent and the patient powers, which is what happens when I'm seeing green.

To repeat, once our memory has stored sufficiently many instances of some kind of sensible object, the soul can group them into one collective unit, as a result of which we may be said to have experience. From experience, the human mind, and only the human mind, is able to grasp the essence of the thing in question. Because knowledge is of many kinds (practical, technical, and theoretical),[4] what is grasped in each case is a different kind of essence. Unfortunately, in *Posterior Analytics* II.19 Aristotle does not provide the details of this transition, i.e., that from experience to the acquisition of essences. Ideally, one would hope to be presented with the details of this transition in

---

2 See also Ferrarin (2006: 105).

3 For an excellent discussion of Aristotle's account of the intellection of essences, see Berti (1978).

4 See *Metaphysics* VI.1 for an extended discussion of the three kinds of knowledge.

the third book of the *De Anima*, partly in the chapter on the imagination and partly in the chapters on thinking, but regrettably the Greek text of the passages where Aristotle talks about imagination is corrupt, and Aristotle's rather cryptic remarks on thinking, which have been the source of much controversy, hardly help to clear the air. What we know for sure is that thinking is impossible without the imagination (*De Anima* III.7, 431a17), in which case the imagination – itself anchored in perception – must somehow mediate between perception and thinking. How it mediates, we don't know.

On this point, Husserl has an interesting story to tell, which is pertinent here not because it is a reading of the Aristotelian text but because, at least in broad strokes, it does not contradict anything Aristotle says. According to Husserl, the imagination allows the free variation (*freie Variation*) of an *eidos*, which in turn enables one to distinguish between what is and what is not essential to the thing in question.[5] For example, one may imagine a house to be this or that color or to be this or that tall, all of which, therefore, must be incidental to being a house. However, one may not imagine a house without walls and a roof, in which case these latter are fundamentally unlike the accidental properties of houses. According to Husserl, then, the acquisition of essences is enabled by what he calls "free variation," which is one kind of imagination. Of course, Aristotle does not say this much, but this is a possible story to tell about how one can proceed from perception to thinking. Nevertheless, how the imagination helps reveal an essence, as a result of which we can *think* the essence, cannot be settled on the basis of the extant texts of Aristotle.

Speculation aside, what we know is that, according to Aristotle, in thinking too there is an agent capacity, which is a power to affect, as well as a patient capacity, which is a power to be affected. Furthermore, here too we can talk about a communion of the agent and the patient powers, and only in the case of such communion does the intellect not only potentially but actually think. This means that perception and thinking are similar on many levels. If we may be allowed briefly to adopt an anachronistic terminology, we may say that for Aristotle objectivity is an agent capacity (power to affect) and subjectivity a

5 As Husserl (1973: 341) puts it, the essence is that "without which the object cannot be intuitively imagined as such." Sokolowski (2000: 179) explains free imaginative variation in the following terms: "We let our imagination run free, and we see what elements we could remove from the thing before it 'shatters' or 'explodes' as the kind of thing that it is. We try to push the boundaries, to expand the envelope of the thing in question. If we can discard some features and still preserve the object, we know that those features do not belong to the eidos of the thing. However, if we run into features that we cannot remove without destroying the thing, we realize that these features are eidetically necessary to it."

patient capacity (power to be affected), both in the case of perception and in that of thinking; however, the *activity* of perception or of thinking is a *communion* of objectivity and subjectivity. To put this in Aristotle's own terms, in perception there is a sensible object (αἰσθητόν), which affects the sensitive soul, just as in thinking there is an intelligible object (νοητόν), which affects the intellective soul. "The soul is *in a way* all existing things" (ἡ ψυχὴ τὰ ὄντα πώς ἐστι πάντα, 431b21; emphasis mine), Aristotle says, because the soul is *potentially* these sensible and intelligible objects, even when these objects aren't *actually* cognized. In a way, therefore, the colors and the sounds and so on are already in the soul, albeit potentially. Likewise, in a way the essences are already in the intellect, albeit potentially (more on this later).

In light of the insights that we have gained from our discussion so far, let us now turn to Plato. This should provide an extra layer of difficulty because Plato seldom provides a straightforward account of his true views; what we have, instead, are dialogues wherein Socrates and his interlocutors argue over some point, but more often than not such debates yield only inconclusive results. Importantly, there is hardly any argument that has been advanced against Plato that Plato did not advance against himself.[6] With these caveats, I would like to suggest nevertheless that the basic picture of the mind we encounter in Plato and Aristotle is not unalike. To begin, the *Theaetetus* distinguishes between two kinds of capacities, a capacity for doing and a capacity for suffering (156a), and mentions an account of perception in which the perceived object serves as an agent capacity, whereas the soul serves as the patient capacity – along the lines of the Aristotelian doctrine that perception is a kind of suffering (τὸ γὰρ αἰσθάνεσθαι πάσχειν τι ἐστίν, 424a1). In the *Sophist*, to which dialogue the *Theaetetus* serves as a prequel, the same conceptual framework is discussed in the famous *gigantomachia* passage, not in the context of perception but this time in the context of knowing, that is to say, in the context of thinking. Here (247e) the Giants first define being as "the capacity to act" (δύναμις εἰς τὸ ποιεῖν) or "the capacity to be acted on" (δύναμις εἰς τὸ παθεῖν), but in the remainder of the dialogue, being is described as the capacity for "communion" (κοινανία) between the patient and the agent, namely, the soul and its object.

It is not insignificant that in the context of the *gigantomachia* passage Plato takes pains to raise the question of *what* serves as the agent and *what* serves as the patient (248d). Several answers are suggested in the ensuing text; however,

6 A well-known example is the "Third Man Argument," which Aristotle advances to criticize Plato's conception of Forms, even though the argument itself dates back to Plato's *Parmenides*.

all but one are refuted. Stephen Menn maintains that the only alternative that remains unrefuted in the text is the one according to which the knower is the patient and what is known is the agent: "the stranger does not try to refute the other option, that [...] knowing is a *paschein* and being known is a *poiein*," so "Plato's implicit conclusion is that the *noêta* cannot undergo *paschein* but can exercise *poiein* and that the soul, in knowing, *paschei* under the influence of the intelligible objects."[7] Lesley Brown argues likewise that this alternative "is left untouched by" the refutations that follow,[8] which in turn permits us "to think that in this case [Plato] could expect the reader to notice a similarity between perceiving and knowing."[9] If this interpretation is on the right track, how the soul relates to its object in Plato is (at least in this respect) similar to how the soul relates to its object in the Aristotelian framework, in which the soul serves as the patient, whereas the object of the soul, regardless of whether it is a sensible or an intelligible object, serves as the agent.

Despite these apparent similarities, one can also raise the question concerning the ways in which the Platonic and the Aristotelian conception of the soul differ from one another. And especially important for our present discussion is the question of whether Plato provides a "static" and Aristotle a "dynamic" account of the intuition of essences. According to Béatrice Lienemann (in this volume), the answer is no, and I agree. Let us raise a simple question: is Aristotle, unlike Plato, an empiricist? More precisely, can we say that it is only in the works of Aristotle (i.e., unlike what we encounter in Plato's dialogues) that the journey undertaken by the human soul commences with perception? Once again, I believe that the answer is no. *If* Aristotle is an empiricist – and I stress the "if" because calling Aristotle an empiricist is an oversimplification on several levels – there is sufficient textual evidence to conclude that Plato too is an empiricist. Let me explain.

We find three forms of imagery in what may be called the middle books of the *Republic* – namely, the analogy of the sun, the analogy of the divided line, and the most famous of them all, the allegory of the cave. For the sake of brevity, I shall leave aside the analogy of the sun and only discuss the analogy of the divided line and the allegory of the cave. The analogy of the divided line is meant to tell us something both about the types of apprehension as well as the types of objects that are apprehended. In the sensible realm (i.e., in the case of perception), there are two types of sensible objects, and corresponding

7 S. Menn (1995: 55).

8 L. Brown (2001: 196).

9 Brown (2001: 200). See also Larsen (2014) for a detailed discussion of these issues (see especially 313ff.).

to these are two types of apprehension; likewise, in the intelligible realm (i.e., in the case of thinking), there are two types of intelligible objects, and corresponding to these are, once again, two types of apprehension.

As we ascend higher up in the divided line, objects have more reality to them, and the modes of apprehension are more truthful. Let us begin, first, with the objects that have the least amount of reality – namely, shadows and reflections – corresponding to which is the mode of apprehension to which we can ascribe the least amount of truth (namely, mere belief). Next, we proceed to the second section of the sensible realm, which contains what most people would characterize as real things: tables, chairs, plants, animals, and so on. Unlike was the case with shadows and reflections, the type of apprehension at work here is not simply belief but true belief. As we move from the sensible realm up to the intelligible realm (i.e., from perception to thinking), we are dealing with intelligible objects, and as has been mentioned, this realm is also subdivided into two, and once again the lower half does not have as much truth and reality as the upper half. The type of apprehension that we have in the lower half of the intelligible realm is "mediated thinking" (διάνοια). Ostensibly, the objects apprehended by means of mediated thinking are mathematical objects, but arguably mathematics is only meant to serve as an example here because all thinking is mediated (that is to say, "discursive"), with the sole exception that we will now discuss. This last kind of thinking is thinking simpliciter (νόησις), and it is the kind of thinking operative in the intuition of essences, which is, again, what Husserl would refer to as *Wesensschau*. This very seeing of essences is *built upon* the foundation of mediated thinking, but it is *itself* immediate. It is the kind of thinking that Aristotle says is not synthetic but simple (ἁπλῶς).[10]

After the analogy of the divided line, Plato introduces his allegory of the cave. This allegory commences with the upwards motion of the unchained prisoner and continues with the downwards motion of the same person, which stands for education. The ascent in question here is from the realm of perception to the realm of thinking, which is, therefore, in a very important respect similar to the kind of ascent that we encounter in the *Symposium*: from "a" beautiful body and then "all" beautiful bodies to the Beautiful itself. But lest we read too much into the ascent that takes place in the *Republic* or the *Symposium*, let us examine some of Plato's other works to see whether perception really serves as the necessary first step for intellection. In the *Phaedo*, Plato writes the following:

---

10 See also Berti (2005).

> Do we think with our blood, or air, or fire, or none of these, and does the brain provide our senses of hearing and sight and smell, *from* which come memory and opinion, and *from* memory and opinion which has become stable, comes knowledge?
>
> 96b

Hardly anyone can deny the terminological similarities between Socrates' remarks and the passages that we encounter at *Posterior Analytics* II.19. In the above quoted passage Socrates says not only that memory and opinion come *from* sensation but also that knowledge comes "*from* memory and opinion" (ἐκ δὲ μνήμης καὶ δόξης). Likewise, in an earlier passage in the *Phaedo*, Socrates raises the following questions:

> Whence have we acquired the knowledge of it [i.e., the knowledge of the Equal itself]? Is it not from the things we mentioned just now [i.e., from sensible particulars], from *seeing* sticks or stones or some other things that are equal we come to think of that other which is different from them?
>
> 74b; emphasis mine

What we can gather from this, therefore, is that we proceed *from* the perception of equal things *to* the equal itself; it is not as if we can forgo perception altogether and begin straightaway with the intellection of Forms. A few lines later Socrates reiterates the same idea, once again in the form of a rhetorical question: "But it is definitely from the equal things, though they are different from that Equal, that you have derived and grasped the knowledge of equality?" (74c). To put it differently, particular equal things are different from the Equal itself, but the grasp of the latter is *derived from* our perception of the former. Socrates then concludes the following: "Then surely we also agree that this conception of ours derives from seeing or touching or some other sense perception, and cannot come into our mind in any other way" (75a). Hence, the grasp of the Equal *cannot* come from anywhere else than from the senses – "from seeing or from touching or from the other senses" (ἐκ τοῦ ἰδεῖν ἢ ἅψασθαι ἢ ἔκ τινος ἄλλης τῶν αἰσθήσεων). In a word, the intellection of the Form of the Equal must proceed from the perception of equal things.

Granted, one can find many passages from Plato's dialogues where it is stated unambiguously that mere perception does not yield knowledge. Note, however, that such passages in no way contradict the point I am trying to make. Perception is not knowledge but is one of its enabling conditions. Perception serves as the starting point from which the soul may learn to approach beings *with a different kind of perspective*. Whereas other animals can intend objects ontically, human animals can also intend objects ontologically; we intend not only equal *things* but also the *equality* of equal things. We apprehend not only just actions but also that which renders just actions just in the first place. We not only *perceive* houses but also *think* what it means to be a house. And there is nothing here that Aristotle would reject. On the one hand, thinking operates with a kind of freedom from the contingencies of sensuous data; on the other hand, it is precisely these contingencies that enable our grasp of essences or Forms. Plato's dialogues are attempts to appropriate these contingencies so as to transcend them; they exemplify the free variation of an *eidos*, whereby the hidden *eidos* is gradually made visible.[11]

3 The "Theory" of Recollection

The opening word of a Platonic dialogue is not insignificant; it often brings into relief the principal topic to be addressed. The opening word of the *Cratylus* is βούλει, which when posed as a question would mean "Do you want?" or "Do you desire?", and the dialogue examines the question of whether names are natural or conventional, the question of whether names are bound by the (arbitrary) wants and desires of the linguistic community. The *Phaedo* begins with the word αὐτός, meaning "itself," and it is in this dialogue that Plato first thematizes the Forms. Likewise, the *Meno* begins with the question ἔχεις; – normally meaning "Do you *have* it?" or, to emphasize it even more, "Do you *already* have it?" – which of course pertains to what comes later: the so-called Meno's paradox and the "theory" of recollection.

In the *Meno*, Socrates suggests that, on the face of it, inquiry appears to be either unnecessary or impossible, for if one already knows what he is looking for, inquiry is unnecessary, but if one doesn't know what he is looking for, inquiry is impossible: "He cannot search for what he knows – since he knows it, there is no need to search – nor for what he does not know, for he does not know what to look for" (80e). Lest inquiry be either unnecessary or impossible, Plato relates a story that purportedly offers a fresh perspective:

> As the soul is immortal, has been born often, and has seen all things here and in the underworld, there is nothing which it has not learned; so it is in no way surprising that it can recollect the things it knew before [...]. As the whole of nature is akin, and the soul has learned everything,

11 Cf. Hopkins (2002).

> nothing prevents a man, after recalling one thing only – a process men call learning – discovering everything else for himself [...], for searching and learning are, as a whole, recollection.
>
> 81c–d

Let me propose that this is an untenable resolution of the original problem and that Plato is well aware of its untenability. If learning is recollecting, each and every learning necessarily harks back to a prior time when what is now being recollected was at that time fresh and contemporary. But if I remember now what I learned in a prior life, how did I learn this thing in the first place, namely, in that prior life? If all learning is recollecting, *that* learning, too, is only a recollecting of something that had been learned even earlier. But how did I learn it then? Then, too, I must be remembering what I learned even earlier. This will go on *ad infinitum*; it leads nowhere. For any such framework to make sense, there must be at some point an original instance of learning that is *not* a recollecting, in which case not all learning is recollecting.

The problem here is that a literal reading of Plato's metaphor ends up causing us to miss his basic point. The basic point of the passage is that *in a way* the soul already contains the Forms. To put it more precisely, in a way, the soul already has the Forms; in a way, it does not already have the Forms. In what way, Plato does not say, but for a possible interpretation let us turn to Aristotle. In *De Anima* III.8, Aristotle writes the following:

> Let us now summarize our results about soul, and repeat that the soul is *in a way* all existing things; for existing things are either sensible or thinkable, and knowledge is *in a way* what is knowable, and sensation is *in a way* what is sensible: *in what way we must inquire.*
>
> 431b20–24; emphasis mine

What Aristotle says immediately after this is that there is a sense in which the sensible and the thinkable are already in the soul, but there is another sense in which they are not already in the soul; they are already in the soul as potentialities but not as actualities:

> Knowledge and sensation are divided to correspond with the realities, potential knowledge and sensation answering to potentialities, actual knowledge and sensation to actualities. Within the soul the faculties of knowledge and sensation are [already] *potentially these objects* [...]
>
> 431b24–27; emphasis mine

In an earlier passage from the *De Anima*, Aristotle had already elaborated on how the distinction between potentiality and actuality plays out in perception and thinking:

> If thinking is like perceiving, it must be either a process in which the soul is acted upon by what is capable of being thought, or a process different from but analogous to that. The thinking part of the soul must therefore be, while impassible, capable of receiving the form of an object; that is, must be potentially identical in character with its object without being the object.
>
> 429a13–16

These remarks provide a possible answer to the question posed (or at least implied) by the opening word of the *Meno*: do you already have the Forms in the soul? In a way, you already have the Forms in the soul, but in a way you don't. The Forms are already in the soul potentially, but only after the soul undergoes a *dialectical* journey does it have the Forms actually. As Socrates puts it in the analogy of the divided line, "Then also understand that, by the other subsection of the intelligible, I mean that which reason itself grasps by the power of dialectic" (511b). But even before the soul undergoes this dialectical journey culminating in the intellection of Forms, the thinking part of the soul, as Aristotle would say, must be potentially its object. Intellection is made possible by the presence of both the intellective part of the soul, which serves as a capacity to be affected, and the intelligible object, which serves as a capacity to affect the intellective part of the soul. To adopt the Aristotelian terminology, the intelligible objects are the essences of things (e.g., the *equality* of equal things, the *justness* of just actions, and so on), which are already in the mind, albeit potentially. The dialectical journey Plato talks about is an attempt to *turn the soul* so as to move beyond the contingencies of sensuous data, which are in constant flux, to the Form of the thing in question, which does not undergo change.[12] One moral of the discussion carried out in the *Theaetetus* is that perception yields not knowledge but belief, and regardless of what we add to belief (make it a *true* belief or even a *justified* true belief), we are not

12 In his commentary on the *Meno*, Jacob Klein writes the following: "Our ability to recollect [...] depends on our power to 'look into ourselves'. We reminisce best when we close our eyes. The myth [of recollection] emphasizes this aspect of our recollecting and learning power: the accumulated 'knowledge' is 'inside', is 'within' the soul" (Klein 1965: 96). If I could rephrase Klein's words, I would say that because knowledge derives from memory (ἐκ δὲ μνήμης), and because memory is within the soul, we have to look inward to grasp ("recollect") the Forms.

going to end up with knowledge of any kind. In a word, there is no knowledge of sensibles; one can only have knowledge of intelligibles. These intelligibles are "recollected" not from our acquaintance with them in a prior life but from a kind of soul that can never be conceived as a *tabula rasa*.

4 A Fabricated Rivalry

Thus far, we have not focused on the differences between the philosophies of Plato and Aristotle, and there are many. Aristotle devotes the last two books of his *Metaphysics* to offer a critique of Plato and his successors, Speusippus and Xenocrates.[13] Perhaps most importantly, Aristotle argues that Plato mistakenly ascribes ontological separation to the Forms, which he believes is the root cause of many of the problems that (allegedly) plague the kind of metaphysics endorsed in the Academy.[14] Plato's Forms are for Aristotle essences that inhere in things; these essences are separate not simply but only in thought. To put it more precisely, Aristotle *also* believes that there are immaterial forms that are separate simply, but the specific Forms that the Academy philosophers talk about are for the most part Forms that can only exist in some matter. What is perhaps curious here is that, in the *Metaphysics*, Aristotle too argues that the form of a material thing is ontologically prior to the matter-form compound, so his disagreement with the Academy philosophers is in fact much more nuanced than it is generally taken to be.

None of this really matters, however, because the philosophers of the Middle Ages have already popularized a crude interpretation of the difference between the philosophies of Plato and Aristotle – an interpretation that is, to put it mildly, loosely based on the actual writings of these philosophers but has seldom been questioned until the 19th century. However, once the damage is done, it has proved itself to be largely irreparable, so even after the 19th century the medieval interpretation of the difference between the two philosophers has continued to hold sway, at least to a significant extent. This is not to say that the medieval philosophers interpreted the Greeks in the same way; quite the contrary, this period is first and foremost characterized by interpretive controversies between commentators who vehemently disagreed with each other on countless issues. But the basic framework was taken for granted, and it is a framework according to which Aristotle ascribes ontological priority to particulars, whereas Plato ascribes ontological priority to

13 For a detailed discussion of Aristotle's critique, see Annas 1988.
14 See also Fine (1984); cf. Unlu (2020).

the Forms; it is a framework according to which Aristotle cherishes the senses, whereas Plato wants to forgo what comes from the senses in favor of the apprehension of the Forms. The popular image of these philosophers is already clear in the *School of Athens*, where Plato, unlike Aristotle, points above to some heavenly realm, to the realm of Forms.

Before proceeding further, let us discuss what Aristotle says in the *Categories*, which for many centuries has been more influential than any other of his works. Aristotle says here that (i) some things are said of a subject but are not in a subject, (ii) some things are in a subject but are not said of a subject, (iii) some things are both in a subject and are said of a subject, and lastly (iv) some things are neither in a subject nor are said of one. To be "in" a subject is to be an accident, whereas not to be "in" a subject is to be a substance. To be "said of" a subject is to be a universal, whereas not to be "said of" a subject is to be a particular. According to this fourfold classification of beings, therefore, there are (i) universal substances, (ii) particular accidents, (iii) universal accidents, and (iv) particular substances. Aristotle adds to this that if it were not for particular substances, nothing else would exist, meaning that particular substances are the ontologically prior beings. Thus, if it were not for this tree and that horse, nothing else would exist.

What happens, then, to the "treeness" of the tree or the "horseness" of the horse? These are now interpreted as universal substances, which are not prime substances but only *secondary* substances (δεύτεραι οὐσίαι), meaning that the ontological status of these (secondary) substances is demoted in the Aristotelian account of being, in contradistinction to what we encounter in Plato's dialogues, where the Forms (purportedly the counterpart to the secondary substances of the *Categories*) enjoy ontological primacy over everything else. Thus, Plato and Aristotle endorse fundamentally different ontologies: Aristotle prioritizes individual substances such as this tree and that horse, whereas Plato prioritizes the Form of the Tree and the Form of the Horse.

The problem is, Aristotelian ontology is characterized first and foremost by its tendency to interpret beings in terms of matter and form as well as in terms of potentiality and actuality, but none of these concepts are to be found in the *Categories*. In the *Metaphysics*, we are told clearly that the matter-form compound (e.g., a particular horse) cannot be a primary being; what is prior, rather, is the form. *For us*, then, there is a tension between what Aristotle says in the *Categories* and what he says in the *Metaphysics*. Furthermore, because the *Categories* is a very short work where these issues are discussed only in passing, in contradistinction to the *Metaphysics*, where the same issues are discussed in excruciating detail, *we moderns* think of the *Categories* as a curious work, i.e., either a spurious work falsely attributed to Aristotle or – this is the more popular view in our age – an early work whose relation to Aristotle's mature thought remains unclear.[15] In contrast, in the early Middle Ages it was the *Categories* that was believed to contain the basic tenets of Aristotelian ontology.

What, then, did commentators in the early Middle Ages think about the *Metaphysics*? Well, they didn't know anything about it; they didn't have the text. Nor were these commentators conversant with the *Physics* or the *De Anima*. They learned their Aristotle, rather, from the *Organon* and, strange though it may seem, from Plotinus' *Enneads*. Especially in the Islamic world, the fourth, fifth, and sixth Enneads were believed to belong to Aristotle, which "work" was known to them as *The Theology of Aristotle* (*Athulujiya Aristu*). But above all, it was the *Categories* – the first part of the *Organon* – that contained the core principles of Aristotelian thought, so it is no wonder that almost everyone in this era wrote a commentary on the *Categories*, in which work it is clearly stated that particular substances are the prior beings.

As has been mentioned already, however, once the damage is done, it has proved itself to be largely irreparable, so even when in the later Middle Ages the lost manuscripts of Aristotle were discovered, commentators were extremely reluctant to revise their interpretation of Aristotelian ontology (based on the *Categories*), which had been ossified in the course of several centuries by the church elders. As Robert Pasnau puts it, "whereas readers today tend to treat the *Categories* as a curious early work of uncertain relation to Aristotle's mature thought, the scholastics regarded it as foundational for the whole corpus" (Pasnau 2012: 674). Important here is to note that hell broke loose when the lost manuscripts were discovered. Pasnau explains this historical context in the following terms:

> A bad-tempered decree from Paris in 1210 demanded that the body of one master be exhumed and reburied in unconsecrated ground, that the works of another be burned, and that 'neither the books of Aristotle on natural philosophy nor their commentaries be read at Paris in public or secret' – under penalty of excommunication. This prohibition was repeated in the above-quoted rules of 1215, this time with the books on metaphysics included [...].
>
> PASNAU 2012: 666

15 There are countless works on the authenticity of the *Categories*, but Frede (1987) is believed by many scholars to have settled the debate.

In 1231 Pope Gregory IX wrote a letter reaffirming the ban in Paris (outlawing the *Metaphysics*), which is hardly surprising, given that the *Metaphysics* contains many passages violating what was then the official church doctrine. In the introduction to *A Companion to the Latin Medieval Commentaries on Aristotle's Metaphysics*, Fabrizio Amerini and Gabriele Galluzzo write that given "the historical role that the *Categories* played in the shaping of medieval thought, it should come as no surprise that few medieval interpreters accepted without qualification the claim [maintained in the *Metaphysics*] that form is primary substance".[16] Long story short, commentators in the early Middle Ages found themselves in the situation of having to fabricate a rather simplistic interpretation of Aristotelian ontology (based almost exclusively on the *Organon* and especially on the *Categories*), and commentators in the later Middle Ages were reluctant to revise these doctrines that had already become ossified through several centuries. Importantly, however, this pseudo-Aristotle needed a rival, at which point a pseudo-Plato was to enter the scene. Note that *none* of Plato's works were available in the early Middle Ages, with the sole exception of the *Timaeus*,[17] and even in the case of the *Timaeus*, only a half of the dialogue was available (as far as 53c), thanks to a somewhat "creative" translation and commentary by Calcidius,[18] which in fact included many things absent from the original text. In a word, then, commentators in the early Middle Ages learned their Plato by and large from secondary sources, which they pitted against Aristotle's *Organon*, meaning that these commentators knew little of the actual texts of the two philosophers that they were nevertheless eager to compare and contrast.

So what can we say about the medieval interpretations of Plato? First and foremost, this "Plato" argued the opposite of what "Aristotle" argued in his works. The pseudo-Aristotle revered what comes from the senses, which made sense because he also believed the primary things to be particulars, espe-

---

16 Amerini and Galluzzo 2014: 13.

17 It is by no means a coincidence, of course, that in the *School of Athens* it is the *Timaeus* that Plato holds in his hand.

18 "Calcidius wrote his *Commentary* in the late fourth century AD to elucidate his Latin translation of *Timaeus* 17a–53c. [...] It served as the primary source for studying Plato throughout the Middle Ages and into the Renaissance. The *Commentary* secured the *Timaeus* a place of unparalleled importance in the history of medieval western philosophy and determined the way in which the dialogue was read, interpreted, and used in teaching. For about a thousand years Plato meant almost exclusively the *Timaeus* and the *Timaeus* meant primarily Calcidius' rendering and commentary" (Somfai 2004: 203). "For centuries the West drew its knowledge about Plato chiefly from this work. During this time Plato's fate lay, so to say, in the hands of Calcidius. [...] Plato chiefly meant the *Timaeus*, and the *Timaeus* as translated and explained by Calcidius" (van Winden 1965: 2).

cially particular substances such as this tree and that horse. In contrast, the pseudo-Plato more or less shunned the senses altogether in favor of the direct apprehension of the Forms. Despite overwhelming evidence that contradicts this simplistic picture, many contemporary scholars, even though their readings are admittedly more nuanced, are still somewhat reluctant to challenge this interpretation entirely. The truth is, *for both Plato and Aristotle*, perception does not yield knowledge, but it is perception that paves the way for the intellection of Forms/essences. If one insists on calling Aristotle an empiricist, Plato too is an empiricist, and to the same extent. To put things more carefully, however, it would be misleading to call any of these philosophers an "empiricist," even though both of them maintain that the path to knowledge commences with the perception of sensible particulars. Perception is a necessary first step in the dialectical journey that the soul must undergo to reach the Forms (Plato) or in the process whereby we grasp the essences of things (Aristotle). And these Forms or essences are *already* in the soul, but only *in a way*, for it is only this dialectical journey that can actualize what already potentially exists in the intellect.

To repeat, what distinguishes Aristotelian ontology are, first and foremost, the concepts of matter and form as well as the concepts of potentiality and actuality, which concepts are not to be found in the *Categories*. According to Aristotle, form is prior to the matter *and also to the matter-form compound*, in which case *this tree and that horse cannot be the ontologically prior beings*; what is prior, rather, are the forms. Therefore, as has already been mentioned, we moderns think of the *Categories* as either a spurious work falsely attributed to Aristotle or a very early work that conflicts with the rest of his corpus. But because in the early Middle Ages commentators had no access to the rest of his corpus, having to content themselves instead with the *Organon*, the *Categories* was believed to contain the basic tenets of Aristotelian ontology, as a result of which this tree and that horse were believed to be the ontologically prior beings.

An ontology that prioritizes particular things would be more likely to call for an epistemology that prioritizes what comes from the senses because it is the senses that enable our acquaintance with particular things. And if such is the kind of epistemology endorsed by Aristotle, and if his main rival is Plato (none of whose works were available for almost a thousand years – with the aforementioned exception of a loose translation of half of the *Timaeus*),[19] it would

19 Incidentally, in *Physics* IV.2, 209b13–15, Aristotle famously argues that the *Timaeus* misrepresents Plato's true views and that in his "unwritten teachings" (ἄγραφα δόγματα) Plato speaks – e.g., about space – "in a different way" (ἄλλον τρόπον).

make sense for Plato to have advanced a different kind of epistemology, one that more or less forgoes entirely what comes from the senses in favor of the direct apprehension of the Forms. What I have tried to explain above, however, is that this is a very crude picture of the difference between the philosophies of Plato and Aristotle; Aristotle too maintains that perception does not yield knowledge, just as Plato too maintains that the intellection of Forms involves a process that commences with perception. The lack of access to the actual texts of these philosophers in the early Middle Ages is at least one of the reasons that explain why it was natural for commentators to have overstated the extent to which the two philosophers disagreed with each other. Granted, since the nineteenth century, scholars have developed a more nuanced understanding of the difference between the philosophies of Plato and Aristotle, but it would not be an overstatement to say that there is still an extent to which Aristotle, unlike Plato, is believed to be an empiricist, even though there is considerable textual evidence showing that their epistemologies are not unalike: the epistemological framework explained in *Posterior Analytics* II.19, in which the journey toward the thinking of essences commences with perception, finds an analogue in several passages in the *Phaedo* and elsewhere, in which the intellection of Forms proceeds from the perception of particulars, just as Plato's metaphor of recollection in the *Meno* finds an analogue in Aristotle's statement in the *De Anima* that the soul is in a way (i.e., potentially) all things. While other parallels may be found, these should suffice to show that Plato's and Aristotle's epistemologies are more similar than is generally believed.

5 Conclusion

Both Plato and Aristotle argue that we proceed *from* perception *to* thinking. Perception is directed toward sensible particulars, which are in constant flux, and as such it does not yield knowledge, but nevertheless perception is a necessary first step toward knowledge. Knowledge is of the Forms or the essences of things, as a result of which non-human animals – despite their acquaintance with sensible particulars – cannot be said to know anything. Nor can human animals be said to have knowledge either, unless they take the necessary steps to *turn the soul* so as to move beyond the contingencies of sensuous data. Again, Plato's dialogues are attempts to appropriate these contingencies so as to transcend them. The grave mistake often committed by Socrates' interlocutors is to conflate the sensible and the intelligible. Plato warns us that the Equal itself is different from equal things, that piousness is different from particular pious actions – recall Euthyphro's misleading statement that the pious is to do what I am doing now (*Euthyphro*, 5d) – and so on. Because it is natural for the *hoi polloi* to confuse the sensible and the intelligible, Plato's works repeatedly remind us to avoid this type of mistake, which is not to deny, however, that the path toward thinking commences with perception.

Neither Plato nor Aristotle is very clear on *how* we move from perception to thinking, even though we find vague statements in Aristotle's texts suggesting that the imagination mediates between the two.[20] If so, the imagination must somehow facilitate the move beyond the contingencies of sensuous data, but in Plato and Aristotle there is no clear counterpart to something like *freie Variation*. Let me propose nevertheless that many of Plato's dialogues exemplify the free variation of an *eidos* (equality, piousness, and so on), whereby the once concealed *eidos* is gradually unconcealed, but the method at work is never made thematic. What we need in such a context is a fresh interpretation of Plato's account of knowledge, one that not only acknowledges the presence of the transition from perception to thinking but also scours Plato's dialogues to uncover the specifics of this transition.

As for Aristotle, future research needs first and foremost to clarify the role of imagination in thinking, in the course of which it also needs to address the questions of why the intellect cannot think without an image and why thinking is nevertheless irreducible to imagining. Because Aristotle's terse remarks in *Posterior Analytics* II.19 provide only the outlines of a model, however, and because the third book of the *De Anima* leaves many questions unanswered, scholars may have to look elsewhere in search of a possible interpretation of how the imagination mediates between perception and thinking.

Bibliography

Amerini, F. and Galluzzo, G. (2014). "Introduction," in F. Amerini and G. Galluzzo (eds.), *A Companion to the Latin Medieval Commentaries on Aristotle's Metaphysics* (pp. 1–18). Boston: Brill.

Annas, J. (1988). *Aristotle's Metaphysics: Books M and N*. Oxford: Oxford University Press.

20 At least one thing we can say here is that thinking transcends the *particularity* of images. As Ferrarin puts it, thinking "uses images as particular examples and [...] disregards precisely what makes the image an image, its particularity" (Ferrarin 2006: 105).

Berti, E. (2005). "Heidegger and the Platonic Concept of Truth," in C. Partenie and T. Rockmore (eds.), *Heidegger and Plato: Toward Dialogue* (pp. 96–107). Evanston, IL: Northwestern University Press.

Berti, E. (1978). "The Intellection of 'Indivisibles' according to Aristotle, *De Anima* III.6," in GER. Lloyd and GEL. Owen (eds.), *Aristotle on the Mind and Senses* (pp. 141–163). Cambridge: Cambridge University Press.

Brown, L. (2001). "Innovation and Continuity: The Battle of Gods and Giants, *Sophist* 245–249," in J. Gentzler (ed.), *Method in Ancient Philosophy* (pp. 181–207). Oxford: Oxford University Press.

Ferrarin, A. (2006). "Aristotle on Φαντασία." *Proceedings of the Boston Area Colloquium of Ancient Philosophy* 21: 89–123.

Fine, G. (1984). "Separation." *Oxford Studies in Ancient Philosophy* 2: 31–88.

Frede, M. (1987). "The Title, Unity, and Authenticity of the Aristotelian *Categories*," in *Essays in Ancient Philosophy* (pp. 11–28). Minneapolis, MN: University of Minnesota Press.

Hopkins, BC. (2002). "Husserlian Transcendental and Eidetic Reductions and the Interpretation of Plato's Dialogues." *Bochumer Philosophisches Jahrbuch für Antike und Mittelalter* 7 (1): 81–114.

Husserl, E. (1973). *Experience and Judgment: Investigations in a Genealogy of Logic* [EU], tr. by J. Churchill and K. Ameriks. London: Northwestern University Press.

Klein, J. (1965). *A Commentary on Plato's Meno*. Chapel Hill, NC: The University of North Carolina Press.

Larsen, JK. (2014). "The Virtue of Power (The *Gigantomachia* in Plato's *Sophist* 245e6–249d5 Revisited)." *New Yearbook for Phenomenology and Phenomenological Philosophy* 13: 306–317.

Menn, S. (1995). *Plato on God as Nous*. Carbondale, IL: Southern Illinois University Press.

Pasnau, R. (2012). "The Latin Aristotle," in C. Shields (ed.), *The Oxford Handbook of Aristotle* (pp. 665–689). Oxford: Oxford University Press.

Sokolowski, R. (2000). *Introduction to Phenomenology*. New York, NY: Cambridge University Press.

Somfai, A. (2004). "Calcidius' *Commentary* on Plato's *Timaeus* and its Place in the Commentary Tradition: The Concept of *Analogia* in Text and Diagrams." *Bulletin of the Institute of Classical Studies* (Supplement) 83: 203–220.

Unlu, H. (2020). "Aristotle on Ontological Priority." *Acta Philosophica* 29: 137–158.

van Winden, JCM. (1965). *Calcidius on Matter: His Doctrine and Sources*. Leiden: Brill.

CHAPTER 7

Theôria as Understanding of First Causes

Abida Malik

1 Introduction

For centuries, knowledge in the sense of justified true belief has been the major focus of epistemological debates. Only in the last two decades has more attention been dedicated to other cognitive states like understanding (see, for example, Kvanvig [2003]). Especially underrepresented in the contemporary debate are forms of knowing that cannot be (fully, or successfully) communicated, i.e. tacit knowing. If we look back to antiquity and especially to Plato, however, we can see that understanding (νοῦς, ἐπιστήμη), taken as a highly demanding[1] and partly tacit form of knowing, is center-stage in his epistemology.

In this paper, I aim to show that contemplation of the Form of the Good is best understood as a two-fold process: contemplation (θεωρία) of this Form leads to (1) direct apprehension of it which, simultaneously, provides us with (2) highly demanding propositional knowledge (being able to give an account of it). In the following, I label (1) and (2) as understanding$_1$ and understanding$_2$, respectively. Understanding$_2$ is an immediate effect[2] of contemplating *theôria*'s highest object (i. e. the Form of the Good) and attainable only by philosophers. Through its directness of apprehension, understanding$_1$ seems to have more similarities with knowledge by acquaintance than propositional knowledge. While propositional knowledge can sometimes be transmitted via testimony (e.g. A passes on a simple fact to B), contemplation of the Forms as the culmination of *paideia* is something every philosopher must experience personally. Although understanding$_2$ Forms as first principles or causes enables philosophers to give a causal account of phenomena and therefore provides them with what we would today call propositional

1 I use the label "demanding" throughout the paper to characterize cognitive states that are hard to acquire or transmit (e. g. because they require a lot of background knowledge). On the issue of hard and easy cases of transmission of propositional knowledge, understanding, and know-how, see Grimm (2020).

2 That is, these two kinds of understanding are conceptually distinct, but not temporally set apart from each other.

 | HTTPS://DOI.ORG/10.1163/9789004747128-009

inferential knowledge, this knowledge is only a further consequence of the immediate understanding$_1$ gained through acquaintance with the Form of the Good and the other Forms.[3] This direct contact is often described as intuitive grasping (ἐφάπτεσθαι, *Rep.* 490b4). So while understanding$_2$, according to Plato, implies being able to give an account, the final step in the process to achieve understanding$_2$ (contemplation of the Form of the Good which leads to understanding$_1$) remains partly tacit. In other words, it involves a direct, non-propositional grasp of the Good.

In the following, I will give (i) a short overview of Plato's terminology of contemplation, (ii) outline and discuss different approaches that have been proposed in the literature on how best to understand the process of contemplation, to then (iii) argue for the thesis that contemplation is best understood as incorporating two concepts of knowledge, which together can be subsumed under the heading *understanding of (first) causes.*[4] To do this, I will show that Plato's account of knowledge of the Forms (= understanding$_{1+2}$) fulfills certain characteristics: it (a) is holistic, (b) concerns the grasp of structures, (c) is explanatory, and therefore propositional and demanding, while at the same time being (d) partly tacit. The apparent contradiction between (c) and (d) can be resolved by differentiating between understanding$_1$ and understanding$_2$: while (a) to (c) are applicable to understanding$_2$, (d) is characteristic of understanding$_1$. These characteristics frequently reappear in the contemporary debate in epistemology about the relation between knowledge and understanding and are usually ascribed to the latter (see, for example, Kvanvig [2003: 202]; Ahlstrom-Vij & Grimm [2013: 341]; Malfatti [2021]). The contemporary interest in understanding can therefore partly be seen as a revival of Plato's notion of *epistêmê*.

---

3 It is controversial whether Plato would have called this kind of understanding knowledge. I agree with Jessica Moss (2021: 122–131) and Whitney Schwab (2016) that we should refrain from calling it knowledge, since Plato seems to restrict *epistêmê* to Forms. Schwab (2016, sect. 6) classifies it as an informed, privileged, authoritative form of *doxa*. Strictly speaking, this constitutes even another type of understanding (understanding$_3$, not only conceptually but also temporally set apart) because it is concerned with particulars. In this paper, I am mainly interested in understanding of the Forms, and especially the Form of the Good.

4 I am using the plural here because *theôria* is not restricted to the Form of the Good. In this paper, however, though at times I talk about Forms in general, I wish to concentrate on the Form of the Good. The idea that *epistêmê* in Plato is often better captured as understanding has already been famously proposed by Burnyeat (see Burnyeat & Barnes [1980: 186f.]) and Nehamas (1985: 25).

2 Terminology of Contemplation

When Plato talks about achieving understanding of the Forms, he uses many different verbs and nouns which mostly seem to refer to either the visual realm ((*apo*)*blepein*, *horan*, *kat*(*idein*), *theasthai*, etc.) or the haptic realm (*lambanein*, (*eph*)*haptesthai*, etc.).

The first list need not surprise us, since Plato lets Socrates[5] – who denies having knowledge (*Rep.* 506d6–e5, 533a1–6)[6] – introduce the allegory of the sun when he tries to describe what the Form of the Good consists in and what contemplation of it involves. The image of the sun persists also in the following allegories of the line and the cave. Thus we have good reason to take Plato's visual language in the context of contemplation as metaphorical: we 'see' with our mind (more specifically, our rational soul-part) what the Form of the Good consists in. Socrates even explicitly states that "we say that the many beautiful things and the rest are visible but not intelligible, while the forms are intelligible but not visible" (*Rep.* 507b9–10).[7] In the following, I will refer to this process of seeing as contemplation or *theôria* interchangeably.[8]

Both groups seem to be perfectly compatible with our contemporary talk of understanding. According to Plato, we "grasp" (ἅψασθαι, ἐφάπτεσθαι, *Rep.* 490b3–4) the nature of the Good with the reasoning faculty of our soul – as in the case of seeing, we are clearly speaking of an intellectual, metaphorical grasping here. Since we use the very same language today when we want to express that we finally understand something, we should not take this terminology (at least not alone) as evidence for some kind of mystical intuition.

In accordance with this metaphorical use, seeing the Forms is equated with knowledge of the Forms: when we see the Form of the Good, we know its essence (*Rep.* 516b6–7 by analogy). If we want to discover what *theôria* of the Forms involves, therefore, we need to pay attention to the intellectual vocabulary Plato uses in these contexts. At first sight, one might suspect he is rather

---

5 When I talk about Socrates in this paper, I always refer to the Platonic Socrates, i.e., Socrates as depicted as dialogue partner in Plato's works.

6 In *Rep.* 517b6–7 he even says that only the god knows whether his explanations are true. The fact that Socrates only talks about an image of the Good (the Sun) and his denial of the possession of *epistêmê* on the matter are often taken as evidence to attribute *doxa* to Socrates (Moss indicates some supporters of this position, see Moss [2021: 180, n. 38]). However, I agree with Moss that there cannot be *doxa* of Forms and that Socrates' cognitive state should better be classified as *dianoia*; see Moss (2021: 180–195).

7 The translation of the *Republic* follows Reeve's revised version of the Grube translation (in Cooper [1997]), if not otherwise stated.

8 For a more detailed explanation of the different visual terms applied, see Ward (2022: 50–52).

careless in his use – we can find many general words for knowledge/knowing (γνῶσις, γνωρίζειν, οἶδα) – but, on the other hand, at least from the Line allegory onwards, he seems to make a kind of distinction between *dianoia* and *nous*/*noêsis* restricting the latter to knowledge of the Forms. As I will try to show in section 4, *nous* (sometimes used synonymously with *epistêmê*) provides us not with piecemeal propositional knowledge, but with highly demanding understanding.

3 State of the Art

Every reader of the *Republic* seems to be confronted with a dilemma once they try to pin down what Plato means when he talks about *theôria* (or grammatically and semantically related forms). On the one hand, Plato's descriptions often seem to suggest some kind of immediate, direct, or even intuitionist kind of knowledge, implying an acquaintance with the Form of the Good. On the other hand, this intuitionist reading has problems in accommodating other text passages where Plato stresses the necessity of being able to give an account of the Good (e.g. *Rep.* 534b8–d1), suggesting a propositional reading. Accordingly, what contemplation of the Forms (or of the Form of the Good, specifically) consists in has been and still is a matter of debate. In the following, I will outline some of the most influential approaches to then suggest that the controversy often arises by confusing two different types of understanding, and that an integrative account could therefore help solve some of the conflicts.

3.1 *The Intuitionist Reading*

Interpreters who support an intuitionist reading (e.g. Halfwassen [2006: 231–233]; Oehler [1962: 112–114]; Szlezák [2011: sect. 10.5]) maintain that we grasp the Form of the Good via some immediate, direct intuition, sometimes similar to divine inspiration and separate from discursive thinking (*dianoia*).[9] Accordingly, they translate *noêsis* (the faculty with which we grasp the Good

9 Oehler, however, expressly separates this intuitive grasping from any kind of mysticism and emphasizes the connection to *dianoia* (Oehler [1962: 112–114]). Horn and Rapp who argue against the intuitionist reading take intuition as a "separate form of knowledge acquisition". It is "independent from experience and leads to a priori knowledge," "evidential experience or accompanied by such," and "knowledge acquired via intuition is in no need of further reasons or justification" (Horn & Rapp [2005: 13], own transl.). In this strong form my reading also rejects intuitionism.

and the Forms in general) as "intuitive thinking" (Szlezák [2011: 160][10]) or "intellectual vision" (Halfwassen [2006: 231]; Krämer [2011: 138]).[11] And indeed, many text passages seem to suggest such a reading, especially if we look at the *Symposion* (210e4–5) or the *Phaedrus* (247b6–e4). And in the *Republic*, although we apply a methodical scientific approach (first we learn arithmetic, geometry, stereometry, astronomy, harmonics to then apply finally the science of dialectic), the final step of how we actually achieve knowledge of the Form of the Good is not sufficiently spelled out. Jens Halfwassen even speaks in this context of a "mystic agnosticism" (2006: 224). However, intuitionism still cannot be the whole story. In *Rep.* VII at the latest it becomes crystal clear that propositional knowledge is central to understanding the Forms.

3.2 *The Propositional Reading*

According to the propositional reading (already supported by Cross [1954]), what we gain through contemplation is full-fledged propositional knowledge, i.e., we are able to give an account of the Forms. Plato goes so far as to claim that in order to attribute knowledge to someone they must be able to give an account (e.g. *Rep.* 534b8–c6). Robert C. Cross, moreover, emphasizes that "throughout the dialogues knowledge, forms, and logos turn up together." (1954: 444)[12] Like the intuitionist reading, the propositional reading certainly must be accommodated by a comprehensive account of the process of contemplation. There might be doubts, however, whether reading the relevant passages exclusively in this sense is sufficient to explain what happens when we finally achieve knowledge of the Good and of Forms in general.[13] It can certainly not be equated with acquisition of piecemeal knowledge.[14] Know-

10 All translations from contemporary texts are my own.

11 Thomas A. Szlezák also entertains the possibility that knowledge of the Form of the Good represents "an own level of knowledge that would surpass the noetic knowledge of ideas just as this surpasses διάνοια." (Szlezák [2012: 140])

12 According to Cross, a Form is "the logical predicate in a logos, not the logical subject. It is what is said of something, not something about which something else is said." (Cross [1954: 447])

13 Cross would probably classify knowledge of the Form of the Good as an exception. At least that is his move when it comes to the Form of Beauty in the *Symposion*. Still, Cross holds that "[t]he knowledge that interests him [i.e. Plato] in his non-enthusiastic moments is the knowledge in which logos is inextricably involved" (Cross [1954: 443]). That the Good is seen as an exception by Cross is also supported by his later commentary on the *Republic* that he published together with Woozley. There, while still stressing the propositional aspect, they do allow for "some sort of immediate awareness" (Cross & Woozley [1964: 253]) when it comes to the Form of the Good specifically.

14 In the sense that we first know the Good and then maybe some other Forms, but not necessarily all of them, like Reeve (2010: 222), who argues against holism, claims.

ledge of the Good (i.e. definitional knowledge in the sense of "Good is X.") is not "stand-alone" knowledge but necessarily connected to other Forms and the phenomena in the world. I will come back to this in section 4.

3.3 *The Pragmatic Reading*

Supporters of the pragmatic position (e.g. Wieland [1999: 159–185, see also 224–236, 261]) identify the knowledge we achieve via contemplation with practical knowledge, or know-how that is non-propositional. Julius M. E. Moravcsik, who talks about understanding in this context, also goes in this direction: his interpretation seems to strictly separate understanding from propositional knowledge while conceding only a minor role to the latter, namely its practical function (1979: 60). Since this approach has its origins in the hermeneutical tradition of the 20th century, it has also been labeled "hermeneutic-pragmatic" (Horn & Rapp [2005: 23; see 23–25]). The main problems of this approach, which have already been recognized by Horn and Rapp, is that it (1) openly conflicts or does not know how to accommodate passages which clearly point to a (at least partly) propositional reading and, connected to this point, (2) with Plato's clear preference for abstract, theoretical knowledge. Philosophers desire to do philosophy without having to engage in practical matters – and there is no problem with this when they have been raised outside of Plato's ideal state (only within *kallipolis* they have to rule for a certain amount of time to give something back to society for the education they enjoyed, see *Rep*. 520a6–c3).

3.4 *The Aporetic Reading*

The aporetic position, represented by Cornelia J. de Vogel (1988: 46–50), maintains that the kind of knowledge we achieve of the Good is very limited. She takes as evidence, among other passages, the statement that the Good lies "beyond being" (ἐπέκεινα τῆς οὐσίας, *Rep*. 509b9), and that it therefore lies on another level than the other Forms. De Vogel concludes that "[t]his transcendency certainly limits its knowability." (1988: 50) This reading seems attractive insofar as Socrates repeatedly denies having knowledge about the Good and emphasizes that the truth of his explanations is uncertain (*Rep*. 506d2–e5, 517b6–7, 533a2–5). At the same time, however, Socrates classifies the Good as an object of teaching (*Rep*. 505a2), and the success of *kallipolis* depends on philosophers who achieved knowledge of the Good (e.g. *Rep*. 505d5–506b2). Although de Vogel acknowledges that "on the level of the intelligible the Good is knowable" and that "[t]he philosopher *must* know it" (1988: 50), she still downplays its knowability by referring to its transcendence. Still, it seems unclear why this fact should be taken as conclusive evidence for any restriction on the knowledge of the Good. If philosophers need exactly this knowledge to be good rulers, it does not make sense to postpone this knowledge to some later point when the philosophers dwell on the Isles of the Blessed (as de Vogel seems to maintain, see 1988: 50).

3.5 *The Atomist Reading*

Horn and Rapp reject the intuitionist and the pragmatic reading; the alternative they propose is an atomist reading. The main idea is that through contemplation we get to know "elementary parts of knowledge, as opposed to their compounds, like knowing letters, as opposed to syllables or words." (2005: 25) This would be knowledge of "indivisible, conceptual units." (2005: 26) While they only provide a rough sketch of this approach, I think developing it further and reconciling it with a very weak intuitionist reading could prove very promising in the context of understanding_1. The idea would be that we understand_1 these elementary parts of knowledge and gain, consequently, understanding_2 of them.

3.6 *The Integrative Reading*

In the following I will argue for an integrative reading that takes seriously the Platonic passages that suggest a direct acquaintance with the Forms as well as passages that clearly support a propositional account.[15] I will try to outline the characteristics that Plato attributes to the knowledge we achieve via contemplation. This is to show that it would be better to talk about understanding rather than knowledge of the Forms because these characteristics coincide with the features we today attribute to understanding. It will then become clear that some characteristics are in line with the propositional account (understanding_2), others with a weak intuitionist or atomist account (understanding_1).

I will start with evidence for understanding_2. It will quickly become clear that Plato gives us a lot more evidence for understanding_2, which, in general, seems to be less mysterious than understanding_1. It is understanding_1 which

15 Julia Annas already seems to support such a reading in her commentary on the *Republic*, when she claims that "[k]nowledge involves an experience which is, in some sense, direct, by contrast with which belief is, or can be, second-hand. (But there is nothing mysterious about this directness; it comes from close 'acquaintance' with the subject-matter in a quite ordinary sense.) Knowledge also involves the articulate ability to say what the known item is, and why it is the way it is; and this involves knowing what makes for its goodness or badness." (Annas [1981: 191f.]) Different from my reading, however, she talks about knowledge and sees understanding as some kind of essential property of it (understanding is "the mark of knowledge", Annas [1981: 193]).

gives rise to aporetic, or intuitionist readings and it seems to openly conflict with understanding$_2$. This conflict, however, only arises if we try to squeeze Plato's statements into one single conception of knowledge (or understanding). In what follows, I will try to show that it makes much more sense to divide up this conception into two different kinds, where understanding$_2$ is the immediate result of understanding$_1$.

4 *Theôria* as Understanding

In this paper I focus on *theôria* as vision or contemplation of the Form of the Good, and in this context at least one thing seems to be clear: this "seeing" represents the culmination of the philosopher's educational journey and seems to occur in a specific moment. So, where the intuitionists get it right is that knowledge of the Good is, in a certain sense, sudden and direct. I suggest taking understanding$_1$ as direct knowledge (by acquaintance) of elementary units, which we become aware of suddenly (after long study of the sciences and especially dialectics), and which immediately provides us with highly demanding propositional knowledge (understanding$_2$). This process need not be at all mystical, but is rather similar to the discovery of a mathematical proof (for the latter see Horn & Rapp [2005: 26]). Bertrand Russell subsumes knowledge by acquaintance under "knowledge of things" (1912: 25) and also allows universals among these "things": "general ideas, such as *whiteness*, *diversity*, *brotherhood*, and so on." (1912: 28) He sees acquaintance, furthermore, as foundational for all other kinds of knowledge (1912: 26). What I am acquainted with are "things immediately known to me just as they are." (1912: 25) If we proceed from these assumptions, knowledge of the Forms (understanding$_1$) seems to resemble knowledge by acquaintance. We do need preparatory studies, however: only after extensive studies of the sciences do we achieve direct awareness of the elementary units of being. In other words, we need to prepare our rational soul-part.

This awareness is not simple, piecemeal propositional knowledge. Seeing the Good entails more than that: by being acquainted with it, we understand$_2$ that this specific conceptual unit is the cause of understanding (ἐπιστήμη) and truth (*Rep.* 508e3–4), we see how it is entangled with other Forms, and how it is related to the world of becoming. If we take the Forms as causes (and the Good as supreme cause), we can therefore legitimately speak of an understanding of causes, i.e. systematic and comprehensive knowledge of the Forms, of which we can give an account (understanding$_2$). In the following, I will give textual evidence for this claim. My aims in this section are to show that (i) knowledge of the Forms is comprehensive and holistic, that (ii) it involves grasping relations, structures or patterns, (iii) that it provides us with explanations, and, connected with the latter, that it is highly demanding and propositional while at the same time (iv) being partly tacit. The tacitness is due to the acquaintance part (understanding$_1$), the propositional aspect a direct consequence of it (understanding$_2$).

4.1 *The Evidence*

4.1.1 Holism

Several passages suggest that knowledge of the Forms is holistic and not – as Reeve (2010: 222) maintains – potentially piecemeal. This, so I argue, is the first striking characteristic that makes knowledge of the Forms a candidate for what we today usually call understanding. Socrates repeatedly emphasizes that the philosopher desires not partial, but complete wisdom (*Rep.* 475b8–9) since, in general, when we love something, we desire all of it and are not satisfied with parts (*Rep.* 475b4–7). In this vein, we want to know the complete *ousia* (i. e. all the Forms) and not just parts of it (*Rep.* 485a10–b8). The same goes for truth: the lover of learning desires the whole truth as far as it is possible (*Rep.* 485d3–4). This topic is prominent on the meta-level of the dialogue: something becomes clear – in other words, we understand it – when we grasp it as a whole (*Rep.* 491c7–9). Reeve argues against holism since he considers knowledge of the Forms as potentially piecemeal in the sense that we first understand what the Good consists in, how it functions as a principle, then maybe gain knowledge of a specific Form (e.g. Beauty) without necessarily gaining knowledge of all the other Forms in this process (Reeve 2010: 222). However, apart from the passages mentioned (which could be read in Reeve's way, if we restrict the completeness mentioned there to the Form of the Good) there is strong evidence that we do not gain isolated knowledge: the philosopher *qua* dialectician is *synoptikos* (*Rep.* 537c7), i.e., philosophers have comprehensive knowledge, they see how all things hang together. This is also underlined by the claim that the philosopher sees all of time and being (*Rep.* 486a8–9: θεωρία παντὸς μὲν χρόνου, πάσης δὲ οὐσίας) and that our souls desire the whole of divine and human things alike (*Rep.* 486a4–6).

There are even stronger passages that suggest holism: Socrates underlines that the philosophers "know each thing that is" (*Rep.* 484d5–6). This is clearly referring to the Forms since only they *are*.[16] Therefore, philosophers do not

16 Whatever that means specifically. I keep the neutral translation of *einai* here as "to be". Famously there is a scholarly debate about mainly three different readings (existential, predicative, veridical; see, for example, Fine [1999: 217–225]), but my argument does not hinge on this question.

know certain isolated Forms, but all of them. This is reinforced by a later passage describing the dialectician who grasps the *logos* of each being (*Rep.* 534b3–4).

These statements are hard to reconcile with the thesis that we might achieve isolated pieces of knowledge of specific Forms without seeing their connections to their "formal" environment. Rather, for Plato, if we understand what a Form consists in, we see how it fits into the whole pattern.

4.1.2 Grasping Relations and Structures

Another aspect (connected to holism) that distinguishes simple, or easy propositional knowledge from understanding is that we do not learn a simple fact, but grasp[17] a whole structure, we understand how different items (Forms and Forms, or Forms and particulars) are related to each other. In *Rep.* 506a4–7 Socrates talks about the connection of particulars (*dikaia* and *kala*) to the Good and emphasizes that we cannot understand this connection without prior knowledge of the Form of the Good. While this passage could still be read in favor of piecemeal knowledge, immediately afterwards it becomes clear that the philosopher needs to be knowledgeable in all of *these* things (ὁ τούτων ἐπιστήμων, i.e., (s)he needs to have knowledge of the Form of the Good and all the other Forms, *Rep.* 506a9–b1). Moreover, when we come into contact with particulars, we relate them to their respective Forms – only this gives us understanding.[18] Plato illustrates this *ex negativo* by referring to people who do not have this understanding:

> Do you think, then, that there's any difference between the blind and those who are really deprived of the knowledge of each thing that is? The latter have no clear model in their souls, and so they cannot – in the manner of painters – look to what is most true [εἰς τὸ ἀληθέστατον ἀποβλέποντες], make constant reference to it [ἀεὶ ἀναφέροντές], and study [θεώμενοι] it as exactly as possible.
>
> *Rep.* 484c6–d1

As a good ruler of the city state in our lawgiving process we need to refer to the Forms. We need to understand how they are related to the world of becoming.

17 I will not enter a discussion of what this grasping specifically involves. This is a major point of controversy in contemporary epistemology (for a useful overview see Baumberger, Beisbart & Brun [2017: 12–17]), but not essential for this paper. I only want to draw attention to the fact that Plato seems to discuss a specific cognitive state here that clearly resembles our contemporary discussion on understanding.

18 As explained in n. 3, Plato would not call this cognitive state *epistêmê*, however.

In other words, we need to have rationally grasped the dependence relations between particulars and Forms.[19]

With regard to knowledge of the Good a different passage suggests that it is not isolated but connected to all other Forms. It is not just the case that we can somehow situate the Good within other Forms. Plato goes much further in claiming that we can situate it within the whole of being (and becoming): we can "distinguish (διορίσασθαι) in an account (τῷ λόγῳ) the form of the good from everything else (ἀπὸ τῶν ἄλλων πάντων)" (*Rep.* 534b8–c1).

4.1.3 Explanatoriness

But if we do not get some mystical, divine insight, what does happen when we "see" the Good? I suggest understanding this process as getting acquainted with the Good (seeing it as a pure element embedded in its formal environment; understanding$_1$) which simultaneously leads to the ability of giving an account of it (understanding$_2$). Next, as a further consequence, there is a grasp of the Good in its function as first cause and enabler of knowledge, of being, and of truth (understanding$_3$; *Rep.* 508e3–4, 509b6–8). Therefore, I follow Myles F. Burnyeat (1981: 102, 107, 107 n. 22) and one influential strand of the contemporary analytic approach to understanding, according to which understanding involves knowing explanations.[20] Acquaintance, in the Platonic picture, is intimately connected to explanation. That is because the Forms themselves constitute (causal) explanations, and it is necessary that philosophers can give evidence of their understanding of the Forms (in an account). Plato's Socrates specifically stresses the ability to give an explanation (in Greek: διὰ τί, i.e., the cause) that distinguishes dialecticians from astronomers and the like (people practicing the lower sciences in the wrong way):

> They [i.e. people practicing harmonics today] seek out the numbers that are to be found in these audible consonances, but they do not make the ascent to problems. They don't investigate, for example, which numbers are consonant and which aren't or what the explanation [διὰ τί] is of each.
>
> *Rep.* 531c1–4

19 In the contemporary debate, too, understanding causes, if we follow Ahlstrom-Vij and Grimm (2013), involves "a grasp of the relevant dependency relations." (2013: 341)

20 Kelp (2015: 3804) calls this strand "explanationism". This conception is radically different from the hermeneutical tradition which opens up a contrast between explaining and understanding (see George [2020: section 2.2], and also Burnyeat's explanations [1981: 107 n. 22]).

The problem of these people is that they stick to their sensory data and do not abstract from them, they do not try to make the ascent to first principles, which would explain the being of the particulars perceived by our sensory apparatus.[21]

But this is exactly what the Forms do. When we gain knowledge of them, we understand how they act as causes in the world of becoming and how they relate to other Forms and contribute to phenomena. This is already foreshadowed shortly after the passage just quoted:

> Moreover, I take it that, if inquiry into all the subjects we've mentioned [i.e. arithmetic, geometry, stereometry, astronomy, harmonics] brings out their association [κοινωνίαν] and relationship [συγγένειαν] with one another and draws conclusions [συλλογισθῇ] about their kinship, it does contribute something to our goal and isn't labor in vain, but that otherwise it is in vain.
>
> *Rep.* 531c9–d4

The "lower" subjects are useful only insofar as they draw the soul upwards: they should incite us to look for first principles, but this is only successful if we focus our efforts on discovering how these subjects hang together, how they are related to each other. From this information we deduce what is their own (οἰκεῖα), what essentially belongs to them.

This deduction is vital. After having successfully made the ascent we get to see/know the Good (in an acquaintance-like way), which at the same time means that we can give an account of it (see *Rep.* 534b8–c5 and below). Further contemplation leads to even more understanding: In *Rep.* 516b6 contemplation (θεάσασθαι) is translated by Grube/Reeve as "study," which is fitting if we understand this process as having direct acquaintance with the Good for some time. Contemplation of the Good is here expressed via analogy with the Sun:

> Finally, I suppose, he'd be able to see [κατιδεῖν] the sun, not images of it in water or some alien place, but the sun itself, in its own place, and be able to study what it is like [θεάσασθαι οἷός ἐστιν].

---

21 See Lienemann (in this volume) who focuses on the role Plato assigns to sense perception in the ascent to knowledge. She also refers to a passage in the *Republic* (523b9–e1), where Plato's Socrates distinguishes between sense perceptions that do not and sense perceptions that do motivate us to inquire further and therefore, eventually, reach *epistêmê*.

> Necessarily so.
>
> And at this point he would infer and conclude [συλλογίζοιτο] that the sun provides the seasons and the years, governs everything in the visible world, and is in some way the cause [αἴτιος] of all the things that he used to see.
>
> It's clear that would be his next step.
>
> *Rep.* 516b4–c3; transl. slightly modified

This further understanding of the function of the Good, that we only seem to gain after some time, could be called inferential knowledge,[22] which also entails the ability to relate particular phenomena to their respective Forms. Seeing the Good necessarily entails this step:

> [...] In the knowable realm, the form of the good is the last thing to be seen [ὁρᾶσθαι], and it is reached only with difficulty [μόγις]. Once one has seen [ὀφθεῖσα] it, however, one must conclude [συλλογιστέα] that it is the cause [αἰτία] of all that is correct and beautiful in anything, that it produces both light and its source in the visible realm, and that in the intelligible realm it controls and provides truth and understanding [νοῦν], so that anyone who is to act sensibly in private or public must see it.
>
> *Rep.* 517b8–c5

This passage is also evidence for the fact that the knowledge we try to reach is very demanding. Contemplation itself seems to involve direct contact with the Form of the Good as the culmination of the philosopher's educational journey. Only in a further (very important) step via deduction do we understand that the Good functions as a cause on the ontological and epistemological level. The necessity of this step is important: I take this to be a nomic necessity (see Shields [2007]) since we are clearly internally compelled, like by laws of logic, to draw this inference. We cannot choose not to deduce, rather, the deduction becomes necessarily clear to us. That this inference represents a further step is backed by the aorist in the passage (*ophtheisa*): first we see the Good, then we make our deductions.

This kind of knowledge is clearly highly demanding and, once the philosophers have it, verbalizable by them – likewise, those who are not able to give an account do not have knowledge (Plato makes this very clear: *Rep.* 531e4–6, 533c1–3, 534b3–d1). What we can give an account of, however, is the outcome of contemplation (the definition of the Good and further deductions

22 Although Plato doesn't allow this as far as particulars are concerned (see n. 3).

about it), not the process of seeing itself. The process of seeing seems to involve a specific grasping component:

> We said that sight tries at last to look at the animals themselves, the stars themselves, and, in the end, at the sun itself. In the same way, whenever someone tries [ὁρμᾷν] through argument [διὰ τοῦ λόγου] and apart from all sense perceptions to find the being itself of each thing and doesn't give up until he grasps [λάβῃ] the good itself with understanding [νοήσει] itself, he reaches the end of the intelligible, just as the other reached the end of the visible.
>
> *Rep.* 532a3–b2

In the following, Socrates explains that it is dialectic which leads us to grasp the Good. But what does this grasping consist in? According to Plato, when we grasp the Good we grasp its *logos*, i.e., its definition:

> Then, do you call someone who grasped [λαμβάνοντα] the definition [λόγον] of each being dialectical? But insofar as he's unable to give an account [λόγον διδόναι] of something, either to himself or to another, do you deny that he has any understanding [νοῦν] of it?
>
> How could I do anything else?
>
> *Rep.* 534b3–7, transl. modified

Understanding$_1$ entails being able to give an account (= understanding$_2$). Understanding$_1$ as direct (not mystical) insight is the precondition for this propositional knowledge. I have modified the Grube/Reeve translation in the first sentence since grasping seems to stick closer to the text (in this first sentence Socrates uses the Greek equivalent for 'grasping' (λαμβάνειν), not for 'giving an account' (λόγον διδόναι)). With this modification, we have the following line of thought:

> (P1) I am dialectical/a dialectician iff I grasp the definition of each being (= understanding$_1$).
>
> (P2) If I'm unable to give an account of X (= understanding$_2$), then I don't have any understanding$_1$ of X.
>
> (C) Therefore, if I am a dialectician (= possess understanding$_1$), I have the ability to give an account (= possess understanding$_2$).

P1 involves the grasping as well as the holistic component (mentioned above). Grasping a definition of X, then, seems to involve understanding$_2$ X. The latter becomes clear in the next passage: Having grasped X's *logos* (i.e., having understood$_1$ it), we can give an account of it, i.e., we have propositional knowledge:

> Then the same applies to the good. Unless someone can distinguish [διορίσασθαι] in an account [τῷ λόγῳ] the form of the good from everything else [ἀπὸ τῶν ἄλλων πάντων], can survive all refutation, as if in a battle, striving to judge things not in accordance with opinion but in accordance with being, and can come through all this with his account still intact, you'll say that he doesn't know [εἰδέναι] the good itself or any other good.
> *Rep.* 534b8–c5

I take the ability to "distinguish in an account the form of the good from everything else" as the ability to give a definitional account of it.[23] Moreover, I take the potential to "survive all refutation" as a safety requirement: the definitional account we are giving survives all tests and objections – in Socratic-Platonic language, all *elenchoi*. Knowledge of the Good is safe.

4.1.4 Tacitness

In spite of the philosopher's ability to give an account, their knowledge remains partly tacit in the sense that they cannot simply pass it on by giving this account. This is part of the evidence for the claim that knowledge of the Forms is a special and demanding kind of knowledge: the impossibility or difficulty in transmitting it via testimony is due to its partial tacitness. Another part of the evidence can be found in the fact that Plato reserves the specific faculty of *nous* and the demanding method of dialectic for its achievement.

Plato more or less consistently[24] applies the terms *nous* or *noêsis* when he refers to cognition of the Forms. It is with this faculty, often translated with 'intuition', 'intuitive understanding', or 'intuitive thinking' (for the latter, e.g.

---

23 Against readings that take this passage as evidence for the indeterminability of the Good (e.g. Halfwassen [2006: 257f., 263f.]). Halfwassen takes the Good to be the "absolute One itself, lying beyond all determination." (2006: 258)

24 A strict and clear line between *nous*/*noêsis* and *dianoia* cannot be drawn since in some passages Plato seems to use them interchangeably (e.g. *Rep.* 511a1, 525c3, 526b2). However, attributing to him a loose language in this respect is not justified either: in many cases, at least, Plato seems to reserve *nous*/*noêsis* to the realm of Forms (e.g. *Rep.* 511c3–e4, 532b1). Halfwassen (2006: 231 n. 29) even maintains that there is a systematic distinction between *noêsis* and *dianoia*.

Szlezák [2011: 160]), that we achieve knowledge of the Forms (*Rep.* 511c3–d5, 525c2–3, 532a5–b2); *noêsis* is described as a *pathêma* in our soul which is set over the objects of the highest realm (*Rep.* 511d6–8). Indeed, it is set over[25] "truth itself" (*Rep.* 526a8–b4). It is distinguished from *dianoia* which seems to be restricted (although not strictly, see n. 24) to the sciences ranging below dialectic (*Rep.* 511d2–5). There has been the tendency to connect *dianoia* to discursive reasoning and *nous* to direct, non-discursive insight (e.g. Halfwassen [2006: 231 n. 29]). Marcelo D. Boeri (2017) argues, however, that such a strict separation cannot be maintained and develops instead a deflationary reading (of the Line Allegory). According to this approach, *dianoia* and *noêsis* are not that detached from another, but simply different levels on which our souls can operate (2017: 17). Boeri argues for a deflationist form of intuitionism that is not mysterious at all. It deserves to be called intuitionism insofar as he maintains that we grasp the Forms directly. We do this via *noêsis* which is a kind of cognition that is non-propositional (2017: 19f., 28, 34, 36). This approach partly coincides with what Horn and Rapp have proposed, since also Boeri (2017: 28) suggests that the objects of our understanding are elementary, basic units. And indeed, although we clearly gain propositional knowledge as a result of understanding the Form of the Good (and, as a direct consequence, of the other Forms), grasping what the Form of the Good consists in – being able to give an account of what it consists in – seems to result from some kind of direct acquaintance with the Form of the Good. And this is what makes easy transmission (via testimony) impossible.

Passages which point to some kind of direct acquaintance are the direct object constructions[26] Plato frequently uses in this context (e.g. *Rep.* 505a5–6): we know the Good, not specific truths about it, at least not in the very first step. Specific truths seem to arise only after having seen/known the Good. This becomes especially clear in the sun allegory:

> Finally, I suppose, he'd be able to see [κατιδεῖν] the sun, not images of it in water or some alien place, but the sun itself, in its own place, and be able to study what it is like [θεάσασθαι οἷός ἐστιν].
>
> *Rep.* 516b4–7; transl. slightly modified

25 This is my translation of *epi* in this sentence. The Grube/Reeve translation renders it differently: "to use understanding itself on the truth itself". Either way, the point seems to be that *noêsis* is applicable to truth, and therefore to the highest realm.

26 Though, taken by themselves, we should not give too much weight to them. In combination, however, with the information that by seeing the Good we can give an account, it makes sense to divide up the process of cognition and to attribute knowledge by acquaintance to the first part (understanding$_1$).

The Grube/Reeve translation renders both *katidein* and *theasasthai* as a direct object construction while in the Greek, strictly speaking, only *katidein* has a direct object (the sun). I think that this distinction is important, since it nicely illustrates the different steps involved in the process: first, we get directly acquainted with the Good. This understanding_1 simultaneously leads to a grasp of its essence (of which we can then give an account): we see the Good itself, which directly leads to an understanding_2 of what it consists in.[27] In a further step, we achieve inferential knowledge (we might call this understanding_3 or, with Schwab (2016), privileged *doxa*)[28] by relating the Good to the other Forms and the world of becoming: by inference we understand that the Good is the supreme cause (*Rep.* 516b9–c2 by analogy). If we look at the passage just quoted, one might ask how understanding_1 could possibly be tacit at all: after all, we are able to give a definition of the Good (this becomes even clearer in *Rep.* 534b8–d2). Although the philosophers have this propositional knowledge, I suggest that it is partially tacit, in the sense that it cannot be passed on successfully (because direct acquaintance which enables us to achieve this propositional knowledge is lacking). Some general remarks on knowledge transmission might be called for to better understand this claim.

Transmission of knowledge is a much-debated topic in contemporary epistemology, and so-called testimony plays a central role in it: in the common view, testimony is conceived as statements or, "tellings generally" (Fricker 1995: 397), which function as a vehicle for knowledge transmission. Knowledge of a speaker (S knows that p) is transmitted to a hearer or a whole audience via testimony if certain conditions are in place (e.g., the speaker needs to be sincere and competent). This view is based on the idea that a lot of our knowledge (or what we today would usually call knowledge) has social sources. We cannot possibly achieve first-person knowledge of the numerous beliefs we maintain and work with on a daily basis.[29] While Plato wouldn't grant these beliefs the status of knowledge,[30] today we wouldn't object to

27 Annas (not focusing on the Good here, but on Plato's knowledge conception in general) calls this "*knowing what*" (1981: 192), but I don't see any reason why we should not simply classify it as propositional knowledge (since typically the answer to a "What is X?" question is propositional).

28 See also n. 3.

29 For an overview on the epistemology of testimony, see Gelfert (2014).

30 At least within the *Republic*, *epistêmê* seems much more demanding. There are also passages in other dialogues which point in that direction. See, for example, *Thet.* 201a7–c3 which can be read as a claim that beliefs achieved via testimony are not knowledge. This interpretation, although very influential, is controversial (see Bjelde [2021: 490–492] who argues against Burnyeat's and Nehamas' claims that *epistêmê* cannot be transmitted (see Burnyeat & Barnes [1980: 179]; Nehamas [1985: 24–29] and Bjelde

knowledge attributions in this context. If the knowledge that we would like to transmit is not very demanding, this process indeed often seems to be successful.[31] However, knowledge of the Forms, as we have already seen, is highly demanding. Although it is teachable (even the Form of the Good, which is called *megiston mathêma, Rep.* 505a2), it is not simply transmitted via testimony from teacher to student. The philosopher who has contemplated the Form of the Good is able to give an account of it, but the student is not able to just take up this account merely by listening to the teacher.[32] On the one hand, this could be explained by a lack of sufficient background knowledge or preparatory knowledge (education in the other sciences like arithmetic needs to be completed first, forced direct contemplation of the Good will not do, see *Rep.* 515e6–516a4). In other words, the philosophers-to-be are not properly primed yet (for the notion of priming see Grimm [2020: 133–135]). On the other hand, even after completion of the curriculum it is not a philosopher-teacher who tells their students what the Good consists in. The philosophers need to go this final step by themselves (*Rep.* 540a4–9). This is a clear indication that part of this understanding is tacit.

Understanding does not need to be tacit and demanding in general though: there are cases where not a lot of work on the side of the recipient is needed (e.g., understanding directions if we speak the language and are adult human beings).[33] In the case of the Forms, however, a lot of work needs to be done by the philosophers. The philosopher kings might tell their students what the Good consists in, but Plato puts special emphasis on the claim that they need to go through the whole curriculum, and therefore need to experience

[2021: 490–494]). See also *Men.* 97a9–b3 which could be interpreted as knowing the road to Larissa only if we actually went down that road ourselves. However, my claim for the demandingness of *epistêmê* does not depend on these (non-conclusive) passages, but is based on the depiction in the *Republic*. Bjelde, who focuses on the *Meno*, takes *epistêmê* to be "the cognitive state that enables one to teach." (2021: 495) This excludes it from being either knowledge or understanding (2021, 495f.).

31 It is successful because we have the relevant background knowledge or preparatory knowledge. Grimm calls this "proper priming" (2020: 135). For examples of transmission of propositional knowledge, which for their success depend on the subject's proper priming, see Grimm (2020: 133f.).

32 If this were the case, Plato could have postulated that future philosophers in *kallipolis* are not required to go through the whole educational curriculum. This would only be necessary for the first generation. There seems to be no evidence for this assumption in the text, however. It rather seems to be essential for every philosopher-to-be to pass through the whole curriculum.

33 See Grimm (2020), who argues against any general claims on the easiness or difficulty of transmission of understanding (and know-how and propositional knowledge). For specific examples of easy understanding, see Grimm (2020: 130).

for themselves the vision of the Good. Acquaintance with the Good is a personal experience (which can then be formalized in a *logos* which the philosophers can give – but this *logos* is not sufficient for transmitting the experience). The best we could hope for if they only get to know of the Form of the Good via this *logos* would be some parrot-like repetition of it, without any real understanding.[34] Understanding$_2$ (propositional knowledge) might theoretically be transmitted, but certainly not understanding$_1$ (knowledge by acquaintance). Since understanding$_2$, however, crucially depends on understanding$_1$ we cannot even gain real understanding$_2$ by listening to the account.

The personal experience, however, is not sufficient by itself, either. If we dragged some random person out of the cave and made her look directly at the Form of the Good, she would not be able to gain any understanding: in the language of the allegory, she would not be able to look at the Good because her eyes are not yet accustomed to the bright light (see *Rep.* 515e6–516b8). It is crucial to gradually increase our preparatory knowledge to be ready to finally look at the Good (*Rep.* 516a5–b2). Socrates emphasizes that one must be experienced in the sciences first, otherwise dialectics (and therefore, vision of the Good) is simply not possible (*Rep.* 533a8–9). Another striking point in this respect is that the philosophers must be led and compelled to look at the Form of the Good – even after being fully educated in the other sciences (*Rep.* 540a6–9). While the final stage (vision of the Form of the Good) can be plausibly constructed as a nomic necessity (see Shields [2007]), for the upward journey itself we seem to need a guide who teaches us in the sciences (which then, once we have mastered them, inherently (logically) compel us to see the Good). We do not achieve this demanding kind of understanding all on our own, at least not usually. In very rare cases, some individuals outside the ideal state might be successful in this, but these are clearly exceptions (see *Rep.* 496a11–e2).[35] This passage also underlines that it is essential that each philosopher experience the vision of the Good on their own. Therefore, there has to be some inherently personal component in *theôria*.

---

34 Duncan Pritchard seems to have in mind such a parrot-like repetition when he distinguishes knowing-why from understanding-why. His young son might have knowledge, according to this account, when he can repeat an explanation of why his house burnt down (because he was told the cause: faulty wiring), but he does not really understand it (see 2009: 38).

35 Another reason could be that philosophers raised outside the ideal state in general seem to be more independent and even slightly better in character. For this (small) distinction in the depiction of the philosophic character in the *Republic*, see Malik (2020, chap. 3.3).

5 Conclusion

In this paper I have tried to show that the debate about the kind of knowledge provided by contemplation of the Forms, and especially the Form of the Good, and how contemplation works in the first place, can partly be resolved if we assume that two different kinds of understanding are involved: first, we get acquainted with the Form of the Good (this acquaintance is a result of the discursive method applied in the context of the sciences like arithmetic, geometry, and so on). This kind of knowledge by acquaintance should better be called understanding$_1$ since this acquaintance is not open to everyone, nor to philosophers when they haven't been properly educated yet. Therefore, it is much more demanding than knowledge by acquaintance, as usually understood. As a result of understanding$_1$ we achieve understanding$_2$, a very demanding form of propositional knowledge, namely a systematic and comprehensive knowledge of causes. This is manifested by our ability to define the cause (give an account of the Form of the Good and the other Forms) and our ability to connect particulars with their respective cause (e.g., an action is morally good because it participates in the Form of the Good). This latter ability, however, would not be a manifestation of *epistêmê*, seeing that it is concerned with particulars, and has therefore also been labeled understanding$_3$ in this paper. The advantage of this integrative reading is that it does justice to text passages which point to an intuitionist position as well as other passages which suggest a propositional reading, and therefore better captures Plato's differentiated conception of understanding. Finally, Plato's conception of *epistêmê*, apart from his strong realist metaphysical assumptions about a realm of ideal entities, strikingly resembles our contemporary debates on understanding. In the central books of the *Republic* where Socrates talks about achieving *epistêmê* of the Forms, and especially of the Form of the Good, aspects that are central to the contemporary debate (like comprehensiveness, a grasp of structures or relations, explanatoriness and partial tacitness) play a decisive role. A further analysis of these similarities is certainly a promising topic for future research.[36]

---

36 This paper has benefited a lot from critical comments and literature suggestions by Anna Pavani, Julia Petz, Moritz Vogel, Denis Walter and other members of our Plato Group where I first presented a draft of this study. Furthermore, I thank the audiences at the Annual Conference of the British Society for the History of Philosophy (BSHP) 2023 in Aberdeen and at the First Middle East Meeting of the International Plato Society (IPS) 2023 in Ankara where I received most valuable feedback. For extremely helpful corrections regarding language and style, I am very grateful to Kirsten Anderson. Last, but not least I thank Refik Güremen for his very useful comments on the paper.

Bibliography

Ahlstrom-Vij, Kristoffer & Grimm, Stephen R. 2013. "Getting it right." *Philosophical Studies* 166: 329–347. https://doi.org/10.1007/s11098-012-0038-x.

Annas, Julia. 1981. *An Introduction to Plato's Republic*. Oxford: Clarendon Press.

Baumberger, Christoph, Beisbart, Claus & Brun, Georg. 2017. "What is Understanding? An Overview of Recent Debates in Epistemology and Philosophy of Science." In *Explaining Understanding. New Perspectives from Epistemology and Philosophy of Science*, edited by Stephen R. Grimm, Christoph Baumberger & Sabine Ammon, 1–34, London: Routledge.

Bjelde, Joseph. 2021. "Knowledge is Teachable." *Mind* 130 (518): 475–502. https://doi.org/10.1093/mind/fzaa092.

Boeri, Marcelo D. 2017. "Diálectica, Pensamiento 'Intuitivo' y 'Discursivo' en Platón." *Tópicos, Revista de Filosofía* 52: 11–42. https://doi.org/10.21555/top.v0i52.767.

Burnet, John, ed. 1901/1903/1978. *Platonis Opera. Tomus II–IV. Tetralogias III–VIII continens*. New York: Oxford University Press.

Burnyeat, Myles F. 1981. "Aristotle on Understanding Knowledge." In *Aristotle on Science. The «Posterior Analytics». Proceedings of the Eighth Symposium Aristotelicum held in Padua from September 7 to 15, 1978*, edited by Enrico Berti, 97–139. Padova: Editrice Antenore.

Burnyeat, Myles F. & Barnes, Jonathan. 1980. "Socrates and the Jury: Paradoxes in Plato's Distinction between Knowledge and True Belief." *Proceedings of the Aristotelian Society (Supplementary Volumes)* 54: 173–191 and 193–206. https://doi.org/10.1093/aristoteliansupp/54.1.173.

Cooper, John M., ed. 1997. *Plato. Complete Works*, Ed., with Introduction and Notes by John M. Cooper, Associate Editor DS. Hutchinson. Indianapolis/Cambridge: Hackett Publishing Company.

Cross, Robert C. 1954. "Logos and Forms in Plato." *Mind* 63 (252): 433–450.

Cross, Robert C. & Woozley, Anthony D. 1964. *Plato's Republic. A Philosophical Commentary*. London: MacMillan & Co.

Duke, EA., Hicken, WF., Nicoll, WSM., Robinson, DB. & Strachan, JCG., eds. 1995. *Platonis Opera. Tomus I. Tetralogias I–II continens*, New York: Oxford University Press.

Fine, Gail. 1999. "Knowledge and Belief in *Republic* V–VII." In: *Plato 1. Metaphysics and Epistemology*, edited by Gail Fine, 215–246. New York: Oxford University Press.

Fricker, Elizabeth. 1995. "Critical Notice. Telling and Trusting: Reductionism and Anti-Reductionism in the Epistemology of Testimony." *Mind* 104 (414): 393–411. https://doi.org/10.1093/mind/104.414.393.

Gelfert, Axel. 2014. *A Critical Introduction to Testimony*. London/New York: Bloomsbury.

George, Theodore. 2020. "Hermeneutics." *The Stanford Encyclopedia of Philosophy* (Winter 2021 Edition), Edward N. Zalta (ed.), <https://plato.stanford.edu/archives/win2021/entries/hermeneutics/>.

Grimm, Stephen R. 2020. "Transmitting Understanding and Know-How." In *What the Ancients Offer to Contemporary Epistemology*, edited by Stephen Hetherington & Nicholas D. Smith, 124–139. New York: Routledge.

Halfwassen, Jens. 2006. *Der Aufstieg zum Einen. Untersuchungen zu Platon und Plotin*, 2., um einen Forschungsbericht erweiterte Aufl. München/Leipzig: KG. Saur.

Horn, Christoph & Rapp, Christof. 2005. "Intuition und Methode. Abschied von einem Dogma der Platon- und Aristoteles-Exegese." *History of Philosophy & Logical Analysis* 8 (1): 11–45. https://doi.org/10.30965/26664275-00801002.

Kelp, Christoph. 2015. "Understanding phenomena." *Synthese* 192: 3799–3816. https://doi.org/10.1007/s11229-014-0616-x.

Krämer, Hans. 2011. "Die Idee des Guten. Sonnen- und Liniengleichnis (Buch VI 504a–511e)." In *Platon. Politeia*, 3rd, rev. edition, edited by Otfried Höffe, 135–153. Berlin: Akademie Verlag.

Kvanvig, Jonathan. 2003. *The Value of Knowledge and the Pursuit of Understanding*. Cambridge: Cambridge University Press.

Malfatti, Federica I. 2021. "On Understanding and Testimony." *Erkenntnis* 86: 1345–1365. https://doi.org/10.1007/s10670-019-00157-8.

Malik, Abida. 2020. *Seelen im Wandel. Eine Studie zum Charakterbegriff bei Platon*. Freiburg/München: Karl Alber.

Moravcsik, Julius ME. 1979. "Understanding and Knowledge in Plato's Philosophy." *Neue Hefte für Philosophie* 15: 53–69.

Moss, Jessica. 2021. *Plato's Epistemology. Being and Seeming*. New York: Oxford University Press.

Nehamas, Alexander. 1985. "Meno's Paradox and Socrates as a Teacher." *Oxford Studies in Ancient Philosophy* 3: 1–30.

Oehler, Klaus. 1962. *Die Lehre vom noetischen und dianoetischen Denken bei Platon und Aristoteles. Ein Beitrag zur Erforschung der Geschichte des Bewusstseinsproblems in der Antike*. München: CH. Beck.

Pritchard, Duncan. 2009. "Knowledge, Understanding and Epistemic Value." *Royal Institute of Philosophy Supplement* 64: 19–43. https://doi.org/10.1017/S1358246109000046.

Reeve, Christopher DC. 2010. "Blindness and reorientation: education and the acquisition of knowledge in the *Republic*." In *Plato's Republic. A Critical Guide*, edited by ML. McPherran, 209–228. Cambridge: Cambridge University Press.

Russell, Bertrand. 1912 [sec. ed. 1998, reissued 2001]. *The Problems of Philosophy*. Oxford: Oxford University Press.

Schwab, Whitney. 2016. "Understanding *Epistēmē* in Plato's *Republic.*" *Oxford Studies in Ancient Philosophy* 51: 41–86. https://doi.org/10.1093/acprof:oso/9780198795797.003.0002.

Shields, Christopher. 2007. "Forcing Goodness in Plato's *Republic.*" *Social Philosophy & Policy* 24 (2): 21–39. https://doi.org/10.1017/S026505250707015X.

Szlezák, Thomas A. 2011. "Das Höhlengleichnis (Buch VII 514a–521b und 539d–541b)." In *Platon. Politeia*, 3rd, rev. edition, edited by Otfried Höffe, 155–173. Berlin: Akademie Verlag.

Szlezák, Thomas A. 2012. "The Idea of the Good as *Arkhē* in Plato's *Republic.*" In *The Other Plato. The Tübingen Interpretation of Plato's Inner-Academic Teachings*, edited by Dmitri Nikulin, 121–142. Albany (NY): SUNY Press.

Vogel (de), Cornelia J. 1988. *Rethinking Plato and Platonism*, 2nd impression. Leiden et al.: Brill.

Ward, Julie K. 2022. *Searching for the Divine in Plato and Aristotle. Philosophical Theoria and Traditional Practice.* Cambridge/New York: Cambridge University Press.

Wieland, Wolfgang. 1999. *Platon und die Formen des Wissens*, 2nd, reviewed and extended edition. Göttingen: Vandenhoeck & Ruprecht.

PART 3

Image Perception, Imagery and Theôria *as Higher Cognition*

CHAPTER 8

Imaging Cognition in Plato's *Republic*

Marco Picciafuochi

1 Introduction

One of the first impressions that even the most distracted reader of Plato's *Republic* gains after having a first glance at the text is that its linguistic shape is very rich in imagery – here referring to the characters' use of rhetorical figures which allow to create "images" by means of language. The use of such a figurative richness is extensive to the point that it turns out to be difficult, even for the most attentive reader, to find in it a single notion not being expressed by means of imagery.[1] One might well ask whether such an abundance of figurative language in the text has a deep philosophical value, rather than being merely a matter of style.

According to these observations, what I intend to do is try to take Plato's deployment of imagery seriously, and to draw attention to how this proves conceptually relevant in the shaping of his notion of cognition[2] in *Republic* 6 and 7.

In doing so, in section 2, I will introduce the exegetical tools I will be deploying in my analyses. Particular attention will be given to the relation that exists between cultural models, conceptual metaphors, language uses, and imagery.

1 It seems that Plato himself was completely aware of this fact. It is noteworthy, indeed, that in *Rep.* VI 487e6 Socrates' way of speaking is explicitly defined by Adeimantus as a "speaking through images" (*di' eikonôn legein*). Moreover, this is an impression shared by Aristotle as well, one of the most authoritative testimonies on Plato's style, who, in *Met.* 9, A 991a22 (but see also *Met.* M 5, 1079b26), refers to Plato's vocabulary of paradigm and participation to express the relationship between Forms and sensible particulars by means of the epithet: "empty phrases and poetical metaphors" (*kenologein esti kai metaphoras legein poiêtikas*). That Plato's style was perceived by Aristotle as rich in similes and comparisons can be maintained even referring to *Rhet.* III 4, 1406b32–1407a1, where Aristotle, in order to explain the notion of analogical similes, recalls three different examples taken from Plato's *Republic* (see esp. *Rep.* V 469e, VI 488e, X 601b).

2 It is important to clarify that when I speak of cognition, I generally refer to the way(s) in which the soul epistemically relates to its objects, as emerges, for example, in *Rep.* V 476c1–480a13 and VI 509d6–511e5. As a mere working assumption, this way of conceiving cognition does not need to be exhaustive, its value being proportional to its utility in identifying specific parts of the text that require closer examination.

 | HTTPS://DOI.ORG/10.1163/9789004747128-010

Finally, I will clarify the assumption surrounding the present paper, according to which the philosophical value of Platonic imagery is to be found in its ability to induce metaphorical *thinking*, both in Socrates' interlocutors and – outside the dialogical fiction – in Plato's readers. In section 3, I will analyse two different structural metaphors for "high cognition": the oneiric metaphor (3.1) and the erotic metaphor (3.2). In section 4, then, I will try to see if they can help us better understand the difference existing between *dianoia* and *noêsis* in *Republic* 6. Conclusions will follow in section 5.

2 Exegetical Tools

Before looking more closely at Plato's text, it is important to introduce the exegetical tools which will be used in the rest of the paper and to briefly clarify what the import is of drawing attention to Platonic imagery.

First, we should introduce the concept of *schema* as it is generally conceived in Anthropology and Linguistics.[3] According to these fields of study, a schema is the organisation of cognitive elements into an abstract mental object shared by the members of a given culture. This schema includes both default values and open slots which can be variously filled in with appropriate specifics.

Another important notion to be introduced is that of *prototype*. A prototype is a typical example of a type of object, often with many properties "chunked" together to form a rich, specific image.[4] In other words, a prototype is the instantiation of a schema.

A schema, either alone or interrelated with other cognitive schemas, can be considered as a *cultural model* in so far as it is used to represent something, to reason with or to calculate from by mentally manipulating its parts.[5] A cultural model is something that permeates the mentality of a certain social group, a socially embedded conceptual grid by means of which the members of that group understand experiences and notions and behave accordingly.

It is important to note that a cultural model may be used for understanding not only those experiences and notions to which it is directly related, and from which it probably originates, but also unrelated ones. It is for the second

3 Most of the definitions offered in this paragraph are taken, with slight variations, from Roy G. D'Andrade, (1995: 179–180); but, on the same notions, see also Mark Johnson (1987: xiv, 19–23); Naomi Quinn and Dorothy Holland (1987: 24–27); George Lakoff (1987); David W. Allbritton (1995).

4 D'Andrade (1995: 179).

5 D'Andrade (1995: 180).

"extended" use of the cultural model that conceptual metaphors are fundamental tools.

A *conceptual metaphor* is, broadly speaking, a form of understanding and, precisely, the understanding of something in terms of something else; it is the cognitive use of a conceptual domain to give structure to another conceptual domain. It always involves a *target domain*, that is the conceptual domain that is talked of, and a *source domain*, that is the domain that provides structures and ideas. Structures and ideas are *mapped from the source onto the target*.[6]

Among all the behaviours affected by conceptual metaphors, it is language to make their presence particularly clear. Conceptual metaphors then, when active on the level of thought, determine in language the use of the so called "metaphorically motivated linguistic expressions,"[7] which may vary from simple prepositions to complex rhetorical figures and narratological structures.[8]

Included in the concept of metaphorically motivated linguistic expression is also the rhetorical notion of *imagery*. The term imagery, if understood in its rhetorical sense, refers to the use of all those figures of speech that allow something to be described by means of a linguistic "representation," or "image," which bears some similarity with it. But to describe a certain content through a linguistic representation means, from a cognitive point of view, to *think* of it or its elements under the terms of the elements of the image. Therefore, whereas, from a rhetorical point of view, imagery is categorised into various figures of speech – such as linguistic metaphors, similes, analogies, allegories,

6 The scholarly literature on conceptual metaphor uses the form X IS Y, in capital letters, for easy reference to single conceptual metaphors, including any subordinate ones. Each term substituting X and Y refers to a conceptual domain. The "IS" must not be understood as a real copula introducing a predicate. Rather, it is a simple way to express the relation of "mapping" existing between the source domain Y and the target domain X. For the notion of conceptual metaphor, a paramount study is George Lakoff and Mark Johnson (1980). After the publication of this seminal study, a huge number of contributions have appeared on the subject. For reason of space, I defer to Alice Deignan (2005: 13–32) for a well-organised discussion of the so-called "Conceptual Metaphor Theories."

7 For the expression, see Deignan (2005: 39).

8 A common example of metaphorically motivated linguistic expression is the preposition "in" in sentences such as "I have something *in* mind." Although "in" is not a linguistic metaphor here, i.e., it is not perceived as metaphorical in the process of understanding the sentence, it can still be seen as a metaphorically motivated linguistic expression, that is, a linguistic realisation of the conceptual metaphor MIND IS A CONTAINER; as for more complex narratological structures, one could refer to Homer's *Odyssey* or Melville's *Moby Dick* and say that they are different literary elaborations of sets of conceptual metaphors, one of the most important of which is LIFE IS A JOURNEY; on the conceptual metaphoricity of empty items of language, such as prepositions and delexalised verbs like *have, do, give, take, make* and so on, see Deignan (2005: 49–52).

and so on – from a cognitive point of view all these figures are understood as linguistic realisations of the same metaphorical mode of thinking.

Moreover, an important fact about imagery is that it is not just an immediate and unconscious reflection of thought on language, as it could be the use of a given preposition according to an implicit conceptual metaphor. As composed of figures of speech, imagery is also part of a communicative strategy by which the sender – be a speaker or a writer – consciously *does* things with words in order to pursue precise *ends* in any given communication process. These ends can vary a lot: a simile, for example, can be introduced to make the discourse more pleasant, or to persuade the receiver. However, insofar as it is based on conceptual metaphors, it may always be the case that a simile is used by the sender to get the receiver to *think* of certain things in terms of certain others, i.e., to *think* metaphorically.[9]

In concluding this paragraph, we could resort to the example of the *commercial transaction* schema to make all these notions more concrete. This schema includes two roles, the buyer and the seller, and a relation of exchange between them. Typically, the exchange foresees the giving of money to the seller for receiving rights over some object. Two prototypes of the *commercial transaction* schema would be the typical actions of doing shopping in a supermarket or online.

In cultures based on capitalistic economies, the *commercial transaction* schema, functioning as a model, is used not only to understand experiences directly related to shopping, but also to represent and calculate, among other things, the value of time. In fact, we do assign a certain monetary value to time when we sign a job contract, and we talk about time as if it were money when we tell someone "Don't *waste* your time" or "I love *spending* my time with you." The conceptual metaphor TIME IS MONEY is clearly active behind these sentences, and the words "waste" and "spending" seem to be motivated by it, even if the linguistic expressions themselves may not be perceived as metaphorical during communication.[10]

However, a hypothetical speaker could also wish to use the TIME IS MONEY metaphor to produce certain effects in her listeners by using expressions

---

9 Think, for example, of the frequent use of similes and linguistic metaphors in education for pedagogical purposes: see Thomas G. Sticht (1993: 621–632).

10 It is quite easy to identify conceptual metaphors in the common-sense understanding of abstract notions, as evidenced by the numerous examples in Lakoff and Johnson (1980). It is however equally frequent that metaphorical understanding works at sophisticated levels of theorisation. On the frequent use of metaphors in science, see Thomas S. Kuhn (1993: 533–542).

which voluntarily capture their attention. To this end, probably, she will resort to imagery which guide her audience to the mode of thinking she desires.[11]

In this perspective, then, if one is interested in understanding the nature of certain concepts which, in particular cultures or authors, seem to be metaphorically structured, one should look not only for metaphorically motivated linguistic expressions, but also, and above all, for imagery. Indeed, linguistic metaphors, similes, analogies, and so on, cannot be taken neither as simple stylistic elements, nor as immediate linguistic reflections of deeper thought structures. Rather, they should be seen as proper *conceptual instructions* imparted by the sender to the receivers through spoken or written language in a communicative exchange, in order to guide the processes of understanding by means of given conceptual metaphors. It is clear, then, that the study of imagery can, with all due caution, guarantee a privileged access to the metaphorical dynamics that give shape to thoughts according to the sender's intentions.[12]

Now that the full range of exegetical tools has been laid out, it is possible to state the purpose of the operation I wish to undertake in this paper more explicitly. In short, I want to see what a cognition-oriented analysis of Plato's imagery can reveal about his epistemology of cognition in books 6 and 7 of the

11 Consider, for example, Benjamin Franklin's *Advice to a Young Tradesman* [21 July 1748], in which the TIME IS MONEY metaphor is introduced at the very beginning of the article with the sentence: "Remember that time is money," where the linguistic metaphor "time is money" and the example which follows in the text are used to induce the readers, especially the young readers, to think of the use of time in any of their activity as spending or earning money, with the aim of persuading them to minimise the amount of non-working time; for an edition of Franklin's paper, see Leonard W. Labaree (1961: 304–308).

12 This is especially true of Plato, and even more so of the *Republic*, where few ideas are expressed without recourse to similes and linguistic metaphors, or, in general, to anything Plato seems to group under the label *eikôn*. Some useful comments on what Plato means with the term *eikôn* can be found in Elizabeth E. Pender, (2000); but see also Suzanne Saïd (1987); more extensively Elsa Grasso (2003) and Christoph Poetsch (2019). Even if it is somewhat different from what I argue here, see what Nicholas D. Smith maintains about the epistemological and pedagogical roles of "images" in the *Republic*: cf. Nicholas D. Smith (2019: 10–22); see also the useful insights in: Pender (2000: 43–60); Pierre Destrée and Radcliffe G. Edmonds (2017); Elsa Grasso (2023); Franco Ferrari (2023); Francisco J. Gonzalez (2023). All these scholars share the idea that Platonic "images" are rhetorical means of inducing the reader to think in a certain way. From the perspective of this paper, such an induced way of thinking is, more precisely, metaphorical in nature, so that analysing the use of imagery in the *Republic* means examining the ways in which Plato *chooses* to use *language* rhetorically to provoke metaphorical *thinking* in his readers, in order to make them grasp certain notions that would otherwise be difficult to understand, or even to conceive.

Republic. Such an analysis will consist in identifying and examining some of the conceptual metaphors that motivate Plato's linguistic behaviours against the background of his own cultural models and prototypes. In doing so, particular attention will be paid to imagery, insofar as it can reveal the conceptual metaphors which Plato wants his readers to reason with in order to shape and understand his concept of cognition.[13]

3 Imaging Cognition in *Republic* 6–7

With this exegetical apparatus, we are now ready for analysing some of the passages of *Republic* 6 and 7 in which Plato employs imagery to describe and

13 Even if, at least at first glance, the use of a similar set of exegetical tools may appear unusual, it is possible to show that it is completely consistent with other interpretative approaches which try to relocate Plato's thought in its own cultural horizon. One of the most significant attempts in this direction is the one proposed by Andrea W. Nightingale in her seminal (2004) book. What Nightingale does indeed is to identify the existence of a precise cultural model in the Athenian culture of the fifth and fourth centuries B.C., which she names "traditional *theôria*," and to examine the ways in which philosophers from Plato onwards have manipulated this model to shape different prototypes of philosophical *theôria*. In addition, Nightingale brilliantly shows how these prototypes gradually contributed to the formation of a distinct notion of intellectual cognition within the Greek culture of the time. The appearing of such a notion occurred essentially thanks to the functioning and the gradual crystallisation of specific conceptual metaphors derived from the overall cultural model of traditional *theôria*. Related to this last point, moreover, is one of the many virtues of Nightingale's analyses, i.e., their ability to stress the impact of cultural models in the narratological and linguistical choices of the philosophers she considers. This is particularly evident with Plato: Nightingale's main conclusions on the Athenian philosopher derive precisely from a detailed analysis of his use of imagery in the *Republic*, and, more exactly, from an account of the "theoric" structure and dynamics underlying the image of the cave in book 7. In doing so, however, Nightingale programmatically declares to be more interested in placing the philosophical foundation of *theôria* within its historical and cultural context, than in offering a philosophical analysis of fourth-century epistemologies: see Nightingale (2004: 7). On the contrary, what I propose here is, on the one side, to follow the path opened by Nightingale, but, on the other, to develop it in the epistemological direction she discards. This is the reason why I prefer to draw attention not so much to Plato's dependence on particular cultural models, but, rather, to the conceptual metaphors at work behind his use of imagery and language which actually allow him to use those same models for his theoretical ends. Similarly, Giulia Cervato, in this volume, sees the references to the Eleusinian Mysteries in the *Phaedrus* (250b–c) as Plato's way of using a cultural model in order to promote a metaphorical understanding of cognition. For a more detailed discussion of *theôria* in ancient Greek culture and further references on the subject, see Simone Astorino's paper in this volume.

characterise cognition. Due to space constraints, I will focus only on what might be termed, according to the image of the line, as "high cognition" – encompassing the cognitive states of *dianoia* and *noêsis* – even if a similar interpretative strategy can also be applied to "low cognition", namely, *eikasia* and *pistis*.[14]

3.1 *Expanding the Metaphor of Sight with Dreams*

In a passage of book 7, in dealing with dialectics, Socrates first distinguishes it from other kinds of techniques, then he tries to explain the difference it specifically bears with mathematical disciplines, which, in a few lines, will be identified as forms of *dianoia* (533d6–7). Referring to the latter, Socrates contends:

> SOCR: [...] And as for the rest, I mean geometry and the subjects that follow it, we described them as to some extent grasping what is (*tou ontos ti ephamen epilambanesthai*), for we saw that, while they do dream about what is (*oneirôttousi men peri to on*), they are unable to command a waking view of it (*hupar de adunaton autais idein*) as long as they make use of hypotheses that they leave untouched and that they cannot give any reason of (*mê dunamenai logon didonai autôn*). What mechanism could possibly turn any agreement (*homologian*) into knowledge (*epistêmên*) when it begins with an unknown principle and intertwines the conclusion and the steps in between from what is unknown? GLAU. None.[15]
>
> *Rep.* VII 533b5–c3

The first thing to be noted is that dreaming is not contrasted with being awake in itself, but, more specifically, with the kind of sight possessed during the waking state (*hupar ... idein*).[16] Hence, the passage let us think that the grasp that mathematical disciplines are said to have on "what is" (*tou ontos ti ephamen epilambanesthai*) is conceived in terms of vision as well, even if this kind

---

14 For an overview of the immense secondary literature on the divided line, very useful is Yvon Lafrance (1987). More recently, see Nicholas D. Smith (1996); Kenneth Dorter (2004); Richard Foley (2008); Terry Echterling, (2018).

15 Throughout this paper I used G.M.A. Grube's translation of the *Republic* in Cooper (1997). Most of the time, I slightly modified it.

16 One of the most comprehensive studies on dreams in ancient Greek authors, which also includes a section devoted to Plato, is RGA. Van Lieshout (1980); but see also the classic Eric R. Dodds (1951: 102–134).

of vision is different from the waking view. What distinguishes, though, the grasp of dream-view from that of waking-view?

A first passage we should consider in this regard is contained in *Republic* 5, where a definition of *to oneirôttein* is given:

> SOCR: [...] So look: isn't this dreaming: whether asleep or awake, to believe that a likeness is not a likeness but rather the thing itself that it is like? GLAU. I certainly think that someone who does that is dreaming.
> *Rep.* V 476c3–5

Here, dreaming (*to oneirôttein*) is defined as a condition that can interest both the sleeping (*en hupnôi*) and the waking state (*egrêgorôs*). In fact, dreaming seems to be a sort of misleading belief (*êgêtai*) about everything that bears some resemblance with something else (*to homoion*). When a person dreams, what seems to happen is that he or she considers what resembles something not as simply resembling that something, but as being identical with it (*auto*).

So, if we apply the same schema to our passage in book 7, we see that, in conceiving mathematical disciplines as forms of *oneirôttein*, Socrates is saying that mathematicians, as long as they do not put in question their hypotheses, consider the *homologia*, that is, the coherence they obtain by intertwining premises, intermediate propositions, and conclusions according to certain assumed hypotheses,[17] not simply as a coherent system similar to science, but as something identical with science (*epistêmê*), even if they miss the principle of such a coherence.

In regard of dreams, also the *Timaeus* seems to be particularly rich, since an entire section of it (70d7–72d3) is devoted to the description of the psychophysical dynamics involved in dreaming. In this dialogue, dreaming is conceived as a form of divination (*to manteion*, 71e2).[18] Moreover, Timaeus explains that:

17 On the double meaning of *homologia* in this passage, intended both as the coherence existing among mathematical propositions and as the agreement of opinions among experts, see Franco Ferrari (2019). The ambiguity is certainly intentional, even more so as the two senses of the word seem not to be in contrast, but to imply each other. Indeed, the agreement among experts occurs *if and only if* the mathematical construct is consistent with the assumed hypotheses.

18 On divination and, more generally, on altered states of consciousness in ancient Greek thought, see Dodds (1951: 64–101); Yulia Ustinova (2018); Eva Anagnostou-Laoutides (2023).

A (ὄναρ). *Onar* consists mostly of deceptive visions, that is, *phantasmata* (71a6, c3, e8), *phanenta* (72a3), and *eidôla* (71a5, b5).[19]

B (ὄναρ). Visions, in dreams, are outcomes of a psycho-physical dynamic consisting in the reflection of the power of thoughts coming from the intellect on the liver (71a3–d5).

C (ὄναρ). The dreamer cannot explain his/her visions rationally until he/she wakes up (71e6–72a2).

D (ὄναρ). With the awake, the dreamer becomes interpreter of his dreams, that is, a prophet of divinations (*prophêtai de manteuomenôn*, 72b4–5).

E (ὄναρ). Nothing within the dream can make the dreamer interpreter of his dreams (72a2–6).

If we compare this description of dreams with the passage of *Republic* 7, we can see that several correspondences hold. In fact, similarly to the dreaming state:

A (διάνοια). *Dianoia* consists in an *homologia*, that is, in a coherent set of propositions built upon hypotheses.

B (διάνοια). *Homologia* is the outcome of a constructive procedure consisting in assuming hypotheses and intertwining propositions according to them.

C (διάνοια). The mathematician cannot "give reason" (*logon didonai*) of the *homologia* until he knows the principle (533c8–9).

D (διάνοια). Once the mathematician knows the principle, he becomes a dialectician.

E (διάνοια). Nothing internal to the hypothetical procedures and to the coherent set of propositions they permit to build (*homologia*) allows the mathematician to know the principle and so to acquire science (*epistêmê*).

Putting together what *Republic* 5 and the *Timaeus* reveal about Plato's conception of dreams, the difference between the dream-view of dianoetic thought and the waking view of science can be stated as follow:

- According to points A and B, dream-view of dianoetic thought has no direct access to "what is," which it sees just in the form of appearance, that is, in the form of *homologia*, outcome of a constructive procedure.

19 On the use of such a vocabulary in Plato's dialogues, see Saïd (1987) and, recently, Giorgia Stocchi (2022).

Furthermore, in force of the similarity that every appearance bears with its original, two things occur: (1) as long as it does not put in questions the hypotheses on which it relies, dianoetic thought (even in force of point E) is always deceived by the appearance of "what is," and confuses *homologia* with science; (2) it may occur that some information, even all the information contained in the *homologia*, turns out to be true, but this (according to points C and D) can become manifest only once a direct access to "what is" is acquired. Thus, the only possible "grasp" that *dianoia* has on "what is" is what we would call, in modern terms, a conjectural understanding.

– The waking view of science, guaranteed, as Socrates states a few lines after, only by the dialectical method, has instead direct access to "what is." Thus, even when it deals with appearances, dialectic is never deceived, since it always recognises appearances as something similar but not identical with the original. This means that, with regards to mathematical disciplines, by possessing knowledge of the principle, dialectic gives reason of the hypotheses, and sees the *homologia* occurring between mathematical propositions as simple *homologia*, without confusing it with science.

3.2 *The Erotic Metaphor in* Rep. *VI 490a8–B7*

Another important conceptual metaphor for cognition which quite often appears in the central books of the *Republic* is the one that we might call the "erotic" metaphor, that is, a metaphor by which cognition is structured with reference to desire, sex, and birth.[20]

A passage in *Republic* 6 can be particularly interesting for us. After having introduced the famous image of the ship (487e7–489a2), Socrates and Glaucon continue their inquiry on the nature of the philosopher. In this regard, Socrates states:

> First of all, if you remember, he had to be guided by the truth (*hêgeito d' autô … prôton men alêtheia*) and always pursue it in every way (*diôkein*

---

20 As for every aspect of Plato's thought, a plenty of studies have been published on Plato's concept of *erôs*, especially concerning the *Symposium*; see, for example, Thomas Gould (1963); Gerasimos Santas (1979); Gregory Vlastos (1981); David M. Halperin (1990) and Halperin (2017); for the concept of *erôs* in the *Republic*, see Paul W. Ludwig (2007). Particularly interesting for the subject of asymmetry and reciprocity of the erotic desire in Plato's thought is Bruno Centrone (2009). On Plato's use of the erotic metaphor to understand cognition, some remarks in Monique Dixsaut (2000); more extensively, Laura Candiotto (2016).

> *auton pantôs kai pantê edei*), or else he'd really be a boaster, with no share at all in true philosophy. [...] Then, won't it be reasonable for us to plead in his defence that it is the nature (*pephukôs*) of the real lover of learning (*philomathês*) to struggle toward what is (*pros to on ... hamillasthai*), not to remain with any of the many things that are believed to be, that, as he moves on, he neither loses nor lessens his erotic love (*oud' apolêgoi tou erôtos*) until he grasps the nature of each thing that is, with the part of his soul that is fitted to grasp it (*prin autou ho estin hekastou tês phuseôs hapsasthai hôi prosêkei psuchês ephaptesthai tou toioutou*), because of its kinship (suggenei) with it, and that, in getting near what really is (*plêsiasas ... tôi onti ontôs*), having intercourse (*migeis*) with it, and begetting understanding and truth (*gennêsas noun kai alêtheian*), he knows, truly lives, is nourished, and is relieved from the pains of giving birth (*gnoiê te kai alêthôs zôiê kai trephoito kai houtô lêgoi ôdinos*), but not before?.
>
> *Rep.* VI 490a1–b7

Before analysing the passage in more detail, it is worth recalling the main traits of the cultural model in which it is inscribed. This is what Halperin has defined the "socio-sexual system,"[21] typical of the Athenian culture during classical period. Such a model, strongly male-centred and patriarchal, is characterised by two main features: it is polarised, i.e., it includes two actors playing different roles, and hierarchical. It also includes two variants, or prototypes, of sexual desire, the one, so to say, "masculine," and the other "feminine."

> [I]t was characteristic of the ancient Athenians to regard sexual desire as an appetite and, hence, to construe it (by analogy with hunger and thirst) as an acquisitive passion, a longing for the possession and consumption of a desirable object. [But] such a conception of eros – with its emphasis on the object-directed, acquisitive nature of desire – represented to the Athenians a specifically "masculine" model of erotic dynamics. Female desire, as the Greek constructed it, tended by contrast to be related to the physiological economy of the female body, rather than the mind's desires; hence, it is not aroused by individual objects but is governed instead by the requirements of woman's physical constitution with its generative functions.[22]

21 Halperin (1990: 266).

22 Halperin (1990: 276).

According to Halperin, sexual desire was conceived by ancient Athenians as an appetite, in analogy with hunger and thirst. In this frame, masculine desire (1) is strongly object-oriented and (2) has an acquisitive nature. On the contrary, feminine desire (1) is governed by the requirements of woman's physical constitution and (2) is generative or procreative in kind.

First, we should notice that the main features of the socio-sexual model – and, more specifically, of its masculine variant – can be found already active in the first two lines. Here two actors come to the fore: on the one side, the philosopher, "the beautiful and good man" (*ton kalon te kagathon*, 480e5); on the other, the truth (*alêtheia*). The relationship between the two is conceived in terms of a person following a leader. The truth is indeed said to be guiding (*hêgeito*) the philosopher, who, on his side, must follow it totally and in every way (*dein diôkein pantôs kai pantêi*).

What emerges here is then a polarised and asymmetrical relationship in which the role of dominance is occupied by the truth, that of obedience by the philosopher. Moreover, it is already enough clear that the relationship between the philosopher and the truth is highly object-oriented and acquisitive. The verb *diôkô* is indeed strongly connoted in Plato's language in the sense of "trying to catch," being the same verb that Plato often uses to express the act of chasing the pray in hunting or to pursue the enemy in war.[23]

However, it is in the immediately following lines that the masculine socio-sexual prototype comes explicitly to the fore. Beside the explicit erotic vocabulary here deployed – the philosopher addressed as *philomathês* (literally, the *lover* of learning), animated by an unceasing love (*oud' apolêgoi tou erotos*) – two things should be noted: (i) the first one is Plato's use of the verbs *haptomai* and *ephaptomai* with the clear aim of stressing the acquisitive nature of the desire of the philosopher, in accordance with the masculine prototype; (ii) the second one is the change in actors: from the beginning, Socrates ceases to speak of the truth, that is substituted by *to on*,[24] and, at a certain point, even the *philomathês* is substituted with the part of its soul (*hôi psuchês*) that is akin to what is.

Hence, despite being the prototype of desire still masculine in nature, by substituting a female desiring subject (the soul) and a neutral desired object (*to on*) to the previous male desiring subject (the philosopher) and female

23 See, for example, *Prot.* 310c3–4, *Rep.* II 375a5–7. On the hunting metaphor in Plato's thought, see Francesco Aronadio (2014).

24 The ease with which Socrates moves from speaking of *hê alêtheia* to speaking of *to on* in this passage is most likely due to Plato's conception of truth as fundamentally *ontological* in nature: see on this point Bruno Centrone (2014).

desired object (the truth), Socrates already paves the way, starting from the language, for what will be in a few a dramatic change of perspective.

Indeed, in the last part of Socrates' speech the nature of sexual desire radically changes and moves from being a masculine desire, acquisitive in nature, to being a feminine desire, of generative or procreative kind. The soul, and precisely its akin part to what is, gets progressively closer to what is (*plêsiasias*) until the point of being fecundated by it (*migeis*[25] *tôi onti ontôs*).

The final overturning has as its result that of emphasising the peculiar nature of cognition when it comes to what is. What seems to happen is that, in the same moment in which the philosopher's soul reaches and grasps what really is, the roles are radically inverted, and cognition can no longer occur as an acquisitive act of what is by the male philosopher. It occurs instead as a physiological process, and exactly as a childbirth of a pregnant soul. In this new feminine prototype of sexual desire, it is "what is" to play the active and masculine role, whereas the soul emerges in its passive and feminine nature.

In the very last lines of the passage, Socrates asserts that what originates from the fecundated soul are *noûs* and *alêtheia*. Thanks to such a childbirth, the soul frees itself from the pain, and, in this new state, it knows, lives, and grows up "in the mode of truth" (*alêthôs zôiê kai trephoito*).

What does Socrates mean with these expressions? For understanding this point, the value of the participle *gennêsas* is, in my opinion, crucial. Lots of translators intend it in a perfective way, as if Socrates was referring to a self-contained act, whose outcomes are exactly *noûs* and *alêtheia*.

I think instead that the value of *gennêsas* is durative, that is, it does not express a concluded action, limited in time, but an articulated process, the attention being carried in the context not on its finitude and completeness, but on its internal ongoing dynamic. There are two reasons that bring me to think so:

1) **The aspect of the verb**: the aspect is indeed aorist, not perfect, as it should be when the action is concluded. As usually deployed to conceive actions without any temporal connotation, aorist's value is strongly related to the verbs on which it syntactically depends. In this case, *gennêsas* depends on three verbs: an optative aorist, *gnoiê*, and two optative presents, *zôiê* and *trephoito*. It is however quite easy to see that the gen-

25 It is important to remember that the verb *mignumi* acquires a clear sexual connotation in this context, as Vegetti has rightly pointed out: see Mario Vegetti (2007: 764, n. 24). Plato's choice of terminology is unlikely accidental: *mignumi*, which usually expresses the common action of "mixing" things, is also the technical term of Hippocratic medicine to designate the fecundation dynamic, which, in animals and humans, always provides for the interaction of male and female seeds: cf. e.g., Hipp. *De genitura* 5.4–7 (= VII Littré).

eral aspect of the last phrase depends more on the two presents than on the aorist, since the use of the conjunction *te kai* partially identifies the knowing with the way of life and growing up of a living being. Thus, the expression *gnoiê te kai alêthôs zôiê kai trephoito* conveys a single concept that should be understood, according to the value of the presents, as an ongoing process. Consequently, this is also true for the dependant participle *gennêsas*.

2) **The resulting coherence of the erotic metaphor**: the erotic metaphor, indeed, allows Socrates to implement the masculine and feminine prototypes of sexual desire to conceive the cognitive *evolution* of the soul of the philosopher in its passage from the dimension of belief to that of knowledge and understanding. This passage is described as a progressive approximation that culminates with a sudden transformation, in which the soul moves from sexually acting as a male to acting as a female. In this move, it is *the nature of the soul to be transformed* indeed, and hence the nature of cognition: the cognitive contact with *to ontôs on* lends the soul the structure of a pregnant living being, whose proper way of life is, precisely, giving birth.[26] Thus, understanding and truth must be seen as direct expressions of the proper nature of the soul, of its acquired – or recovered – generative power. In such a new (or renewed) condition, they are not simply outcomes of a single act of generation. They are exactly what generation consists in, the two sides of the same coin, that is the life of the pregnant soul, its *gennân* as ongoing *alêthôs zôein kai trephein*. Coherently, the final relief from pain is not properly "final." It should instead be conceived as an ongoing process as well, as a continuous effort of emancipation, protracted as long as generation is still occurring, that is, out of metaphor, as long as the soul exercises its highest cognitive power.

4 Using Oneiric and Erotic Metaphors to Interpret High Cognition

In the final part of my paper, I would like to see if the oneiric and erotic metaphors can help us deepen our understanding of the distinction that Soc-

26 That the general scope of the cognitive process described is naturalistic in nature is already evident in Socrates' use of the adverb *pephukôs*, which, at the very beginning of Socrates' speech, underlies the natural disposition of the *philomathês* towards what is, as Aronadio (2002: 238).

rates poses between *dianoia* and *noêsis* in the passage of the line. Let's read the first passage:

> Consider now how the section of the intelligible is to be divided. [...] In one subsection, the soul, using as images the things that were imitated before, is forced to investigate from hypotheses (*hêi to men autou tois tote mimêtheisin hôs eikosin chrômenê psuchê zêtein anankazetai ex hupotheseôn*), proceeding not to a first principle but to a conclusion. In the other subsection, however, it makes its way to a first principle that is not a hypothesis, proceeding from a hypothesis and without the images used in the previous subsection (*aneu tôn peri ekeino eikonôn*), using forms themselves and making its investigation through them.
>
> *Rep.* VI 510b2–8

Even though both the subsections of high cognition are introduced, it seems that the main focus of the passage is the association between *dianoia* and the use of images.[27] Precisely, the hypothetical method is presented as somehow strictly related to the soul considering and using the objects of the previous subsections (plants, animals, and artificial objects, see *Rep.* VI 510a5–6) *as* images (*hôs eikosin*) of something else.

Now, the specific feature of images seems to be their complete dependence on the reference to the originals of which they are images. In other words, an object can be considered as "image" only in so far as it is related to the original dimension it reflects or represents, even though this dimension remains inaccessible.[28] This, moreover, seems to be the reason behind the hypothetical method: in order to conceive of sensible objects as images of something else, the soul is forced (*anankazetai*) to *assume*, by hypothesis, that very something else, i.e., the existence and the properties of the objects of which they are images.[29]

---

27 See also *Rep.* VI 510e1–511a2 and 511a7–8.

28 On the relationship between image and original in Plato's thought, and on the positive epistemological value Plato ascribes to sensible objects by conceiving them as images, especially as sacred statues (*agalmata*), cf. Astorino, in this volume. A similar epistemological value of the sensible objects as images of the ontological realm of Forms is recognised by Giulia Cervato (in this volume) who interestingly links the inability to recognise an image *qua* image as the meaning behind the metaphor of blindness as it appears in the *Phaedrus*: cf. Cervato, in this volume.

29 For a similar interpretation of the relationship existing between images and hypothetical method, see Smith (2019: 119–121).

So, as compared to *pistis*, which essentially seems to consist in perceiving concrete sensible objects and in having opinions about them, the use of those same objects as images is positively presented by Socrates as an advantage in cognitive terms. Indeed, conceiving sensible objects as resemblances, similar, but not identical, with a supposed original dimension of more authentic objects, opens up the possibility of cognitively transcending the entire sensible realm, since it forces to assume the existence and the properties of intelligible entities.

But this is not the whole story. It is at this point that it is useful to refer to what the oneiric metaphor has shown us about *dianoia*. As we have seen before, the oneiric metaphor presents *dianoia* as a way of dreaming, and what characterises dreams, according to Plato, is their being composed by deceptive visions (*phantasmata, phanenta, eidola*), that is, by resemblances that pretend to be originals by hiding the difference that separates them from the real originals.

What is interesting about the oneiric metaphor is that to be conceived as deceptive visions are the same things that the mathematicians assume by hypothesis, as a consequence, as we have seen right now, of using as images the concrete sensible objects.

So, the oneiric metaphor shows the limits deriving to *dianoia* from the use of sensible objects as images in trying to see the originals. On the one side, using sensible objects as images opens up the possibility of transcending the sensible realm by supposing the existence and by guessing the properties of a further original dimension, which is not sensible at all. On the other, by only proceeding this way, what can be said of that supposed original dimension will be only conjectural.

The oneiric metaphor seems then to warn about the danger inherent in conjecturing from sensible objects with the aim of seeing their intellectual counterparts, the danger consisting in mistaking as original what is only the result of a construction. The conjectural nature of *dianoia* is indeed similar to what happens in dreaming: something appears to the mind, and it is thought to be real, even if it is only a product of a psychophysical dynamic whose outcomes are *eidôla, phanenta,* and *phantasmata.*

In the last passage I would like to consider, Socrates resumes the discussion of the last subsegment of the intelligible part of the line, that occupied by *noêsis*.

> Then also understand that, by the other subsection of the intelligible, I mean that which reason itself grasps (*autos ho logos haptetai*) by the power of dialectic. It does not consider these hypotheses as first prin-

> ciples but truly as hypotheses, as stepping stones to take off from, enabling it to reach the unhypothetical first principle of everything (*mechri tou anupothetou epi tên tou pantos archên*). Having grasped this principle (*hapsamenos autês*), it reverses and, being in turn contiguous to the things that are contiguous to it (*palin au echomenos tôn ekeinês echomenôn*), comes down to a conclusion without making use of anything sensible at all, but only of Forms themselves, moving on from Forms to Forms, and ending in Forms.
>
> *Rep.* VI 511b2–c2

First, note the reappearing of the same terminology we have seen deployed by Socrates in the erotic metaphor. I refer especially to the twofold use of the verb *haptomai*.

Second, the narratological structure of the passage seems to perfectly reflect the general schema we have seen active in the erotic metaphor. In that occasion, a process of approximation to "what is" was followed by a grasp on "what is" and by a sudden transformation of the soul and its resulting activity.

Even here, what we see is a movement divided in three steps: (i) the approximation to the principle of everything; (ii) the moment of the grasp on the principle; and (iii) a final moment in which the discourse reverses and comes down to a conclusion.

Both elements lead us to suspect that the erotic model could be somehow active in the present passage as well. Let see if, by keeping in mind such a model, a better understanding of the passage results.

If we read the present lines this way, step (i) should be understood according to the masculine prototype of desire. This would mean that the *logos*, in this phase, should be animated by an acquisitive pulsion towards its object (the principle of everything) that would still lie beyond its limits.

Then, the action expressed in step (ii) by the verb *haptomai* should not be intended as a simple grasp, but as the first grasp occurring in a sexual intercourse where the involved partners are the *logos* and the principle of everything.

Finally, according to the feminine prototype of desire, two things should occur: (1) the female pole of the sexual dynamic, in getting pregnant, should observe a change *in its own nature*; and (2) its new nature should be expressed in the form of an *ongoing generative process*, as fulfilment of its new procreative desire.

So, in step (iii), the descent of the *logos* towards a conclusion, what is here described as its use of Forms, its proceeding through them, and its ending in them, should be understood as a generative process, expression of the new nature that the *logos* gets after having been fecundated by the principle of everything. The Forms, in this sense, are not something external to the *logos*, lying beyond its limits and to which the *logos* tries to approximate, but something concerning its very nature, its structure, and dynamic articulation.

To further support this last point, we can highlight Socrates' use of the expression *palin au echomenos tôn ekeinês echomenôn*. Firstly, we should note that the participle *echomenos* seems to express a condition rather than an action, in particular the condition of being "spatially contiguous" to something.[30] Moreover, the anaphoric character of the aside, due to the repetition in the polyptoton of *echomenos* and *echomenôn*, seems to emphasise, from a stylistic point of view: (1) the isomorphism between the *logos* – i.e., the *echomenos* – and the *echomena* – which I would tend to interpret as a reference to the Forms explicitly mentioned right after;[31] and (2) the common "contiguity" of the Forms and *logos* to the principle of everything (*ekeinês*, to read *archê*).

In other words, after grasping the principle, the *logos* somehow receives the structure of the domain of the Forms, so that its usual action of descending towards a conclusion, that was typical also of the hypothetical method, is now a direct expression of that "eidetic" structure that, after the grasp, has become its own.

5 Conclusions

To conclude, I would like to summarize the main traits of "high cognition" as it is imaged by Plato in *Republic* 6 and 7.

1) The oneiric metaphor has shown how the distinction between *dianoia* and *noêsis* is conceived on the basis of two forms of vision, one directed to resemblances, the other to the originals. In this distinction, the oneiric metaphor emphasizes the limits of *dianoia* as compared to *noêsis* by stressing its constructive nature and the merely conjectural character of the conclusions to which it can come. Moreover, on the basis of the erotic metaphor, *dianoia* is conceived as an "acquisitive" phase of cog-

30 Lots of occurrences in this sense may be found in Herodotus and Thucydides, where the participle *echomenos* seems to be part of a geographical terminology: see, for example, Thucid., *Hist.* 2.96.4, 5.67.2, 8.90.5, Herod., *Hist.* 2.12.1, 2.93.4, 2.170.1–2, 4.176.1; similar uses in Isocr., *Pan.* 96.4, Xenoph., *Hell.* 3.2.25, 4.4.9, 6.5.16, 7.1.20. It is really interesting that, in the dialogue *Parmenides* (148e9), Plato recurs to the expression *tên echomenên chôran* to make more explicit the concept of "contact," already expressed in the passage by means of the verb *haptomai*.

31 Similar considerations in Aronadio (2002: 259–260).

nition, during which the real object transcends the cognitive limits of the soul.

2) The oneiric metaphor associates *noêsis* with awake vision, and therefore carries the idea of the presence of the object in front of the observer. In this way, the object is no outcome of any construction, and it offers itself to the cognitive subject in a completely independent way. So, the cognitive subject, in the form of *logos*, cannot do anything else than conform to it. Even more emphatically, the erotic metaphor suggests that the character of noetic cognition is that of the sexual union followed by generation. This seems to mean that it is the very nature of cognition that changes by virtue of the presence, within its boundaries, of its object. Out of metaphors, therefore, it seems that the main feature of noetic cognition is its strong dependence on the nature of its cognitive object, addressed in various ways as "truth," "what is," or "principle of the whole." The dependence is such that noetic cognition is conceived not only as the presence in the soul of thoughts whose content reflects what is found in an external reality, but as the appearance of reality itself in the concrete exercise of thought, once reality has returned to be what it has always been: the nature (as origin and as true essence)[32] of the *logos*.[33]

Bibliography

Allbritton, David W. 1995. "When Metaphors Function as Schemas: Some Cognitive Effects of Conceptual Metaphors." *Metaphor and Symbolic Activity* 10 (1): 33–46.

Anagnostou-Laoutides, Eva. 2023. "Man before God: Music and Silence as Induction to Altered States of Consciousness from Plato to Clement of Alexandria." In *Later Platonists and Their Heirs among Christians, Jews, and Muslims*, edited by Eva Anagnostou-Laoutides and Ken Parry, 25–60. Leiden: Brill.

Aronadio, Francesco. 2002. *Procedure e verità in Platone: Menone, Cratilo, Repubblica.* Napoli: Bibliopolis.

---

32 On the concept of *phusis* in the Presocratics and Plato, see Dietrich Mannsperger (1969); Gérard Naddaf (2005).

33 I would like to express my sincere gratitude to all the participants and the organiser, Prof. Refik Güremen, of the First Middle East Meeting of the International Plato Society, held in Ankara in May 2023, for their thorough comments and stimulating criticisms. I would also like to thank Prof. Bruno Centrone for carefully reading and discussing with me several important aspects of the present paper. Special thanks go to Prof. Francesco Aronadio, for his rich and meticulous insights and for the invaluable support and wise guidance he offered me throughout the entire elaboration of this paper.

Aronadio, Francesco. 2014. "La 'Caccia' Dialettica in Alcuni Luoghi Platonici." In *KALLOS KAI ARETE*. *Bellezza e Virtù. Studi in onore di Maria Barbanti*, edited by R. Loredana Cardullo and Daniele Iozzia, 67–77. Acireale: Bonanno Editore.

Candiotto, Laura. 2016. "Nous e phren: conoscenza intellettuale, razionalità discorsiva e saggezza erotica in Socrate e Platone." *Methodos* 16.

Centrone, Bruno. 2009. "Il Problema Della Reciprocità Nell'Eros Platonico." In *Metamorphoses of Love: Metamorfosi dell'amore*, 33–49. Pisa: ETS.

Centrone, Bruno. 2014. logica, ontologica in Platone." *Méthexis* 27: 7–23.

Cooper, John M., and DS. Hutchinson, eds. 1997. *Plato: Complete Works*. Indiana: Hackett Publishing Company.

D'Andrade, Roy G. 1995. *The Development of Cognitive Anthropology*. Cambridge: Cambridge University Press.

Deignan, Alice. 2005. *Metaphor and Corpus Linguistics*. Amsterdam-Philadelphia: John Benjamins Publishing Company.

Destrée, Pierre, and Radcliffe G. Edmonds. 2017. "Introduction: the Power – and the Problems – of Plato's Images." In *Plato and the Power of Images*, edited by Pierre Destrée and Radcliffe G. Edmonds, 1–10. Leiden: Brill.

Dixsaut, Monique. 2000. *Platon et la question de la pensée. Études platoniciennes I*. Paris: Vrin.

Dodds, Eric R. 1951. *The Greeks and the Irrational*. Berkeley: University of California Press.

Dorter, Kenneth. 2004. "The Divided Line and the Structure of Plato's 'Republic'." *History of Philosophy Quarterly* 21 (1): 1–20.

Echterling, Terry. 2018. "What Did Glaucon Draw? A Diagrammatic Proof for Plato's Divided Line." *Journal of the History of Philosophy* 56 (1): 1–15.

Ferrari, Franco. 2019. "Homologia e dialettica in Platone." *Antiquorum Philosophia* 13: 23–44.

Ferrari, Franco. 2023. "La Relation Insaisissable: Images Et Métaphores Dans L'Ontologie de Platon." In *Platon et la pensée de l'image*, edited by Anca Vasiliu and Elsa Grasso, 63–83. Paris: Vrin.

Foley, Richard. 2008. "Plato's Undividable Line: Contradiction and Method in Republic VI." *Journal of the History of Philosophy* 46: 1–23.

Gonzalez, Francisco J. 2023. "La République Comme Image Trompeuse." In *Platon et la pensée de l'image*, edited by Anca Vasiliu and Elsa Grasso, 85–103. Paris: Vrin.

Gould, Thomas. 1963. *Platonic Love*. London: Routledge & Kegan Paul.

Grasso, Elsa. 2003. *Copie, simulacre et vérité chez Platon*. PhD diss., Aix-Marseille 1, Université de Provence.

Grasso, Elsa. 2023. "L'Image Contre L'Image. Le Triptyque de République Vi Et Vii: Soleil, Ligne, Caverne." In *Platon et la pensée de l'image*, edited by Anca Vasiliu and Elsa Grasso, 17–59. Paris: Vrin.

Halperin, David M. 1990. "Why Is Diotima a Woman? Platonic Eros and the Figuration of Gender." In *Before Sexuality. The Construction of Erotic Experience in the Ancient Greek World*, edited by Froma I. Zeitlin, John J. Winkler, and David M. Halperin, 257–308. Princeton: Princeton University Press.

Halperin, David M. 2017. "L'Erôs Platonicien Et Ce Qu'On Appelle «Amour»." In *Érotique et politique chez Platon. Erôs, genre et sexualité dans la cité platonicienne*, edited by Luc Brisson and Oliver Renaut, 19–69. Sankt Augustin: Academia Verlag.

Johnson, Mark. 1987. *The Body in the Mind: The Bodily Basis of Meaning, Imagination, and Reason*. Chicago: University of Chicago Press.

Kuhn, Thomas S. 1993. "Metaphor in Science." In *Metaphor and Thought*, edited by Andrew Ortony, 2nd ed., 533–542. Cambridge: Cambridge University Press.

Labaree, Leonard W. 1961. *The Papers of Benjamin Franklin*. Vol. 3. New Haven: Yale University Press.

Lakoff, George. 1987. *Women, Fire, and Dangerous Things: What Categories Reveal about the Mind*. Chicago: University of Chicago Press.

Lakoff, George, and Mark Johnson. 1980. *Metaphors We Live By*. Chicago: University of Chicago Press.

Lafrance, Yvon. 1987. *Pour interpréter Platon: la ligne en République VI, 509d–511e. Bilan analytique des études, 1804–1984*. Paris: Bellarmin.

Ludwig, Paul W. 2007. "Eros in the Republic." In *The Cambridge Companion to Plato's Republic*, edited by GRF. Ferrari, 202–31. Cambridge: Cambridge University Press.

Mannsperger, Dietrich. 1969. *Physis Bei Platon*. Berlin/Boston: De Gruyter.

Naddaf, Gérard. 2005. *The Greek Concept of Nature*. SUNY Series in Ancient Greek Philosophy. New York: SUNY Press.

Nightingale, Andrea W. 2004. *Spectacles of Truth in Classical Greek Philosophy: Theoria in Its Cultural Context*. Cambridge: Cambridge University Press.

Pender, Elizabeth E. 2000. *Images of Persons Unseen: Plato's Metaphors for the Gods and the Soul*. Sankt Augustin: Academia Verlag.

Poetsch, Christoph. 2019. *Platons Philosophie des Bildes: Systematische Untersuchungen zur platonischen Metaphysik*. Frankfurt-am-Main: Vittorio Klostermann.

Quinn, Naomi, and Dorothy Holland. 1987. "Culture and Cognition." In *Cultural Models in Language and Thought*, edited by Dorothy Holland and Naomi Quinn, 3–40. Cambridge: Cambridge University Press.

Saïd, Suzanne. 1987. "Deux noms de l'image en grec ancien: idole et icône." *Comptes rendus des séances de l'académie des inscriptions et belles-lettres* 131 (2): 309–330.

Santas, Gerasimos. 1979. "Plato's Theory of Eros in the Symposium." *Noûs* 13 (1): 67–75.

Smith, Nicholas D. 1996. "Plato's Divided Line." *Ancient Philosophy* 16 (1): 25–46.

Smith, Nicholas D. 2019. *Summoning Knowledge in Plato's Republic*. Oxford: Oxford University Press.

Sticht, Thomas G. 1993. "Educational Uses of Metaphor." In *Metaphor and Thought*, edited by Andrew Ortony, 2nd ed., 621–632. Cambridge: Cambridge University Press.

Stocchi, Giorgia. 2022. "Il Lessico Dell'Apparenza in Platone: Phasma, Phantasma E Phantasia." In *Platone e l'esperienza. Sette studi lessicologici*, edited by Francesco Aronadio, 27–67. Napoli: Bibliopolis.

Ustinova, Yulia. 2018. *Divine Mania: Alteration of Consciousness in Ancient Greece*. London: Routledge.

Van Lieshout, RGA. 1980. *Greeks on Dreams*. Leiden: Brill.

Vegetti, Mario, ed. 2007. *Platone. La Repubblica*. Milano: BUR.

Vlastos, Gregory. 1981. "the Individual as Object of Love in Plato." In *Platonic Studies: Second Edition*, 3–42. Princeton: Princeton University Press.

CHAPTER 9

Seeing the Divine Through Its Visible Images: Plato's Philosopher and the *Theôros*

Cesare Simone Astorino

1 Introduction

In classical Athens, *theôria* referred to a journey undertaken by private individuals or official ambassadors, involving the contemplation of something most often religious in content. Scholars have increasingly recognised Plato's indebtedness to this tradition, on which he drew in moulding the figure of the philosopher, as it appears from the so-called middle dialogues. In the *Symposium*, the *Phaedo*, the *Republic*, the *Phaedrus*, and partially in the *Timaeus* and the *Laws*, the philosopher's portrayal is highly reminiscent of that of the *theôros*, i.e. the delegate in charge of the "theoretical" journey. In order to justify this idea, a number of reasons have been recalled. For instance, it has been observed that, like the *theôros*, the philosopher must metaphorically abandon the ordinary dimension in which he lives and embark on a journey to broaden his knowledge, or that both strive towards a visual-contemplative knowledge that is usually (though not exclusively) set over the divine realm.[1]

On the other hand, the dialogues offer several examples where the terminology of *theôria* and various visual verbs are used to describe the sensible world, providing us with instances of Plato's positive account of it, as has already been observed.[2] In this paper, I shall analyse the topics of recollection and philosophical madness in the *Phaedrus* to explore some of these examples. I shall argue that sensible objects play a crucial role in Plato's process of cognition. My thesis is that the descriptions of the sensible world as an image (*eikôn*) and cult statue (*agalma*) of the domain of the forms suggest that the former can lead to the latter. The sensible realm is not a distraction or an obstacle to understanding the intelligible one, as many scholars believe, but rather a *pathway* to it. Like cult statues, sensible objects constitute a means of

1 Andrea Wilson Nightingale (2004: 100–113); Ian Rutherford (2013: 324–338); Julia K. Ward (2021: 76–80). On the comparison between the *theôros* and the Athenian of the *Laws*, cf. Julia Pfefferkorn (2023: 19–24).
2 Nightingale (2004: 139–141).

HTTPS://DOI.ORG/10.1163/9789004747128-011

relationship between the human and the divine. Building upon this characterisation, I also attempt to make the more ambitious claim that sensibles are not just functional, but *necessary* for humans to start their quest towards the divine domain. As statues make it possible for us to initiate interaction with the deities they represent, likewise sensible objects are necessary to start rediscovering the forms.

The paper is divided as follows. After a brief presentation of the institution of *theôria* and the role of the *theôros*, I turn to the *Phaedrus*, focusing on three aspects. Firstly, I shall argue that Plato frames the discussion of recollection within a sharp distinction between events occurring in the soul's discarnate state and in its embodied condition. While in the former scenario, the soul has direct access to the forms, in the latter, since forms are separate, the access is indirect – if not always, at least for a significant and prolonged amount of time. Both scenarios, however, employ visual lexica and the *theôr/ia/eô* family of words. I shall analyse some of these occurrences and show that there are three uses of *theôria* in the *Phaedrus*, as this word can indicate the contemplation of (1) the forms, (2) the sensible objects, or (3) the forms through the sensible objects. In the story, the first kind of *theôria* is available to the discarnate soul, while the other two pertain to the soul during its embodied existence. Since the paper is devoted to the positive role that sensible objects can play and their connection with traditional *theôria*, I shall mainly focus on the second and the third cases.

In order to see how sensible objects can lead one to the forms, I shall then emphasise the idea that they are throughout described as images and cult statues (*agalmata*). The description of these entities as *agalmata* is noteworthy for two reasons. First, it provides insight into how the relationship between the sensible and the intelligible might be understood in Plato's ontology. Second, it suggests that, at least in the *Phaedrus*, the sensible world plays the same role with respect to the intelligible as the statue with respect to the divine it represents. By analysing the passage devoted to recollection (*anamnêsis*), I shall show how the sensible world constitutes a means for humanity to get in contact with the divine. By exploring the philosopher's madness, I shall then contend that sensible objects not only help us interact with the divine but also make this communication possible. Just as the *theôros* can only engage with the gods through their material symbols, the philosopher can only behold Beauty itself through a visible image, and not directly (with few *possible* exceptions). The frustration resulting from the possibility of engaging with the divine only in a mediated way leads to madness.

Finally, I shall seek to explain why people are usually not led to the forms by simply gazing at sensible objects, although sometimes contemplating them proves virtuous. To do this, I shall draw on Jaś Elsner's distinction between "naturalistic visuality" and "ritual-centered visuality" and suggest a further element of connection between the philosopher and the *theôros*.

2 *Theôria*

Traditional *theôria* was a revered practice in ancient Greece, encompassing social, political, economic, and religious aspects. In the classical period, it involved the polis participating in extra-territorial cults, festivals, and sanctuaries by sending delegates (*theoroi*) – although a *theôros* could also undertake such visits on his own and thus engage in a private rather than a civic *theôria*.[3] More specifically, this practice implied that *theoroi*, usually on behalf of the city, detached themselves from their usual life in the polis to embark on a journey or pilgrimage outside the ordinary spaces to reach places consecrated to cults or sites where religious festivals occurred.

Nightingale claims that *theôria* was a kind of pilgrimage involving a triadic and circular pattern.[4] The first step involved detachment from the city, i.e. distancing oneself from the social and cultural rules, conditions, and practices of one's own city. The second consisted in entering a 'liminal' phase, by transitioning through a realm or dimension that had little or none of the features of either the past or the future state. In this phase, the viewer encountered the Otherness, both in the social sense of meeting new people and visiting foreign places and in the religious sense of engaging with the god who presides over a particular festival or shrine. This phase culminated in the witness of events, spectacles, or the god himself. At last, and closing the circle, the *theôros* returned to his city and his ordinary life.

The *theôros* performed functions and was concerned with matters that extended beyond spectatorship, including socio-economic and political activities. Nonetheless, the act of spectating appears to be the most prominent and distinctive aspect of *theôria*: "[…] the festivals that were the primary destination of these delegates were regarded as spectacles in which the default form of participation was by *watching*."[5]

---

3 Following Nightingale (2004: 3 note 2), I shall adopt the masculine throughout the paper when referring to the *theôros* and the philosopher "since *theoria* (both traditional and philosophical) was predominantly – thought not exclusively – practised by males."

4 Nightingale (2004: 42–44). On the importance of circular motion, see Ward (2021: 28–49, 52–59).

5 Rutherford (2013: 5, my italics). Also, cf. Ward (2021: 17) and Nightingale (2004: 44–47).

At least on journeys to consult an oracle or to observe statues in shrines and sanctuaries, however, the *theôros* did not merely "observe." By contrast, on these occasions, the *theôros* engaged in a unique form of viewing, "a sacralized mode of spectating" (Nightingale) or a "heightened perception of something sacred" (Ward). Following Jaś Elsner, this specific mode of spectating can be described more precisely as "ritual-centered visuality," for it "denies the appropriateness [...] of interpreting images through the rules and desires of everyday life. It constructs a barrier to the identifications and objectivations of the screen of discourse and posits a sacred possibility for vision."[6] This means that, in the ritualised context, the *theôros* activated a new kind of sight, "a sacred possibility of vision which is by definition more significant [than the ordinary one] since it opens the viewer to confronting his or her god."[7]

The exact nature of this relationship with the god remains obscure though. For sure, according to the accounts of traditional *theôria*, *theoroi* used not to see the gods, but only their sacred images or statues.[8] These images, usually called *agalmata*, were sculpted statues or other representations of the deities to whom the sites were dedicated and had a votive significance. Although *theoroi* only contemplated these objects, they could likely "see" the gods behind their material representations when engaging in the aforementioned religious kind of viewing. Thus, they performed a kind of *indirect* contemplation. But why indirect? This is because (or so I believe), according to a pivotal tenet of Greek religion, humanity was not allowed to see the gods directly and clearly. As Vernant has put it, "Seeing a god face-to-face [...] is far more than human strength can bear. [...] The body of the gods shines with such an intense brilliance that no human eye can bear it."[9] Specifically, encountering a god directly was reckoned to lead the human perceiver to metamorphoses ranging from changing in physical form to getting disabled, especially blinded, or, even worse, to death.[10] Communicating with the divine, therefore, required intercession through images or disguises, such as statues, which could *conceal* and *contain* its overwhelming power.[11] Thus, if the *theôros* could engage in the

6 Jaś Elsner (2000: 62).
7 Elsner (2000: 62).
8 Rutherford (1995: 283) has rightly emphasised that "For Greek *theoroi*, the culmination of a pilgrimage is contemplation of a statue or sacred image."
9 See Jean-Pierre Vernant (1991: 44) and Deborah T. Steiner (2001: 80–81).
10 Correctly, Steiner (2001: 81): "Cult [...] requires statues that make accessible the immortal presence," and images were believed to answer that need.
11 Again Steiner (2001: 81): "Concealment and containment thus form two dimensions of the single enterprise: in assuming a form or body not his or her own, the god simultaneously masks and contains an untenable force."

divine in a highly ritualised context and through a statue that straddles the human and the divine realm, he could not leave this same statue behind in the perfect and complete grasp and vision of the god himself.

3 Recollection

The palinode contains Plato's most detailed account of the story of the soul in his dialogues. The account is divided into two parts, each corresponding to a different temporal period. The first part narrates the experiences of godly and human/non-human souls before their incarnation, while the second deals with the adventures of human souls once they have descended into bodies. These two temporal moments, interrupted by the souls' descents into the bodies, correspond to two different ways of relating to the forms and three distinct types of contemplation.[12]

Socrates begins illustrating the otherworldly kind of *theôria*, referring to the contemplation of the forms by the godly souls. The description of their vision contains the three key elements of *theôria*: detachment, spectating, and return. Transported by circular revolutions and driven by obedient horses, they depart from their habitual place to go out of the universe and gaze upon (*theôrousi*, 247c1) the forms (247b–c). There, they witness the divine forms (*idousa*, *theôrousa*, 247d3–4; *kathorai*, 247d5, 247d6, 247d6), as these were spectacles or holy objects (*makarian opsin te kai thean*, 250b6–7). They enjoy an effortless 'vision,' as they are entirely passive (*dexasthai*, 247d3),[13] and attain a complete and synoptical knowledge, as they metaphorically rise above the surface of the celestial vault, from where they can behold the entire realm of the forms. After having "seen" all the forms and having "feasted on" them (*theasamenê kai estiatheisa*, 247e3), they return to the heavens and go home.

While the godly souls relish in a comprehensive vision of the forms, the non-godly ones trample and strike one another, trying to get ahead of the others. Most non-godly souls, pulled in different directions by the horses, fail to view the forms and must content themselves with mere beliefs (248a–b).

---

12 As for the way I read Plato's myths, I follow Christopher Gill (1993: 56): "It is characteristic of what we call Plato's 'myths' that they present, usually in narrative or descriptive form, accounts of subjects of which we do not, and indeed cannot, know the truth [...]. The reasons why the truth is not available vary [...]. In all these cases [...] the myths are, to this extent, 'noble lies', false in the form which they are presented, though ones which are made, presumably, 'as like as true as possible' (*R.* 382d2–3)."

13 This may indicate that the soul assumes a feminine role in the process of knowledge. Picciafuochi discusses this topic in this volume, with a particular focus on *Rep.* 490a1–b7.

Only the future human souls, i.e., those who succeeded in following their gods, can gaze at the forms, although to a lesser extent than the gods. Unlike the gods, they struggle to look beyond the surface, and can only do so briefly and intermittently (*mogis*). Despite this, the persistent use of the same visual verbs employed to describe the contemplation enjoyed by the gods (cf. *kathorôsa, eiden, theaomai*)[14] suggests that human souls, when freed from bodily constraints, can attain a cognitive state like that of the gods. In this way, Plato seems to offer a description of what it means to possess knowledge of the forms in the full sense, while also highlighting the difficulty of acquiring this epistemic level not only in life, but also after death.

The introduction of recollection marks a turning point in the story, indicating the transition from what occurs in the *otherworldly* dimension to what occurs in the *worldly* one and from one kind of *theôria* to the other:

> For a human being must understand a general conception formed by collecting into a unity by means of reason the many perceptions of the senses; and this is a recollection of those things which our soul once beheld, when it journeyed with God and, lifting its vision above the things which we now say exist, rose up into real being.[15]
>
> 249b6–c4

By contrasting what happened "then" and "now," Socrates emphasises the distinction between the past time just described, in which souls could directly engage with the forms, and the present time, in which the forms, being separated and therefore not available to the senses, can only be remembered from their sensible instances (and then progressively discovered through memory and *logos*).[16] How this process works is not straightforward, but it is overall

---

14 In the whole section, we find several visual verbs used to refer both to the godly and non-godly vision of the forms: e.g., θεατὴ (247c7); ἰδοῦσα (247d3); θεωροῦσα (247d4); καθορᾷ (247d5, 247d6, 247d6, 248a4); θεασαμένη (247e3); καθορῶσα (248a4); εἶδεν (248a6); θέας (248b4); κατίδῃ (248c3); ἰδεῖν (248b6), ἴδῃ (248c6); ἰδοῦσαν (248d2); ἰδοῦσα (249b6); εἶδεν (249c2), θεάομαι (249e5).

15 This passage is notoriously complex, partly due to the corruption of the Greek text (which has been variously amended) and partly because it offers a very concise summary of Plato's "theory of recollection." Given that this passage is not essential for the purposes of this paper, I shall simply quote Fowler's (1925) translation. Unless otherwise indicated, translations are my own.

16 It is a matter of much controversy whether Plato believed that the soul, even during its discarnate existence, could have direct and non-propositional knowledge of the forms. I shall not enter this debate here. However, it seems to me that, at least within the *Phaedrus*, Plato is drawing a sharp distinction between two different modes of cognition

clear that while the forms cannot be "seen" anymore, they can be recollected starting from the perception of sensible objects.

The contrast between the two lives of the soul and the two corresponding methods of dealing with the forms is further emphasised immediately afterwards. Here, Socrates underscores the exclusive capacity of the philosopher to regain his wings by engaging with the forms in the earthly realm through the memory (*mnêmêi*) and the perception of sensible objects capable of awakening such memory (249c). What the philosopher can, therefore, do through life is to approach the forms through his memory and try to make them emerge gradually.[17] At 250c, Socrates draws this same contrast for the third time, claiming that, while *before* we could see the forms in pure light as we were

related to the two lives of the soul – the discarnate and the embodied. Although the usage of *verba videndi* can be interpreted in different ways, I believe that Plato's extensive use of this kind of verbs in these pages suggests that, in its discarnate condition, souls have a (more) direct approach to the forms. Moreover, while it may be true that "[t]he fundamental conception of the Forms is, from the beginning, linguistic rather than visual in its orientation," I think Nightingale is right in claiming that "the descriptions of the philosophical soul gazing on the Forms capture something essential about the experience and nature of knowledge as Plato conceived it" (Cf. Nightingale [2004: 110] against Charles Kahn [1996: 354–355]).

17 Cf. Herman L. Sinaiko (1965: 78): "[...] literally they [*sc.* humans] are incapable of that instantaneous motion to the realm of Being where they can find their proper nourishment through the direct contemplation of the ideas. This is the negative side of the human condition; on the positive side is the phenomenon of 'memory' or 'reminiscence.' [...] Reminiscence thus serves as the last tenuous link connecting the human souls in its present pathological condition with its former, natural state." On the role of memory as an intermediary between humans and the forms, see also 253a, where the philosopher is said to succeed, through memory (μνήμῃ), in coming into contact with the god he has followed after death, and in assuming his attitudes and occupations so far as possible for a human being. The view that the method of investigating the forms is different in the disembodied and embodied states can be interpreted in various ways. I shall mention two possibilities. One is to assume that there is a *quantitative* distinction between the knowledge of the forms attainable after death and during life and to claim that *post-mortem* knowledge is only more comprehensive than the one that can be accessed during life. On the other hand, one can suppose a *qualitative* difference: whereas the cognitive modality available to the soul in the disembodied state is direct, the one available to it in the incarnated state is mediated. This sets insuperable limitations to the knowledge of the forms during life (cf. Franco Trabattoni [2016]). I shall not side here with any of these readings. However, I think that, even if we attribute to the philosopher the possibility of achieving a perfect and complete knowledge of the forms during his lifetime, the path to this knowledge is extremely uneven and undermined by several factors, such as the ἀμυδρὰ ὄργανα (250b3–4), i.e., our bodily impediments to the perfect knowledge of the intelligible beings, cf. *Phd.* 65a–67d. Cf. Bruno Centrone (1998: 150–151, note 123).

pure ourselves, we are *now* buried in a body, locked in it like an oyster in a shell.[18]

Due to this cognitive divide, the reader can expect that the objects of contemplation are now mostly the sensible ones. Indeed, Socrates says that the fourth kind of *mania* stems from seeing (*horôn*, 249d5) a beautiful object down here that reminds us of the Beauty itself. Moreover, he later mentions the possibility of being led from down here to the intelligible world while contemplating what is named after the Beauty itself (*theômenos autou tên têide epônumian*, 250e3), i.e., its visible manifestations. In these cases, Plato seems to allude to a kind of *theôria* devoted to sensible objects *qua* sensible objects. Elsewhere, however, the term *theôria* appears to be set over sensible objects, but it actually refers to the forms that they represent. This indirect *theôria* of the forms is explicitly mentioned by Socrates when he argues that only a few people, when approaching *images*, can contemplate the *kind* [or the form: *genos*] *of what they imitate* (*oligoi epi tas eikonas iontes theôntai to tou eikasthentos genos*, 250b4–5). This means that there are sensible entities that allow us to see beyond them if approached properly, like an opaque glass. On this reading, sensible objects do not obstruct but rather facilitate our knowledge (in a general sense) of what they reveal.[19]

As a result, we have three kinds of *theôria*: a direct *theôria* of the forms, a *theôria* of sensible objects, and a *theôria* of sensible objects that results in an indirect *theôria* of the forms. Viewed from this perspective, the observation and study of sensible objects is not, *ipso facto*, a misguided and blameworthy path of inquiry that focuses on the wrong entity, but it can also be a virtuous method of directing attention to the forms.

18 One may object to this interpretation that, at 254b4–6, Socrates claims that, when the memory of the charioteer is "borne back to the true nature of beauty", "he sees it again (πάλιν εἶδεν) standing with modesty on a pedestal of chastity" (Fowler [1925] trans. slightly modified). However, even in this case, it does not appear to be a case of direct "vision." It is only the *memory* that is led to a vivid remembrance of an object that it has already seen directly (ἡ μνήμη πρὸς τὴν τοῦ κάλλους φύσιν ἠνέχθη). Thus, the soul does not "see" Beauty itself as before, but, again, only through the *medium* of memory. I thank Mehmet Erginel for bringing this passage to my attention.

19 In this volume, Cervato presents a positive interpretation of the senses and sensible objects in relation to the Eleusinian Mysteries (and, more generally, an analysis of the role these practices play in the background of Plato's *Phaedrus*).

4 Images and Statues

The idea that one can glance, behind sensible entities, at the *genos* of what they imitate may initially seem at odds with Plato's supposed distrust of the senses and the physical world. Yet, the *Phaedrus* consistently maintains that sensible objects can serve as a conduit to the forms. The dialogue emphasises the image-like nature of sensible objects, which, as such, can offer a glimpse of what they represent. This is suggested by the use of several terms that underscore the similarity between sensible and ideal objects[20] and reinforced by the repeated depiction of the sensible realm as a cult statue (*agalma*).

Plato refers to the visible domain and its inhabitants with the word *agalma* twice in the *Phaedrus* (251a7, 252e1) and once in the *Timaeus* (37c8).[21] In order to fully appreciate this characterisation, some preliminary observations are needed. In classical Athens, *agalma* referred to the cult statue or sacred image of the deity, rather than a general image (cf., e.g., *Laws* 931a1). Every religious event in ancient Greece, both in the archaic and classical periods, prominently featured *agalmata* as sacred statues. These statues were typically housed within temples or other ritual sites because they were believed to make visible the unseen forces that govern our world by physically embodying them in a specific form and setting.[22] Statues could either be aniconic or iconic, but in classical Athens most statues were iconic. This means that the gods were portrayed in a highly technical and realistic manner to represent how the divine was likely to appear.[23] At the same time, as mentioned earlier, the statue was believed to provide human beings with a means to communicate with the divine sphere.[24]

20 Cf. 250a6: ὁμοίωμα, 250b3: ὁμοιώμασιν, 250b4: εἰκόνας, 250d5: εἴδωλον, 251a3: εὖ μεμιμημένον. Plato typically uses these words to describe images that resemble their model or, at most, images in a general sense, without any inherent negative connotation. For more information on *eikôn*, cf. David Ambuel (2007: 75): "if pressed to distinguish εἰκών from the other two words for 'image,' [...] it may be fair to say that we are looking at the same picture from two perspectives: εἰκών emphasizes the resemblance that an image bears to the original [...]." By contrast, εἴδωλον can refer either to the image in general or to the negative side of the images. Examples of the former include *Rep.* 516a, 532b, 532c, *Tht.* 191d, 208c. Instances of this term denoting the misleading facade of the image can be discerned in *Gorg.* 436d, *Rep.* 586b–c, *Tht.* 150b–c and *Pol.* 303c. On the other hand, the use of εἴδωλον in the *Phaedrus* and *Rep.* 443c exemplifies its capability to show the essence of the imitated object, indicating the positive facet of the image.

21 Also, cf. *Charm.* 154c, *Symp.* 215a–b, *Rep.* 517d. On this topic, cf. Gábor Betegh (2022: 239–248).

22 Vernant (1991: 153) and Nightingale (2004: 162–164).

23 Cf. Steiner (2001: 89–95).

24 See Nightingale (2004: 163–164) for further information and references.

The network of these notions features throughout this section of the *Phaedrus*. From an *ontological* viewpoint, the comparison of the beloved to an iconic statue of a god suggests that the sensible world, of which he is a part, should be viewed as a statue or *representation*, rather than as a mere reflection.[25] Like an *agalma*, the sensible realm attempts to represent the intelligible, present it somehow, and trace it back to space down here. Conversely, the intelligible world constitutes the sphere of the divine, which needs to be "contained and concealed."[26] Additionally, the sensible world, like an iconic statue, bears some resemblance to that which it expresses, without bridging the deep ontological gap that separates it from its model. The statue and the sensible world are material and corruptible, while the divine is incorporeal, eternal, and changeless.[27]

However, the purpose of the cult statue was not just to represent the divine but also to establish true *communication* and authentic contact with the sacred power; its ambition was to make this power present *hic et nunc* and make it available to humans.[28] The *agalma* served as a tool of connection and relation between the beholder and the divine, as something capable of bringing them closer to each other. Thus, it seems possible that Plato considered the sensible world as a pathway to our interaction with the forms in the same way that the statue relates to the divinity it represents. In brief, as sensible objects are the ontological representations of the forms, they have the power to provoke

25 Therefore, I side with those scholars who interpret the notion of *mimêsis* as "representation," instead of "imitation." For a recent *status quaestionis*, cf. Julia Pfefferkorn and Antonino Spinelli (2021: 9–19).

26 It is not possible to justify this claim here. The idea that the forms constitute the sphere of the divine seems generally accepted and has been recently defended by Betegh (2022: 241). It should be noted that there are some passages in which personal characteristics that are more appropriate to a god are ascribed to the forms. Cf., for example, *Phd.* 81a5 (where τὸ θεῖον is described as immortal and wise) and *Phdr*. 246d8–e1 (where τὸ θεῖον is described as beautiful, wise, good, and everything such). Here, Plato does not appear to differentiate between the traditional gods and the forms. Notably, *Phd.* 80d7 states that our soul, on its way to Hades, reaches παρὰ τὸν ἀγαθὸν καὶ φρόνιμον θεόν. In the *Republic*, the contemplation of the forms is referred to as *theôria* of the gods or as divine *theôria* (517d4–5: θείων θεωριῶν); furthermore, the philosopher's ascent from the cave to the form of the Good is presented as a journey from Hades to the gods (521c2–3: ἐξ Ἅιδου [...] εἰς θεοὺς).

27 Vernant (1991: 153): "In its attempt to construct a bridge [...] that will reach toward the divine, the idol must also at the same time and in the same figure mark its distance from that domain in relation to the human world."

28 Vernant (1991: 153).

a vertical movement from one dimension to another.[29] Socrates appears to allude to this epistemological aspect when he describes *anamnêsis* as the activation of the capacity to perceive the divine behind their sensible instances: "Only a few, turning to images, behold the *genos* of the original they imitate [...] But beauty [...] shone among those visions; and since we came to earth, we grasp *it* shining most clearly through the clearest of our senses" (250b4–d3).[30]

Finally, Plato's use of the word *agalma* to describe sensible objects may suggest that they not only act as gateways to what they represent, but also *make this communication possible*, much like statues function as means for interacting with the gods. Previously, I noted that the use of statues in religious worship was based on the belief that gods could only be encountered through a *medium*, not face-to-face. I believe that Plato drew on this tradition and argued along similar lines. To confirm this, I now examine the passages that discuss philosophical madness.[31]

5 Need for Images

The passage on recollection shows that, while the disembodied souls can enjoy a direct acquaintance with the forms, humans can only remember them through their images in life. From this viewpoint, sensible objects trigger the

29 Cf. Proclus, *in Tim.* 246, 7–9: "it is possible to see (θεωρεῖν) the paradigms in the images (ἐν ταῖς εἰκόσι τὰ παραδείγματα), and to make a transition to those through these (καὶ διὰ τούτων ἐπ' ἐκεῖνα μεταβαίνειν)" (my translation). On this topic, cf. Laura Candiotto (2021). Candiotto has rightly observed that μίμησις provides the metaphysical ground for the epistemic possibility of ἀνάμνησις. However, while I agree with this general idea, I am less convinced by her view that the *difference* from the original arouses the desire for the forms and triggers the philosopher's epistemic journey towards them. In contrast, I believe that, in the *Phaedrus*, the *similarity* between sensible objects and the forms they represent triggers recollection. Candiotto also disputes that sensible objects are faithful copies of the model, citing the tenth book of the *Republic* in support of her thesis (Candiotto [2021: 111, note 28]).

30 ὀλίγοι ἐπὶ τὰς εἰκόνας ἰόντες θεῶνται τὸ τοῦ εἰκασθέντος γένος· [...] περὶ δὲ κάλλους [...] μετ' ἐκείνων τε ἔλαμπεν ὄν, δεῦρό τ' ἐλθόντες κατειλήφαμεν αὐτὸ διὰ τῆς ἐναργεστάτης αἰσθήσεως τῶν ἡμετέρων στίλβον ἐναργέστατα. See Ward (2021: 127): "[...] it is evident that the perception of a sacred statue and the perception of a sensible quality, such as beauty in a face, play analogous roles in traditional and Platonic *theoria*, perhaps in the sense of providing causes in the process of knowing."

31 Ward (2021: 127) writes that "For Plato, then, the role of the perceived sensible *agalma* in relation to the epistemological ascent of the knower cannot be underestimated, for without the visual perception of the *agalma*, the soul does not ascend to the intelligible form." However, she does not offer any explanation for this fact. I shall try to justify it in the next paragraph.

recollection of the forms, thus facilitating our commerce with them. However, two passages – where philosophical madness is discussed – seem to go further and may shed more light on the description of sensible objects as cult statues (*agalmata*).

Socrates describes the fourth kind of madness as occurring when someone, seeing our beauty down here (*to têide tis horôn kallos*), is reminded of true and perfect Beauty. For, by facing the likeness of what they saw up there (*hotan ti tôn ekei homoiôma idôsin*), they are astounded and driven mad (249d–250a). Scholars have sometimes attempted to explain this disruptive experience by claiming that madness is prompted by seeing earthly things as likenesses of the forms, and, thus, by recollection. However, Francisco Gonzalez has rightly noted that recollection can only partially explain the feeling of madness and alienation experienced by the philosopher. As a matter of fact, in the text, Socrates goes on to say that when souls see sensible objects that resemble those up there, they are startled and no longer in control of themselves (*ekplêttontai kai ouket' hautôn gignontai*, 250a6–7) and adds that they are unable to understand what they feel as they cannot adequately "perceive" what shines through their sensible imitations (*dia to mê hikanôs diaisthanesthai*, 250b1). Madness, therefore, results from *two* factors. The first is the perception of sensible entities, which, as images of the forms, can prompt the recollection of these entities: in this way, the philosopher realises the presence of a metaphysical surplus beyond the sensible images. However, this acknowledgement alone would not constitute true madness if the philosopher could leave the images behind and grasp the transcendent reality directly. For madness to arise, the philosopher must have access only to the *images* of the forms through which alone he can study them (at least for the most significant part of the quest).[32] To support his argument, Gonzalez relies on Socrates' recognition that the vision of the form is not precise, *mê hikanôs diaisthanesthai*, because it is a vision through the glass of sensible objects. But another factor can bolster this view. At the outset of his speech on the fourth kind of madness, Socrates credited the philosopher with the ability to restore the wings lost during incarnation by looking at sensible objects and recollecting the forms. Yet, he hastened to add that, although the philosopher has wings and wants to fly (*pterôtai te kai anapteroumenos prothumoumenos anapthesthai*), *he cannot do so* (*adunatôn de*) (249d6–7). Here, Plato alludes to a condition of paradox and frustration, like possessing wings but not being able to fly. What drives the philosopher mad is the desire and the feeling of transcending human limits, insofar as he can glance at the form behind sensible objects, and the

32 Francisco Gonzalez (2017: 53–54).

simultaneous awareness of the impossibility of fully overcoming them (he can see the form through its images, or he has wings that testify to his divine nature, but he cannot use them to leave this world and "see" the other).[33]

Why does Plato insist on madness and the impossibility (or extreme difficulty) of directly seeing the divine while alive in the *Phaedrus*? As I have shown above, at 250c, Socrates claims that while our discarnate soul contemplated the forms in the pure light, being ourselves pure (*en augêi katharai, katharoi ontes*, 250c4–5), now we must rely on *amudra organa* and the memory, as we are entombed in the body. The mention of purity suggests a comparison with a passage from the *Phaedo*, where it is asserted that the forms can only be completely and perfectly known after death, for it is not permitted to the impure to touch and attain the pure.[34] Without pressing this too far, one can at least recognise that Plato sometimes points to the idea that the vision of intelligible realities is neither an easy nor a quick task to achieve but requires the employment of a *medium*, at least to get used to the brightness of these realities; without habit or protection, one is liable to get blinded in the soul. In the *Phaedo*, Socrates compares his quest for the causes and true realities to that undertaken by those who wish to witness a solar eclipse. He suggests that just as the latter uses the water to observe the eclipse without being blinded, he, too, turned to images to investigate the causes of the generation and being of things. The moral to be drawn from this seems the following: as the forms are separate from the sensible world, it is impossible to approach them directly and immediately with our eyes. Therefore, the relationship between us and them – at least for an extended part of the search – must be mediated by images (first by the sensible ones and then by reasonings, *logoi*). Otherwise, like those

33 Apparently, this is at odds with *Phdr.* 249c6–8, where we read that "a man who employs such means of remembrance in the right way, always being initiated into perfect mysteries, alone becomes truly perfect." See, e.g., Franco Ferrari (2019: 46–60). To ward off the possible contradiction, two strategies may be adopted. The first consist in arguing that one "may possess 'pieces of knowledge', on a temporary basis," as has been maintained byChristopher Rowe (2005: 222, note 26). Another way to counter this objection is by relying on the context. Immediately before claiming that the philosopher can achieve a perfect condition through recollection, Socrates asserts that "only the philosopher's thought (διάνοια) becomes rightly feathered, for it is always, so far as possible (κατὰ δύναμιν), turned to those things [*sc.* the forms], through memory (μνήμῃ)." The perfection the philosopher can attain, thus, seems hardly the same as the gods'; by contrast, it can be interpreted as a "second-best" perfection, one that human beings can achieve, *kata dunamin*, through the lens of memory.

34 *Phd.* 67a6–b2: καὶ οὕτω μὲν καθαροὶ ἀπαλλαττόμενοι τῆς τοῦ σώματος ἀφροσύνης, ὡς τὸ εἰκὸς μετὰ τοιούτων τε ἐσόμεθα καὶ γνωσόμεθα δι' ἡμῶν αὐτῶν πᾶν τὸ εἰλικρινές, τοῦτο δ' ἐστὶν ἴσως τὸ ἀληθές· μὴ καθαρῷ γὰρ καθαροῦ ἐφάπτεσθαι μὴ οὐ θεμιτὸν ᾖ.

who attempt to contemplate the eclipse of the sun directly, one would find himself blinded in his soul (99c–d).[35] Likewise, in the *Laws* (897d–e), having acknowledged the inherent limitations of our human perception (*thnêtois ommasin*, 897d10) in understanding the nature of Intellect and its movement, the Athenian suggests viewing them through an *image* rather than attempting a direct approach (which would "produce darkness at noon").[36] Finally, in the *Republic*, illustrating the prisoner's ascent out of the cave, Socrates explains that if one were to immediately take the prisoner outside and let him see the sunlight, he would get blinded and would not be able to see any of the forms. In fact, it requires time for him to adapt before he can see things in the world above. Only ascending from one image to the next, he will eventually be able to look at the Sun, not images of it in water or some other place, but the Sun itself, in its place, and to study it (516a–c).[37]

The passages from the *Phaedo* and the *Laws* underscore a distinction between direct and indirect knowledge. While indirect knowledge is not ideal, it appears necessary because of our human mortal nature.[38] Moreover, both texts abound in visual verbs and refer to the images through which we are meant to examine the divine realities using the word *eikôn*, "likeness" or "faithful copy." As a result, the inquiry into the forms and the other metaphysical principles, e.g., the Intellect, seems to require a *medium*, which can be either the sensible reality or *logos*. The *Republic* presents a slightly different view, suggesting that it may be possible to directly see the Sun at the end of an arduous ascent. However, it is notable how Socrates emphasises the importance of using images and the inability to attain the truth directly and instantly without proper training, as this would blind the prisoner. Overall, these passages suggest that sensible objects are not merely an additional element that helps us connect with the intelligible world, but a *fundamental* element of the quest for the forms.[39] Without the visible image, attempting to reveal the essence is impossible. The image does not reveal the essence, but it indicates

35 Cf. Franco Trabattoni (2023: 142–158).

36 *Laws* 897d8–e2: Μὴ τοίνυν ἐξ ἐναντίας οἷον εἰς ἥλιον ἀποβλέποντες, νύκτα ἐν μεσημβρίᾳ ἐπαγόμενοι, ποιησώμεθα τὴν ἀπόκρισιν, ὡς νοῦν ποτε θνητοῖς ὄμμασιν ὀψόμενοί τε καὶ γνωσόμενοι ἱκανῶς· πρὸς δὲ εἰκόνα τοῦ ἐρωτωμένου βλέποντας ἀσφαλέστερον ὁρᾶν.

37 For a different interpretation of blindness in the *Phaedo*, the *Phaedrus* and the *Republic*, cf. Cervato in this volume.

38 Compare the ἀμυδρὰ ὄργανα of the *Phaedrus* with the θνητὰ ὄμματα mentioned in the *Laws* and the insistence on sight and *theôria* in the passages from the *Phaedo* (σκοπῶν, θεωροῦντες καὶ σκοπούμενοι, τὰ ὄμματα, σκοπῶνται, βλέπων [...] τοῖς ὄμμασι, σκοπεῖν).

39 I am assuming that, in the *Republic*, the images the soul contemplates before seeing the Sun directly are sensible objects *qua* images of the forms. Cf. Nicholas D. Smith (1996: 25–45).

it in some way. It functions as a sign, an icon, or a meaningful representation.[40]

Again, the description of the sensible world as *agalma* can provide additional insight. As noted, it was a fundamental tenet of ancient Greek religion that humanity could only encounter the divine through images or disguises of the gods – and that statues were part of these attempts to conceal and contain the divine power. This principle may also extend to Plato's philosophers, at least in the *Phaedrus* and the dialogues mentioned. I suggest that Plato viewed direct access to divine realities as unachievable, or only attainable at times, and that his invitations to approach them through images (*Phaedo*, *Republic*, *Laws*) or cult statues (*Phaedrus*) represent a philosophical adaptation of the religious prohibition against seeing the gods face-to-face. Significantly, both Plato's works and certain Greek myths consistently mention blindness. In the myths, blindness is often depicted as a punishment inflicted by the deities on humans for their visual transgressions.[41] This is because, as we read in the *Iliad*, "the gods are dangerous to look upon in full clarity" (*Il.* XX, 131). Similarly, Plato argues that direct observation of the forms or the motion of the Intellect would blind the soul. To prevent being blinded, we are to study them through their images. The comparison between the two types of blindness is further underpinned by Plato's view that the forms (and Intellect) are not *like* divine beings but *are* divine entities.[42]

---

40 Here, I am paraphrasing one of Palumbo's (2008: 274, note 101) key observations and adapting it in translation. This is the original quote, which I reproduce in full: "[…] nessuna immagine, a rigore, può permettere una visione diretta delle idee. Tuttavia senza l'immagine diviene impossibile anche ogni tentativo di rivelazione dell'essenza. È possibile affermare allora che l'*eidolon* non rivela l'essenza ma in qualche modo la indica, esplica la funzione di segno, di icona, di rappresentazione significativa, come il sole in *Resp.* VI 509A è l'icona del Bene". Also, cf. Sinaiko (1965: 82): "In one sense the human concerns of the philosopher are the *necessary* temporal means through which he strives for his divine and eternal ends" (my italics). As previously mentioned, this type of knowledge is not considered the most optimal by Plato. He believed that cognitive states using visible images are inferior to those that do not rely on them. Plato's views on this are evident in the contrast between *dianoia* and *noêsis* drawn at the end of Book 6 of the *Republic*. For a discussion of the merits and drawbacks of cognition based on visible images see Picciafuochi in this volume.

41 Cf. Richard GA. Buxton (1980: 30–32). Here, he presents several examples, such as that of Epizelos losing his sight after encountering a φάσμα during the battle of Marathon or that of Erymanthos getting blinded after witnessing Aphrodite bathe following her union with Adonis. Also, Tiresias famously lost his sight when he saw Athene bathing and Anchises, upon awakening to Aphrodite's presence, averted his gaze, fearing punishment for his visual transgression.

42 Betegh (2022: 241).

If I am right, the philosopher of the *Phaedrus* and the *theôros* both confront the divine indirectly through statues or sensible objects because the divine they seek is not directly visible (at least to embodied souls).[43] However, it must be observed that they are also very different in this regard. While *theôria* aligns with the Delphic religion's belief that humans should not exceed their inherent limitations, Plato urges us to become godlike (cf., e.g., *Tht.* 176a–b, *Ti.* 90b–d). According to Plato, the ideal of godlikeness does not involve any hybris, as it is grounded on the notion that the best part of our soul, the rational one, is *akin* to the divine.[44] Therefore, although human beings seem to be prevented from attaining a full and direct knowledge of the forms during their lifetime,[45] unlike *theoroi* they should aim to ascend and establish themselves in the divine realm.

6 The Philosophical Ritualised Visuality

A further key point links the philosopher to the *theôros* and provides some valuable clues to understanding the role of sensible objects in Plato's *Phaedrus*. If my previous claims are correct and sensible objects can lead to the forms, it becomes unclear why not everyone can recall them and why this

43 While it may prove difficult to show how the quoted passage from the *Laws* is helpful for the comprehension of the *Phaedrus*, it is difficult to deny the consistency of the *Phaedrus* with the *Phaedo*. As Betegh (2022: 251) has noticed, "[…] at least in the *Phaedo* and the *Phaedrus*, the full encounter with the forms happens outside the sphere of ordinary experience: in the realm of the 'unseen' (Hades) or in the 'region above the heavens'". In both dialogues, moreover, recollection appears as the means that human beings have at their disposal during life, and is based on the previous knowledge of the forms by the discarnate soul.

44 Gabor Betegh (2006: 636): "What counts as the utmost *hubris* in the traditional conceptions became the normative program of human life in the philosophical doctrines."

45 This does not necessarily imply that Plato did not express a different or more moderate stance in other dialogues. My reading, moreover, leaves open the possibility that the soul sometimes comes to a different kind of knowledge even while in life. Although in ancient Greek religion the direct and clear vision of the gods was usually regarded as a harbinger of danger and harm, which could even lead to death, there was a place for epiphany as well, namely for the direct manifestation of the divine to the living person without causing him any injury. Additionally, as Ferrari (2019: 58) notes, in the *Symposium*, the person who can fully use his rational component is described as *theophiles* (212a). From this point of view, Plato's philosopher seems to be the heir of the archaic 'divine man:' like the latter, he is an "exceptional figure who seems to transcend the limits of humanity to skim the divine sphere" (the translation is mine). This interpretation is undoubtedly a possible reading of the text and may find some resonance in the *Phaedrus* in the passage discussed above (note 33).

does not occur consistently. Plato seems to acknowledge this difficulty and addresses it by emphasising the significance of one's approach to sensible objects. I shall argue that the distinction Plato draws echoes one already found in the cultural institution of *theôria*.

In the *Phaedrus*, Plato presents two ways of beholding the beautiful beloved based on the soul's ability to remember past experiences. When seeing a god-like face or a bodily form that has captured Beauty well, philosophers initially feel overwhelmed and unable to fully comprehend the Beauty they see. However, as they continue to gaze, a profound reverence befitting a god (*hôs theon sebetai*) takes hold, recognising the beloved's divine status (251a). This encounter, this moment of awe, leads to a profound inner transformation in the philosopher. In stark contrast, non-philosophers, instead of reverently beholding the beloved (*hôst'ou sebetai prosorôn*), indulge in pleasure and pursue physical desires like four-footed beasts, driven by the urge to procreate (250e). Their souls do not swiftly recollect the reality they witnessed before incarnation (250a), and they fail to move from here to there and neglect the image's fundamental nature. They regard the physical world as the ultimate and sole reality, turning it from an icon to an idol and blocking their path to the divine.[46] Hence, the distinction between philosophers and non-philosophers in the *Phaedrus* does not seem to amount to the difference between those who possess complete and perfect knowledge of the forms and those who do not. Neither one nor the other seems capable of dispensing altogether with images and looking directly at the forms, or at least to do that without a long and strenuous study. Instead, the difference lies in the (in)ability to approach the sensible images correctly, for only those who use them as reminders of the forms (*hupomnêmasin orthôs chrômenos*), and not as ends in themselves, can be initiated at the Highest Mysteries (249c).[47]

Earlier, I observed that *theôria* at religious sanctuaries was marked by a specific kind of viewing referred to as "ritual-centered visuality" by Elsner. This form of observation considers sacred objects, such as altars and cult statues, by focusing on their function in religious rituals. In other words, it centres around the features relevant to the role of the statues in religious rituals and ignores unrelated qualities, such as aesthetic ones. By contrast, "naturalistic"

46 Cf. Suzanne Saïd (1987: 330): "L'idole fait du visible, qui est tout son être, une fin en soi. Elle arrête le regard qui s'abîme en elle et lui interdit d'aller plus loin. L'icône au contraire porte en elle son propre dépassement."

47 This distinction between philosophers and non-philosophers depends on other features, such as the amount of time that has passed since "initiation," the extent to which the soul has "seen" in its disembodied state, the quality of its memory, the strength of its desire to gaze.

visuality deals with the aesthetic qualities of the objects, primarily viewing the statues as artistic manufactures. Building upon Elsner's framework, I now suggest interpreting Plato's contrast between philosophical and non-philosophical viewpoints on images as a contrast between "ritual-centered visuality" and "naturalistic visuality." Thus interpreted, non-philosophical souls perceive a beautiful boy in a naturalistic manner, potentially leading to *agalmatophilia*, or sexual attraction to a statue.[48] On the contrary, philosophers have developed the ability to engage with the sensible world *ritually*, viewing it as an image (*eikôn*) or a statue (*agalma*) of the Beauty itself, recognising it as a pointer to the intelligible. Consequently, they disregard the aesthetic aspects of the beautiful boy and instead delve deeper to capture the divine essence beneath the surface.

7 Conclusions

In this chapter, I have attempted to underscore the centrality of *theôria* in Plato's philosophy. I have noted that Plato uses the *theôria* family of words to describe the contemplation of the forms and the sensible world. I have then delved into why Plato considers the sensible world worthy of contemplation and study, highlighting the positive function of sensible objects in his epistemology. Additionally, I pointed out further similarities between Plato's philosopher and the traditional figure of the *theôros*.

Firstly, I have examined the institution of *theôria* and the role of the *theôros*, noting that he communicated with the divine in the sanctuaries only indirectly, through statues representing the gods. This is due to the Greek religious belief that seeing the gods face-to-face is either impossible or highly perilous for human beings. I have then focused on recollection and philosophical madness in the *Phaedrus*. After reconstructing the context in which the theory of recollection arises, I have analysed the description of sensible objects as *agalmata*, i.e., cult statues. I have attempted to elucidate why Plato employs this term by pointing out its possible ontological and epistemological grounds. On the one hand, I have observed that this description suggests that sensible objects are to be conceived of not as reflections but as representations of the forms. On the other hand, I have sought to show that sensible objects not only straddle the way between the human and the divine spheres but also make it possible for human beings to rediscover the forms. Since forms are separate and divine, direct observation of them is impossible (at least without careful and long prep-

48 Radcliffe G. Edmonds III (2017: 67, 73–80).

aration); as a result, an intermediary is required. This intermediary is provided by sensible objects. Then, I have suggested that Plato may have transposed to the philosophical level the religious idea that seeing the divine directly is beyond the human's reach. This is the first sort of continuity that I have endeavoured to trace between the traditional *theôros* and the platonic philosopher. I have also tried to show a second feature of continuity: the view that one can approach a statue (*agalma*) in two different ways. While philosophers, like *theoroi*, engage in the sensible objects as statues in a *ritual* way, thus considering them as images of the forms, non-philosophers approach them in a *naturalistic* manner, neglecting their power to lead humans towards the divine realm.

Overall, this study has aimed to cast a new light on Plato's conception of the visible world and the profound influence of *theôria* on his thought, by presenting an account that may seem unconventional but hopefully not less interesting and historically plausible than the traditional reading of Plato.

Bibliography

Ambuel, David. 2007. *Image and Paradigm in Plato's Sophist*. Athens: Parmenides Publishing.

Betegh, Gábor. 2006. "Greek Philosophy and Religion." In *A Companion to Ancient Philosophy*. Edited by Mary Louise Gill and Pierre Pellegrin. Malden: Blackwell Publishing Ltd.

Betegh, Gábor. 2022. "Plato on Philosophy and the Mysteries." In *The Cambridge Companion to Plato*. 2nd edition. Edited by David Ebrey and Richard Kraut. Cambridge; New York: Cambridge University Press.

Buxton, Richard GA. 1980. "Blindness and Limits: Sophokles and the Logic of Myth." *The Journal of Hellenic Studies* 100: 22–37.

Candiotto, Laura. 2021. "Mimesis and Recollection." In *Platonic Mimesis Revisited*. Edited by Julia Pfefferkorn, Julia and Spinelli, Antonino. Baden-Baden: Academia Verlag.

Centrone, B. 1998. Platone. *Fedro*. Rome: Laterza.

Edmonds III, Radcliffe G. 2017. "Putting Him on a Pedestal: (Re)Collection and the Use of Images in Plato's *Phaedrus*." In *Plato and the Power of Images*. Edited by Pierre Destrée and Radcliffe G. Edmonds III. Leiden/Boston: Brill.

Elsner, Jaś. 2000. "Between Mimesis and Divine Power: Visuality in the Greco-Roman World." In *Visuality before and beyond the Renaissance: Seeing as Others Saw*. Edited by Robert S. Nelson. Cambridge: Cambridge University Press.

Ferrari, Franco. 2019. *La via dell'immortalità. Percorsi platonici*. Turin: Rosenberg & Sellier.

Gill, Christopher. 1993. "Plato on Falsehood – Not Fiction." In *Lies and Fiction in the Ancient World*. Edited by Christopher Gill and Timothy P. Wiseman. Liverpool: Liverpool University Press.

Gonzalez, Francisco. 2017. "The Power and the Ambivalence of a Beautiful Image in Plato and the Poets." In *Plato and the Power of Images*. Edited by Pierre Destrée and Radcliffe G. Edmonds III. Leiden/Boston: Brill.

Kahn, C. 1996. Plato and the Socratic Dialogue. Cambridge: Cambridge University Press.

Nightingale, Andrea Wilson. 2004. *Spectacles of Truth in Classical Greek Philosophy.* Theoria *in its Cultural Context*. New York: Cambridge University Press.

Palumbo, Lidia. 2008. μίμησις. *Rappresentazione, teatro e mondo nei dialoghi di Platone e nella Poetica di Aristotele*. Naples: Loffredo.

Plato. 1925. *Plato in Twelve Volumes*. vol. 9. Translated by Harold N. Fowler. Cambridge, MA: Harvard University Press.

Pfefferkorn, Julia. 2023. *Platons tanzende Stadt. Moralpsychologie und Chortanz in den Nomoi*. Leiden/Boston: Brill.

Pfefferkorn, Julia, and Antonino Spinelli. 2021. "Revisiting Mimesis in Plato: an Introduction." In *Platonic Mimesis Revisited*. Edited by Julia Pfefferkorn and Antonino Spinelli. Baden-Baden: Academia Verlag.

Rowe, Christopher. 2005. "What Difference Do Forms Make for Platonic Epistemology?" In *Virtue, Norms, and Objectivity. Issues in Ancient and Modern Ethics*. Edited by Christopher Gill. New York: Clarendon Press.

Rutherford, Ian. 1995. "Theoric Crisis: The Dangers of Pilgrimage in Greek Religion and Society." *Studi e Materiali di Storia delle Religioni* 61: 276–292.

Rutherford, Ian. 2013. *State Pilgrims and Sacred Observers in Ancient Greece. A Study of* Theōriā *and* Theōroi. New York: Cambridge University Press.

Sinaiko, Herman L. 1965. *Love, Knowledge & Discourse in Plato: Dialogue and Dialectic in Phaedrus, Republic, Parmenides*. Chicago: Chicago University Press.

Smith, Nicholas D. 1996. "Plato's Divided Line." *Ancient Philosophy* 16: 25–45.

Steiner, Deborah T. 2001. *Images in Mind. Statues in Archaic and Classical Greek Literature and Thought*. New Jersey: Princeton University Press.

Saïd, Suzanne. 1987. "Deux noms de l'image en grec ancien: idole et icône." *Comptes rendus des séances de l'Académie des Inscriptions et Belles-Lettres*, 131(2): 309–330.

Trabattoni, Franco. 2016. *Essays on Plato's Epistemology*. Leuven: Leuven University Press.

Trabattoni, Franco. 2023. *From Death to Life. Key Themes in Plato's Phaedo*. Leiden/Boston: Brill.

Vernant, Jean-Pierre. 1991. *Mortals and Immortals*. Edited by Froma I. Zeitlin. Princeton: Princeton University Press.

Ward, Julia K. 2021. *Searching for the Divine in Plato and Aristotle. Philosophical* Theoria *and Traditional Practice*. Cambridge; New York: Cambridge University Press.

CHAPTER 10

Vision, Blindness, and the Eleusinian Mysteries in Plato's *Phaedrus*

Giulia Cervato

1 Introduction

The famous allegory of the winged chariot (*Phdr.* 246a–250c) could be considered, to borrow Andrea W. Nightingale's words, one of the most fascinating "spectacles of truth."[1] As known, in this beautiful scene gods and worthy souls are described as *seeing* "reality" and beholding "absolute justice, temperance, and knowledge" (*Phdr.* 247d), which manifest themselves in the splendid "plane of truth" (*Phdr.* 248b).[2] Not coincidentally, the verb *theōréō*, as well as its synonym and possible cognate *theáomai*,[3] appears several times in this section.[4]

The vocabulary of the dialogue and its setting highlights the particular intertwining of meanings of the verb *theōréō*: as many scholars famously underlined,[5] from its original meaning of perceptual observation, it soon extended to indicate the attendance at religious festivals, and hence, to intellectual contemplation of intelligible contents.[6] In the first place, this scene refers to the most recent[7] and philosophical connotation of *theōréō*, which

1 Nightingale (2004).

2 From now on, the *Phaedrus*' translations employed will be Fowler's.

3 Rutherford discusses the possible etymologies of *theōría* and *theōréō*: they could derive either from *theá*+**wōros*, meaning "sight-watching" or from *theós*+**wōros*, meaning "god-watcher." See Rutherford (2013: 5).

4 *Phdr.* 247c1, 247d4 for *theōréō*; 247e3, 249e5, 250b5, 250e3 for *theáomai*. The noun *theéōría*, conversely, never appears in the dialogue.

5 See, among others, Kerényi (1962: 141–154), Nightingale (2004), Bénatouïl-Bonazzi (2012), Rutherford (2013), Ward (2022).

6 As Julie K. Ward suggests (see Ward [2022: 1]), this coalescence probably depends on the original observational value of *theōréō*, which might refer more specifically to "seeing something highly significant." The eminence of the object observed through the senses later allows this verb to be extended to sacred spectacles and, hence, to subsequently hint at a kind of vision that is so lofty that it can dispense from senses themselves.

7 According to Nightingale, the transformation of traditional *theōría* into a form of intellectual contemplation is an "invention" of fourth-century BC. philosophers (see Nightingale [2004: 72]).

 | HTTPS://DOI.ORG/10.1163/9789004747128-012

has to do with the apprehension of eternal truths through intellectual and epistemic gazing. In the second place, as will be shown, it hints at sacred rituals, explicitly linking the theoretical contemplation of intelligible truths to religious practices involving traditional deities and cults. In third place, it is deeply concerned with the act of seeing, which – as already mentioned – represents the initial and most general value of *theōréō*.

Still, tracking the plain and simple coexistence of these meanings does not capture entirely the finery and the philosophical implications of the dialogue: not only do the meanings of *theōréō* co-exist and overlap, but they are also shown in their mutual relationship and progressive development. In other words, in the *Phaedrus*, Plato stages the evolution of sight from blindness to its sensory connotation until it reaches its epistemic and incorporeal form, describing a sort of *phenomenology* of seeing. Moreover, he does that precisely through the lexicon of religious and traditional *theōría*, with particular reference to the mysteries of Eleusis, in the context of which this peculiar progressiveness was enacted in the most emblematic way.

In light of the above discussion, the objective of the present analysis is threefold. The first aim is to demonstrate that Plato's *Phaedrus* may well have borrowed some elements from the Eleusinian cult imaginary in structuring its philosophical contents. Second, this analysis intends to prove that this metaphoric loan – far from being merely stylistic – keenly showcases a certain *continuity* of vision, which is at the core of Socrates' palinode. The third objective is to discuss the role of blindness as a meaningful stage of this process.

2 The *Phaedrus* and the Eleusinian Mysteries

The connection between the Platonic work and the lexicon of mysteries has been the subject of a wide debate and provoked numerous hypotheses and conjectures.[8] The atmosphere of secretiveness which, by definition, surrounded those rituals, resulted in the lack of textual sources openly discussing them, making this Platonic question particularly difficult to solve. Nevertheless, Plato's contemporaries already hinted at a possible connection between Platonism and the mysteries. In the *Clouds*, for example, Aristophanes mock-

8 See Boyancé (1962), Riedweg (1987: 1–69), Burkert (1987: 66–88), Des Places (1981: 83–98), Dinkelaar (2020).

ingly describes Strepisades' admission to Socrates' *phrontistērion* as an initiation ceremony.[9]

More importantly, references to practices of initiation (*teletaí*) can be found in Plato's work itself. They occur several times in the dialogues[10] but, as already mentioned, especially assume a pivotal role in the allegory of the winged chariot of the *Phaedrus*.[11] Here, these rites are evoked along with other religious terms and practices that shape the scene itself. Preliminarily tracking all the possible hints at those practices will prove essential in analyzing Plato's philosophical resemantization:

> But at that former time, they saw beauty shining in brightness, when, with a blessed company – we following in the train of Zeus, and others in that of some other god – they saw the blessed sight and vision and were initiated into that which is rightly called the most blessed of mysteries, which we celebrated in a state of perfection, when we were without experience of the evils which awaited us in the time to come, being permitted as initiates to the sight of perfect and simple and calm and happy apparitions, which we saw in the pure light, being ourselves pure [...].[12]

The scene describes the moment at which the soul's ascension in the *hyperouránion* peaks. The mystic atmosphere is pervasive and is created by several elements that shall be examined in detail. In the first place, the intensification of technical terms generally linked to religious festivals must be underlined: the apparition of beauty is described as the most blissful *teletē*, which

9 "τοὺς τελουμένους ἡμεῖς ποιοῦμεν" (Aristoph., *Cl.* 258). Adkins proposes linking other elements of the play to the mysteries: firstly, Socrates' disciple's injunction to consider his revelations as *mystēria* (Aristoph., *Cl.* 143) and, secondly, the clouds' words as they appear (Aristoph., *Cl.* 298–313), which may be regarded as "a catalogue of the Attic religious observances, prominently including the – presumably Eleusinian – mysteries" (Adkins [1970: 13–14]).

10 See, *e.g.*, *Phd.* 69c, *Symp.* 202e, 215c, *Resp.* 2.365a–366a.

11 See Riedweg (1987: 40–41).

12 "Κάλλος δὲ τότ'ἦν ἰδεῖν λαμπρόν, ὅτε σὺν εὐδαίμονι χορῷ μακαρίαν ὄψιν τε καὶ θέαν, ἑπόμενοι μετὰ μὲν Διὸς ἡμεῖς, ἄλλοι δὲ μετ'ἄλλου θεῶν, εἶδόν τε καὶ ἐτελοῦντο τῶν τελετῶν ἣν θέμις λέγειν μακαριωτάτην, ἣν ὠργιάζομεν ὁλόκληροι μὲν αὐτοὶ ὄντες καὶ ἀπαθεῖς κακῶν ὅσα ἡμᾶς ἐν ὑστέρῳ χρόνῳ ὑπέμενεν, ὁλόκληρα δὲ καὶ ἁπλᾶ καὶ ἀτρεμῆ καὶ εὐδαίμονα φάσματα μυούμενοί τε καὶ ἐποπτεύοντες ἐν αὐγῇ καθαρᾷ, καθαροὶ ὄντες [...]" (*Phdr.* 250b–c).

is accompanied by ritual celebrations,[13] and accomplished in a state of absolute purity.

Moreover, initiated souls are said to have experienced a blessed destiny in the otherworldly life, when they were preserved from any form of evil. These motifs, evoked in a strikingly similar fashion in the *Phaedo*,[14] refer to a promise of eternal salvation[15] that was common to several cults. Interestingly, the same words and tones are employed by the *Homeric Hymn to Demeter*, in which a happy life is said to await those who have been initiated in and have assisted (ὄπωπεν) sacred rituals.[16]

Very significantly, this text represents the most important pre-Platonic witness of the Eleusinian mysteries. As is known, the text tells the story of Persephone's rape by Hades and how her mother, Demeter, desperately wandered through the earth. It ends precisely with the description of the rituals devoted to Demeter at Eleusis, where the goddess found shelter during her quest. Probably introduced in the Mycenean age,[17] these rituals were, without any doubt, widely spread in Socrates' and Plato's times, as testified by literary sources[18] and by Alcibiades' condemnation of blasphemy.[19] At the time, they were probably celebrated in two phases, as Plato himself testifies elsewhere: the greater

13 In the passage quoted above, Plato uses the verb *ōrgiázō* (ὠργιάζομεν). The noun *órgia* appears in the *Hymn to Demeter* (HH 2.476). Mylonas associates this term with the celebrations that were held in the *Telestērion* at Eleusis. See Mylonas (1961: 237).

14 See *Phd.* 69c–d: "κάθαρσίς τις τῶν τοιούτων πάντων καὶ ἡ σωφροσύνη καὶ ἡ δικαιοσύνη καὶ ἀνδρεία, καὶ αὐτὴ ἡ φρόνησις μὴ καθαρμός τις ᾖ. καὶ κινδυνεύουσι καὶ οἱ τὰς τελετὰς ἡμῖν οὗτοι καταστήσαντες οὐ φαῦλοί τινες εἶναι, ἀλλὰ τῷ ὄντι πάλαι αἰνίττεσθαι ὅτι ὃς ἂν ἀμύητος καὶ ἀτέλεστος εἰς Ἅιδου ἀφίκηται ἐν βορβόρῳ κείσεται, ὁ δὲ κεκαθαρμένος τε καὶ τετελεσμένος ἐκεῖσε ἀφικόμενος μετὰ θεῶν οἰκήσει. εἰσὶν γὰρ δή, ὥς φασιν οἱ περὶ τὰς τελετάς, 'ναρθηκοφόροι μὲν πολλοί, βάκχοι δέ τε παῦροι:' οὗτοι δ'εἰσὶν κατὰ τὴν ἐμὴν δόξαν οὐκ ἄλλοι ἢ οἱ πεφιλοσοφηκότες ὀρθῶς. ὧν δὴ καὶ ἐγὼ κατά γε τὸ δυνατὸν οὐδὲν ἀπέλιπον ἐν τῷ βίῳ ἀλλὰ παντὶ τρόπῳ προυθυμήθην γενέσθαι." Despite the evident similarities, these lines refer to a blessed destiny in the after-life, while the passage quoted from *Phaedrus* mainly refers to a pre-birth state.

15 See Mylonas (1961: 283–284); Nightingale (2004: 85–87). In Plato's revision of these rituals – as Nightingale underlines – the salvific element is associated with an epistemic one: a blissful life is granted only to those who have seen and known the divine principles.

16 HH 2.480–481. For a wider discussion on the intertextual connections of these motifs, see Ustinova (2009: 226–229).

17 Mylonas (1961: 14–15). Basing on literary sources and archeological findings, Mylonas concludes that the cult may have been introduced in Greece in the fifteenth century BC.

18 See, *e.g.*, Pind. *Isth.* 7.3–5, Soph., *Ant.* 1115–1121,

19 Thucydides (Thuch. 6.28.1) and Plutarch (Plut., *Alc.* 19.1) report that Alcibiades was accused of having parodied the Eleusinian mysteries and later condemned *in absentia*.

mysteries and the lesser ones.[20] Referring to the same partition, Plutarch informs us that the lesser mysteries were celebrated in Agra,[21] which, quite significantly, is said to be close to the place where the *Phaedrus* is set.[22]

More importantly, in the aforementioned passage, Plato hints at another internal partition of these rites. As the Hermias of Alexandria testifies, the two participles *myoýmenoi* and *epopteýontes* are used "as in the Eleusis mysteries."[23] Particularly, Plato refers to two progressive phases in the rites, to which he reduces the path of philosophical research *tout court*. The first category – *mýēsis* – involves those who are newly initiated, while the second – *epopteía* – refers to those who reached the ultimate stages of the rites. This same vocabulary is used by Plato in the *Symposium*, where Diotima explicitly says that Socrates may well have succeeded in completing the early stage of his initiation to Love mysteries (κἂν σὺ μυηθείης) but doubts that he can face the terminal ones (τὰ δὲ τέλεα καὶ ἐποπτικά).[24]

The characterization of these two stages is very controversial. The main concern regards the interpretation of the first one. As known, the verb *mýō* means "close, be shut" and – as Kerényi underlines – surely "implies the element of secret."[25] However, it is not clear if it must be referred to one's lips or eyes. Consequently, it could either hint at the gag order and prohibition of disclosure that were certainly in place[26] or at a sort of propedeutic blinding.

20 See *Grg*. 497c. Here Socrates reproaches Callicles for having been initiated to the greater mysteries before the lesser mysteries.

21 See Plut., *Demetr*. 26.1–2. In this passage, Plutarch describes Demetrious Poliorcetes' exceptional initiation to the Eleusinian mysteries (302 B.C), According to him, the lesser mysteries were celebrated in the month Anthesterion in Agra, a suburb of Athens, while the great mysteries were held in Eleusis during the month of Boëdromion. For this topological reference in the *Phaedrus*, see Riedweg (1987: 8) and Motte (1995: 37).

22 See *Phdr*. 229c: "No, the place is about two or three furlongs farther down, where you cross over to the precinct of Agra; and there is an altar of Boreas somewhere thereabouts." Velardi also suggests that the scared images consecrated to the Nymphs and Achelous that Socrates and Phaedrus find under the plane tree (*Phdr*. 230b–c) could be those of a relief found in 1759 in Athens and now exposed at the Berlin Staatliche Museum. It represents the Nymphs, Pan, and Achelous as well as Demeter and Persephone (*LIMC* I 202).

23 "Το δε μυούμενοι καὶ ἐποπτεύοντες ὡς ἀπὸ τῶν τελετῶν τῶν ἐν Ἐλευσῖνι λέγει", "He uses 'initiated into' and 'observing' as in the rites at Eleusis" (Herm. 178.19), transl. by Baltzly and Share.

24 See *Symp*. 209e–210a.

25 Kerényi (1962: 46).

26 The interdiction of spreading information about the Eleusinian rituals is originally expressed in the *Hymn to Demeter* (HH 2. 468–469). Moreover, as is known, Diagoras of Melos was allegedly exiled from Athens because he mocked and divulged the Eleusin-

Tzeztes leans towards the first interpretation and so do other more recent scholars.[27]

Nevertheless, despite not denying a possible reference to both meanings, the majority of scholars[28] tend to support the second option. The reason must be found in the evident visual nuance of the terminal phase of the rites, *epopteía*, which is connected to the verb *epopteúō* that literally means "to watch" or "to see from above." As Clinton remarks,[29] it is then more likely to suppose that both terms refer to sight. The correlations between these two phases may lie in the progression from deep darkness to full luminosity and clearness of sight.

Indeed, several textual sources describe the Eleusinian mysteries as an alternation between moments of gloom and moments of brightness. Patera insightfully underlines[30] that this trait can already be recognized in the *Hymn to Demeter*. The initial gloomy and deadly atmosphere (expressed, for example, by the dusky coat that she throws over her shoulders)[31] is first brightened by the light torch that illuminates her path and, later, by the divine appearance of the goddess herself in her true height and splendor.[32] From a practical point of view, many scholars suggest that the initiation could start with a ritual veiling (*mýēsis*)[33] and continue with the intermediate phase of the so-called *pannychía*, a ritual procession of torch-bearers that took place in the middle of the night. This procession is clearly testified by Aristophanes' *Frogs*[34] and more vaguely evoked by Sophocles' *Oedipus at Colonus*.[35] Finally, the rites probably

---

ian mysteries (see Athenag., *Legatio* 4.1). This episode is evoked also by Aristophanes in the *Birds* (Aristoph., *Av.* 1369).

27 Chantraine (*s.v.* μύω) and Dowden (1980: 414) consider both possibilities as plausible.

28 See Kerényi (1962: 46), Ustinova (2009: 230), Clinton (2003: 50).

29 See Clinton (2003: 50–51).

30 See Patera (2010: 261–262).

31 HH 2.40–42. Significantly, the term κυανόπεπλος becomes an epithet for Demeter (HH 2.319, 360, 374, 442).

32 See HH 2.188–190.

33 See Clinton (2003: 50), Nightingale (2004: 85), Ustinova (2009: 230). This hypothesis is also supported by some iconographic sources: the Lovatelli Urn and the Torre Nova sarcophagus both display mystery rituals where the initiates have their head veiled. On this, see below.

34 See Aristoph., *Fr.* 445–448: "Onward!/In the sacred round dance of the goddess, in the flower-bearing grove/sporting with all who partake in festival dear to the goddess/I will go with the women and girls/where they dance all night for the goddess, to bring the sacred torch (οὗ παννυχίζουσιν θεᾷ, φέγγος ἱερὸν οἴσων)", transl. by Dillon.

35 See *Soph.*, *OC* 1049–1053.

culminated in the display of sacred objects, followed by the manifestation of a great light or a great fire (*epopteía*).[36]

Many other literary sources provide evidence for a similar sequence.[37] More generally, it is possible to state that "the mystic ceremony featured the movement from darkness to light."[38] This progression probably indicates that, to quote Ustinova, "various degrees of spiritual enlightenment were intrinsically connected with darkness and light."[39] This transition may have provided Plato with a useful metaphorical tool, through which he describes not only the ascension of souls to their original metaphysical homeland but also the dynamic through which it happens, as will be shown.

3 The Continuity of Vision: *Pannychía* and *Epopteía*

As already mentioned, one of the first attempts[40] to recognize the appropriation of the mystery metaphors by Plato is due to the fifth-century AD. Neoplatonist philosopher Hermias of Alexandria, who authored a well-known *Commentary on the Phaedrus*. When describing the souls' *theōria* of beauty, in addition to directly associating Plato's vocabulary with Eleusis, he expounds the functioning of *mýēsis* and *epopteía*:

---

36 See Mylonas (1961: 273–275): "The climax of the celebration was attained when the Hierophant standing in front of the *Anaktoron* and in the midst of a radiant light exhibited the *Hiera* to the initiates."

37 In the already mentioned *Frogs*, the prize for the initiates is said to be "the sun and the happy light" ("ἥλιος καὶ φέγγος ἱλαρόν", Aristoph., *Fr.* 455–457, transl. by Dillon, slightly modified). Similarly, Dio speaks of the alternation between light and darkness in the rites (Chr. 12.33), while Hippolytus reports that the ceremonies culminated in the display of sacred symbols (Hippol., *Ref.* 5.8.3).

38 Nightingale (2004: 85).

39 Ustinova (2009: 230).

40 Before Hermias, a fragment of Plutarch testifies the awareness of the intermingling between Plato's description and the traditional rituals. Plutarch fr. 178, *On the Soul* reads "And so the verbs *teleutân* (die) and *teleîsthai* (be initiated), and the actions they denote, have a similarity. In the beginning there is straying and wandering, the weariness of running this way and that, and nervous journeys through darkness that reach no goal (πλάναι τὰ πρῶτα καὶ περιδρομαὶ κοπώδεις καὶ διὰ σκότους), and then immediately before the consummation every possible terror, shivering and trembling and sweating and amazement (φρίκη καὶ τρόμος καὶ ἱδρὼς καὶ θάμβος). But after this a marvelous light meets the wanderer, and open country and meadow lands welcome him (φῶς τι θαυμάσιον)", transl. by Sandbach. As Bonazzi underlines, the description of this ceremony is highly dependent on Plato's *Phaedrus* (Bonazzi [2021: 277]).

> As we have already stated, it has been called a rite from its rendering the soul perfect [...]. One should be aware that a rite is one thing, preliminary initiation another, full initiation, [yet] another. The rite corresponds to the preparation, or to the purification and the like. The preliminary initiation, a term that is derived from the phrase 'to shut the eyes', is more sacred for to shut the eyes is this: no longer take in those sacred mysteries [up there] through the senses but with purified soul itself. And [the goal of] full initiation is to be ensconced among them [sc. the sacred mysteries] and become an observer of them.[41]

Hermias' words recapitulate the stages of the rites and their value. He first proposes an explanation for the noun *teletē* that is coherent with the state of perfection of Plato's souls. Then he confirms the visual character of the *epopteía* and the connection between blinding and *mýēsis*. The conclusions of this analysis, however, are quite unexpected: the epoptic moment seemed to be overlooked and, conversely, the *mýēsis* is assigned a more sacral role. Hermias assumes that, in Plato's view, sensible blindness guarantees a sharper intellectual vision, for senses are considered obstacles to a proper metaphysical sight.

This idea is genuinely present in Plato's work. For example, in the *Phaedo*, with a mirrored image, Socrates expresses the fear that he may become blind in the soul by observing material things with his physical eyes.[42] More importantly, this idea is suggested by another – and possibly the most renowned – philosophical journey: the allegory of the cave. There, Plato recognizes two kinds of blindness: an upward blindness, caused by the shift from darkness to light, and a downward blindness,[43] caused by the shift from light to darkness, "and the same thing – he adds – happens to the soul, too."[44] To use Nightingale's words, in the *Republic* "Plato's theoretical philosopher, in sum, has blind spots: the vision of the theorist is not panoptic, since he sees differently – and different objects – in each realm."[45] The *Republic* thus presents a

---

41 "«Τελετή» ἐκλήθη, καθώςἤδη εἴπομεν, παρὰ τό τελέαν τήν ψυχήν ἀποτελεῖν [...]. Εἰδέναι δὲ δεῖ ὅτι ἄ̨λλο ἐστί τελετή, ἄ̨λλο μύησις, ἄ̨λλο ἐποπτεία. Ἡ μέν οὖν τελετή ἀναλογεῖ τῇ προπαρασκευῇ ⟨καί τοῖς⟩ καθαρμοῖς καί τοῖς ὁμοίοις· ἡ δὲ μύησις, ἥτις παρὰ το «μύειν» ἐλέχθη τοὺς ὀφθαλμούς, θειοτέρα ἐστί· τὸ γὰρ μύειν τοὺς ὀφθαλμουὺς τοῦτό ἐστι, τὸ μηκέτι αἰσθήσει λαβεῖν ἐκεῖνα τά θεῖα μυστήρια, ἀλλὰ αὐτῇ ψυχῇ καθαρᾷ· τὸ δὲ τῆς ἐποπτείας τὸ ἐνιδρυθῆναι αὐτοῖς καὶ ἐπόπτην αὐτῶν γενέσθαι" (Herm., 178. 10–27, transl. by Baltzly and Share.).

42 *Phd.* 99e. For more on the paradoxical value of this image, see Aronadio (2020a: 76).

43 For these definitions, see Reeve (2012: 157).

44 *Rep.*, 7.518a: "ἀλλ᾽εἰ νοῦν γε ἔχοι τις, ἦν δ᾽ἐγώ, μεμνῇτ᾽ἂν ὅτι διτταὶ καὶ ἀπὸ διττῶν γίγνονται ἐπιταράξεις ὄμμασιν, ἔκ τε φωτὸς εἰς σκότος μεθισταμένων καὶ ἐκ σκότους εἰς φῶς. ταὐτὰ δὲ ταῦτα νομίσας γίγνεσθαι καὶ περὶ ψυχήν." The translation is Shorey's.

45 Nightingale (2004: 104).

radical opposition between sensible sight and intellectual sight, and temporary blindness marks the discontinuity between these two capacities.

Nevertheless, *contra* Hermias, this does not seem to be the case in the *Phaedrus*. Rather, in this dialogue, sensible and perceptual sight plays a fundamental role in triggering philosophical research and starting the journey of the *theōrós*. As will be argued, this change in tone and focus in Plato's argumentation is perfectly coherent with the dynamics of the Eleusis rituals as illustrated so far, which, on the contrary, does not fit Hermias' interpretation.

In fact, the *Phaedrus* presents physical eyes as the medium through which the effluence of beauty first enters the soul[46] and awakens it. The soul's conversion and recollection can start precisely by virtue of sensible perception, which has the power to re-orient it. The deceptiveness elsewhere associated with senses is significantly lessened[47] or at least framed in a different context. By alluding again to sacred ceremonies, Plato explains that the "newly-initiated" (ἀρτιτελής) can move back by fixing his gaze (προσορῶν) on a god-like *bodily* form (τινα σώματος ἰδέαν).[48] While being potentially harmful[49], corporeal images are necessary to start the soul's journey and, paradoxically, can be grasped better by those who have more recent contact with ideal beauty.

In a similar vein, the conversion process can be triggered when the newly incarnated soul sees (ἴδῃ) a face that "aptly imitates beauty" (πρόσωπον κάλλος εὖ μεμιμημένον).[50] Even in this case, something that is elsewhere famously deemed deceptive – the activity and the products of *mímēsis*[51] – is evaluated in a positive light.[52] The accuracy of this representation does not produce illusory effects, but, on the contrary, *reveals* its model – beauty itself – and helps the soul to reach it. On the one hand, such trustworthiness depends on the fact that the newly initiated has a fresh, vivid memory of celestial beauty,

46 See *Phdr.* 251b. The effluence of beauty is said to reach the lover "*dià tôn ommáton.*"

47 In Nightingale's analysis, this aspect of the *Phaedrus* is contrasted also with the *Phaedo* (see Nightingale [2004: 165]).

48 *Phdr.* 250a.

49 See *Phdr.* 256c.

50 *Phdr.* 251a. Only in this case, the translation is mine.

51 For the negative value of *mímēsis*, see in particular *Resp.* 10.595c–498b, *Soph.* 236b–c. For a wide discussion on *mímēsis* and its sub-genres – icons, idols, and *phantásmata* – see Desclos (2000).

52 Such a positive outcome of sensible images is particularly underlined by Simone Astorino. In the *Phaedrus* – he states – the process of recollection of embodied souls can only be triggered by "empirical objects" that "facilitate" their "commerce" with forms (see Astorino, in this volume).

which implies that he is able to distinguish[53] it from its corporeal image and recognize the ontological gap between the two of them.[54] On the other hand, however, he can identify it as beauty – although in a diminished status – because of the *continuity* between its corporeal and its ideal form.[55]

As the *Phaedrus* explains, the reason for this consistency is the particular nature of beauty. In fact, it happens to be the *ekphanéstaton* ("most clearly *seen*")[56] of all the celestial forms and is the only one to possess the privilege of *shining through* until the world of becoming. The eminent visibility of beauty corresponds, *a parte subiecti*, to the excellence of sight, which is defined as the most perspicuous of the perceptive faculties (τῆς ἐναργεστάτης αἰσθήσεως) and the most acute sensations (ὄψις γὰρ ἡμῖν ὀξυτάτη τῶν διὰ τοῦ σώματος ἔρχεται αἰσθήσεων).[57] This mutual exceptionality makes it possible to mend the tear between sensible seeing and intellectual seeing.[58] Beauty uniquely reaches sensible sight by offering perspicuous *images* of itself (ἑαυτῆς ἐναργὲς εἴδωλον) even in the mortal world.

The worthy soul will thus be able to follow the traces[59] of beauty as flashes of trembling light until it reaches again the status of the divine contemplation of beauty that it experienced in its incorporeal life. As the torches of the *pannychía* guided the initiates through the night,[60] the corporal likenesses of beauty drive the philosophical soul by virtue of their homogeneity with their principle: this characteristic – Plato says – allows worthy souls to grasp *theoretically* (θεᾶσθαι) the genre (τὸ γένος) of which they are images. In both cases, the flickering and partial apparition of sources of light end with the manifestation of a full luminosity. More accurately, both processions end with *epopteía*:[61]

53 Very poignantly, Palumbo remarks that an image can assume a positive value when it explicitly declares itself as an image and does not try to pass itself off as the model it represents (Palumbo [2009: 69–70, 296]).

54 As known, the more time passes, the more this capacity of discernment deteriorates, for the memory of ideal beauty becomes less and less vivid (see *Phdr.* 250e).

55 This ambiguous "similar dissimilarity" characterizes all images as such, as Socrates points out in the *Cratylus* (see *Crat.* 432c).

56 My emphasis.

57 *Phdr.* 250d–e.

58 See *Phdr.* 250c–d: "περὶ δὲ κάλλους, ὥσπερ εἴπομεν, μετ᾽ἐκείνων τε ἔλαμπεν ὄν, δεῦρό τ᾽ἐλθόντες κατειλήφαμεν."

59 The idea of seeking tracks is used when Socrates describes the lovers' research of a potential beloved who resembles the god they were following in their celestial existence (see *Phdr.* 253a).

60 Quite significantly, in the *Timaeus* Plato says that eyes are made up of a pure not-burning fire, which is "akin of that of day" (see *Tim.* 45b, transl. by Heinemann).

61 Betegh (2022: 242–244) underlines the climatic structure of the mysteries and Plato's reformulation of it.

the close observation of a spectacle of lights and fire for the traditional one and the magnificent appearance of theoretical beauty in the *hyperouránion* for the philosophical one. The heuristic role of this continuity of sight explains why, in the *Phaedrus*, the manifestation of beauty, however splendid it may be, does not provoke any form of blindness, unlike in the *Republic*.

As already seen, the *Phaedrus* describes this moment very poignantly. Beauty (κάλλος) appears "in pure light" (ἐν αὐγῇ καθαρᾷ) and can be caught by equally purified souls. It is bright (λαμπρόν) and shining, and it possesses a unique light (φέγγος). As Riedweg underlines, this description closely resembles the possible performance of the *epoptiká* at Eleusis, with the sudden opening of the *anáktoron* by the hierophant and the emergence of a great burning fire.[62] Moreover, he underlines the strong analogies between Plato's lexical choices and the terms used in the *Hymn to Demeter*, when the goddess was discovered in her true identity by Metanira, queen of Eleusis, and showed herself in her divine radiance.[63]

The name of the ceremony itself provides further information about the souls' contemplative activity. The noun *epopteía* and the verb *epopteýō* normally indicate a kind of vision that benefits from a privileged perspective:[64] they mean seeing closely or from above.[65] This value is hinted at by another passage early in the text, in which the soul is described as *hyperidoŷsa*[66] – that is, looking at reality from the top. The purely cognitive connotation of this kind of *theōreîn* can be better grasped with this in mind: not only could philo-

---

62 See Riedweg (1989: 50). As testified by some epigraphic sources, the ceremony presumably took place in the *telestērion*. For a more recent reconstruction of the epoptic ceremonies, see Clinton (2003).

63 HH 2.275–280: "ὣς εἰποῦσα θεὰ μέγεθος καὶ εἶδος ἄμειψε/γῆρας ἀπωσαμένη: περί τ'ἀμφί τε κάλλος ἄητο:/ὀδμὴ δ'ἱμερόεσσα θυηέντων ἀπὸ πέπλων/σκίδνατο, τῆλε δὲ φέγγος ἀπὸ χροὸς ἀθανάτοιο/λάμπε θεᾶς, ξανθαὶ δὲ κόμαι κατενήνοθεν ὤμους,/αὐγῆς δ' ἐπλήσθη πυκινὸς δόμος ἀστεροπῆς ὥς" ("When she had so said, the goddess changed her stature and her looks, thrusting old age away from her: beauty spread round about her and a lovely fragrance was wafted from her sweet-smelling robes, and from the divine body of the goddess a light shone afar, while golden tresses spread down over her shoulders so that the strong house was filled with brightness as with lightning", transl. by Evelyn-White).

64 This hypothesis is discussed also by Nightingale: "The *epoptēs* might have seen additional sights or perhaps had a closer look at the sacred objects" (see Nightingale [2004: 85]).

65 This meaning results from the combination of the preposition *epí* ("on", "upon") and *opteýō* ("to see"). As Aronadio (Aronadio 2020b: 121) underlines, it is used in Pindar (Pind., *Ol.* 7.11), Aeschylus (Aesch., *Ag.* 1270, 1279, *Coeph.* 1, 583, 985) and Aristophanes (Arist., *Fr.* 1126, 1138, 1143) to describe the gods observing men form the top of the Olympus.

66 See *Phdr.* 249c.

sophical souls then watch ideal beauty freed from their bodies,[67] but they were also able to do so while considering and contemplating the world of becoming at a glance from the viewpoint of first principles.

In other words, the transition from the sensible realm to the pristinely theoretical one entails a shift in perspective: from a bottom-up vision – in which each sensible manifestation must be anamnestically reconnected to their prime cause – to a top-down one, in which first principles can be grasped as such and immediately caught in their ontological excellence. In the first case, the preliminary appearances of sensible beauty are essential to stimulate the theoretical gaze of the soul; in the second case, the ontological priority of ideal beauty is made consistent with its chronological precedence. At the climax of the purely noetic *theōreîn*, images of beauty are immediately seen in the light of their model, which occupies the first position in the act of intellection and does not need to be re-discovered at the end of empirical experience.

In this light, another fundamental lexical choice becomes clearer: as already seen, the spectacle of beauty consists of "perfect and simple and calm and happy *phásmata*." This term – cognate with *phaínō* and thus associated with the idea of appearing – is quite singular: coherently with the mysterial atmosphere of the scene, it refers to celestial and divine phenomena,[68] but it also indicates illusory spectral entities and phantoms,[69] as Plato's work itself testifies.[70] At first, one could wonder why the "truly existing essence" (οὐσία ὄντως οὖσα)[71] at the core of the noetic vision is paradoxically expressed by a term indicating a certain evanescence and a weakened ontological status. Nevertheless, the aforementioned top-down perspective of the noetic gaze offers a straight answer: here, *phásma* is not used in a relative sense and does not designate a relationship. To put it clearly, one could say that, in this perspective, it does not express the appearance *of* something; rather, *phásma* indicates here luminosity and appearance themselves.

---

67 See *Phdr.* 250c: "We saw in the pure light, being ourselves pure and not entombed (ἀσήμαντοι) in this which we carry about with us and call the body (σῶμα)." In this passage, Plato uses the well-known pun involving the words *sēma* ("tomb") and *sōma* ("body"). The philosophical discussion of this phonological analogy is discussed in the *Cratylus* (*Crat.* 400 b–c) and in the *Gorgias* (*Grg.* 493a). For a wider analysis of this parallel, with a specific focus on its mysterial origin, see Gatti (2012: 265–278).

68 See Aesch., *Ag.* 146 and, for Plato, *Plt.* 268. For more on the value of *phásma* as connected with divine apparitions, see Vernant (1990).

69 See Hdt. 6. 69, Aesch., *Ag.* 274.

70 See, *e.g.*, *Tht.* 155 and *Symp.* 179d, where it is used to indicate the ghost of Eurydice.

71 *Phdr.* 247c.

Led by their gods, blessed souls get to these apparitions, that is to say, to experience the proper moment of noetic *theōria*. These apparitions are said to be serene (εὐδαίμονα), perfect (ὁλόκληρα), simple (ἁπλᾶ), and unmoving (ἀτρεμῆ). The light of ideal beauty is firm, stable, and constant, and, from the part of the viewer, it corresponds to a kind of *theōreîn* that is equally stable and serene. This steadiness characterizes the exclusively cognitive vision of forms, which, after being intermittent and hardly conquered during their prenatal elevation, will be threatened again with reincarnation in the world of becoming.

In the *Phaedrus*, every phase of the soul's "journey outside"[72] – which, as Nightingale underlines, represents an essential element of traditional *theōría* – corresponds to a different visual activity. Moreover, the soul's journey becomes a journey of sight. Sight is shown in all its potential developments and evolutions: from a view from below to an overlooking gaze, from an intermittent eyeful to the serene contemplation of eternal forms, and from sensible watching to its purest theoretical expression. The Eleusis mysteries provide Plato with a semantic scheme that he adapts to his philosophical purposes to show all the possible articulations of vision in a sort of phenomenology of sight, which represents one of the main philosophical cores of the dialogue. Analyzing vision and – even more importantly – *educating* it are portrayed as the tools through which recollection and the unification with first principles are possible.

4 *Mýēsis* and Blindness

Despite not being used to mark the break between sensible and intellectual sight, several references to the experience of blindness appear in the *Phaedrus*. The analysis of blindness – the complete absence of sight – could not but occupy an important place in the philosophical project just described. Moreover, in this case, too, Plato chooses to employ the imaginary of mysteries to thematize this last area of investigation.

As already stated, in the Eleusinian mysteries, blindness was probably at the centre of the preliminary phase of initiation – *mýēsis*. This rite likely

72 See Nightingale (2004: 158). The author often underlines that the practices of *theōria* involved a pilgrimage or a journey abroad in order to attend a religious performance or show. In Plato's philosophy, this trait is transfigured under the experience of leaving the sensible order to encounter a radical metaphysical alterity. She observes that in the *Phaedrus* this aspect is hinted at by the location of the dialogue, which is set outside the walls of Athens, along the banks of the Ilissus.

included a ritual veiling, similarly to the way in which the scenes of mystic initiations are depicted on the Lovatelli Urn and the Torre Nova sarcophagus.[73] Very significantly, the *Hymn to Demeter* presents a similar scene: after covering her shoulders with a dark blue cloak, the goddess veils her head[74] as a sign of mourning.[75] Several scholars suggest that this gesture could represent an *aítion* for this ritual practice.

As Cairns remarks,[76] in Greek culture, veiling could indicate a sort of liminality or a shift in one's condition. Such an observation allows two further considerations. Firstly, this value perfectly fits the exoteric and salvific context of the mysteries of Eleusis. Secondly, and more importantly, blindness is religiously significant not as such, but because it allows a sudden change from darkness to light at the moment of one's unveiling, which could represent the acquisition of a renewed and more mindful gaze. If this is true, *mýēsis* did not aim to deny the value of sight, but to make vision sharper by virtue of its re-establishment after complete darkness.

Now, as is renowned, the *Phaedrus* features an eerily similar scene: when pronouncing his first speech, Socrates chooses to talk with his "head wrapped up" (ἐγκαλυψάμενος ἐρῶ)[77] because of the shame that it provokes in him. This feeling of shame derives from the fact that Socrates is urged by Phaedrus to reformulate Lysias' thesis, which he does not share, and, by doing so, to dishonor Love.[78] In Socrates' words, the gesture of veiling has two functions: first, from a metaphorical point of view, lifting his cloak over his head will allow him to "run faster" in his speech and conclude it as soon as possible;

73 See Mylonas (1961: 205–208). The Lovatelli urn (first century BC., *LIMC* 145) represents Heracles' initiation to the mysteries of Eleusis. The hero has his head covered with a thick cloak and, next to him, a celebrant wields the *líknon*, often used in ritual purifications. The Torre Nova sarcophagus (second century AD., *LIMC* 146) depicts a sacred procession: a feminine figure, probably Persephone, holds a torch. Even in this case, an initiate has his head veiled.

74 HH 2.182–183: "And she walked behind distressed in her dear heart,/with her head veiled and wearing a dark cloak which waved about the slender feet of the goddess (ἣ δ'ἄρ'ὄπισθε φίλον τετιημένη ἦτορ στεῖχε κατὰ κρῆθεν κεκαλυμμένη)", transl. by Evelyn-White.

75 For an analysis of the general connection between grief and veiling, see Cairns (2009: 37–54).

76 See Cairns (2001: 20–21, 24–25).

77 *Phdr.* 237a. The verb *enkalýptō* is also used in the *Phaedo*, when Socrates veils his head and uncovers it just before dying (see *Phd.* 118a). For more on this, see Arabatzis (2003: 120–121).

78 As is known, Lysias argued that the lover is inferior to the non-lover and, therefore, that it is better to give in oneself to the latter. Socrates will reprise only the first half of Lysias' thesis.

secondly, having his eyes covered will let him avoid Phaedrus' gaze, which would emphasize his embarrassment.[79]

After delivering the first speech, which he considers blasphemous and false, Socrates regains his sight by uncovering his head. In this passage, the dynamic between sight and its privation is linked again with a mysterial element. Knowing that he dishonored *Érōs*, Socrates looks for a way to purge his guilt. He then says that he knows an ancient ritual of *katharmós* ("purification") that he wants to perform immediately in order to prevent any possible punishment from the god. In saying this, he introduces some elements of novelty:

> Now I, my friend, must purify myself; and for those who have sinned in matters of mythology there is an ancient purification, unknown to Homer, but known to Stesichorus. For when he was stricken with blindness for speaking ill of Helen, he was not, like Homer, ignorant of the reason, but since he was educated, he knew it and straightway he writes the poem:
>
> "That saying is not true;
thou didst not go within the well-oared ships,
nor didst thou come to the walls of Troy"
>
> and when he had written all the poem, which is called the recantation, he saw again at once. Now I will be wiser than they in just this point: before suffering any punishment for speaking ill of Love, I will try to atone by my recantation, with my head bare this time, not, as before, covered through shame.[80]

---

79 See Velardi (2006: 146–147). As the scholar remarks, this interruption of mutuality in gazes must be understood in light of Socrates' following words, in which this reciprocity will prove an essential driving force in arousing erotic mania (see *Phdr.* 250c–253a).

80 "ἐμοὶ μὲν οὖν, ὦ φίλε, καθήρασθαι ἀνάγκη· ἔστιν δὲ τοῖς ἁμαρτάνουσι περὶ μυθο- λογίαν καθαρμὸς ἀρχαῖος, ὸν ὅμηρος μὲν οὐκ ᾔσθετο, Στησίχορος δέ. τῶν γὰρ ὀμμάτων στερηθεὶς διὰ τὴν Ἑλένης κακηγορίαν οὐκ ἠγνόησεν ὥσπερ ὅμηρος, ἀλλ' ἅτε μουσικὸς ὢν ἔγνω τὴν αἰτίαν, καὶ ποιεῖ εὐθὺς·
οὐκ ἔστ' ἔτυμος λόγος οὗτος,
οὐδ' ἔβας ἐν νηυσὶν εὐσέλμοις,
οὐδ' ἵκεο Πέργαμα Τροίας
καὶ ποιήσας δὴ πᾶσαν τὴν καλουμένην Παλινῳδίαν παραχρῆμα ἀνέβλεψεν. ἐγὼ οὖν σοφώτερος ἐκείνων γενήσομαι κατ' αὐτό γε τοῦτο· πρὶν γάρ τι παθεῖν διὰ τὴν τοῦ Ἔρωτος κακηγορίαν πειράσομαι αὐτῷ ἀποδοῦναι τὴν παλινῳδίαν, γυμνῇ τῇ κεφαλῇ καὶ οὐχ ὥσπερ τότε ὑπ'αἰσχύνης ἐγκεκαλυμμένος" (*Phdr.* 243a–b).

When mentioning his need to be purified of his guilt, Socrates evokes another's purification and – even more significantly – another's blinding – namely, those of Stesichorus. Legend has it that Stesichorus wrote a poem in which he blamed Helen for causing the War of Troy. He was then deprived of his eyesight by the goddess as a punishment. He regained it only when he pronounced a new poem, the so-called palinode, in which he withdrew these charges. Like the poet, Socrates dishonored a divine entity and now needs to repair this injury by retracting, reformulating, and re-chanting (πάλιν ἀοιδᾶν) his words. Socrates' palinode is thus compared to Stesichorus' one.[81] As Demos puts it, "Socrates' blindness is metaphorical as well as real when he first discusses love", and, after unveiling his head, "he simultaneously regains his figurative sight into the true nature of love."[82] In both cases, blindness is a symptom of the falsehood of Stesichorus' and Socrates' *lógoi*[83] and can be healed only when this falsehood is recognized and amended.[84]

Nevertheless, a closer look into the dynamics of this play of references could offer new clues about Stesichorus' role in this section. As already mentioned, Stesichorus' fault in stating that Helen was to be held accountable for the Trojan Conflict was the same as Homer's. In Plato's view, unlike Homer, Stesichorus succeeded in redeeming his fault by endorsing another version of the myth, in which Helen never arrived at Troy. According to this variant (which will be later dramatized in Euripides' *Helen*), a mere phantom – an *eídōlon* – replaced the heroine in Troy, while her true self found shelter in Egypt.[85] Stesichorus' rendition of the myth is also evoked in the *Republic*,

81 The story of Stesichorus' palinode is also reported with similar terms by Isocrates (Is., *Hel.* 10.64).

82 Demos 1997, 238–239.

83 For an interpretation of this scene that considers Stesichorus' role in Plato's reflection on speech-making technique, see D'Alfonso (1994: 167–175).

84 Capra (2020: 254–253) suggests that here Socrates may be re-enacting Stesichorus' performance when reciting the palinode.

85 The papyrus of Oxyrhynchus bears evidence of two different palinodes by Stesichorus (Stesich. fr. 193, 2–16 Page) and mentions the beginning of one of them, referring to the Aristotelian philosopher Chamaeleon's *On Stesichorous*. This source testifies to Stesichorus' adoption of the *eídōlon* variant of the myth and mentions the real Helen being carried to Egypt by Proteus. The two palinodes are said to reproach two different poets. The first one's target, as has been said, is Homer; the second one's target is Hesiod, but the reason for his guiltiness is not clear. Cingano hypnotizes that Hesiod may have followed a storyline in which Helen never arrived at Troy but still sailed to Egypt with Parides, committing adultery (see Cingano [1982: 32]). For the relationship between this source and the passage of the *Phaedrus*, see Woodbury (1967: 157–176).

where Plato discusses the false pleasures of the multitudes.[86] Here, illusory leisures become the disputed object of furious and pointless fights, "as Stesichorus says the wraith of Helen (τὸ τῆς Ἑλένης εἴδωλον) was fought for at Troy[87] through ignorance of the truth (ἀγνοίᾳ τοῦ ἀληθοῦς)."[88]

The motifs of this passage offer several elements that enable a better interpretation of the scene of the *Phaedrus*. The senseless souls in the *Republic* mistake an *eídōlon* of pleasure for true pleasure, just like the Trojans and the Acheans mistook the phantom of Helen for the real one. However, upon closer inspection, Stesichorus seems to have done the same thing: he, too, confused the phantom of Helen for the actual woman. His punishment, then, was anything but accidental. His blindness symbolized his *actual* visual impairment, which led him to consider an image of reality as reality itself. Symmetrically, he recovers his sight when he learns to distinguish these two ontological levels and discern the imitation from the thing being imitated.[89] As Plato says in the *Sophist*, the ontological status of an *eídōlon* lies precisely in the relation of resemblance it establishes with its model: it is true "in a way", which is namely *the* way according to which "it is *really* a likeness."[90] Concealing this relationship inevitably produces falsehood.[91]

True blindness – Plato suggests – lies precisely in this: the inability to recognize an *eídōlon* – that is, a likeness – as such. Another detail that is quite trivial but potentially revealing must be considered: as the aforementioned

86 For an insightful analysis of the Platonic fallout of the story, with particular reference to its Euripidean retelling, see Napolitano (2007: 137–233).

87 The pointlessness of the War of Troy is evoked by Euripides as well: "Menelaus: 'She is not; I was tricked by the gods/and had in my arms the baneful image of a cloud (νεφέλης ἄγαλμα λυγρόν)'/Messanger: 'What are you saying? We suffered in vain for the sake of a cloud?'" (Eur., *Hel.* 704–707).

88 *Rep.* 9. 586c, transl. by Brunert. Quite significantly, in the previous section illusory pleasures are also compared to "phantoms of true pleasures" (εἰδώλοις τῆς ἀληθοῦς ἡδονῆς) and "illusions of scene-paintings, so colored by contrary juxtaposition" (ἐσκιαγραφημέναις), (*Rep.* 9. 586b).

89 Tellingly, analogous characteristics play a main role in Marco Picciafuochi's analysis of high cognition in the *Republic* and the *Timaeus*. Namely, the "waking view" guaranteed by *noēsis* can be distinguished from the "dreaming view" of dianoetic thought because the former "has [...] direct access to 'what is'. Thus, even when it deals with appearances, dialectic is never deceived, since it always recognizes appearances as something similar but not identical with the original" (Picciafuochi, in this volume).

90 *Soph.* 240b, my emphasis. For more on the complex lexicon and distinctions in Plato's reflections on images, see the previously mentioned Desclos (2000: 301–328), Palumbo (2009), and Vernant (1990).

91 Picciafuochi makes similar remarks when commenting on the role of *eídōlon* in Plato's account of oneiric visions (see Piacciafuochi, in this volume).

passage of the *Republic* testifies, in Greek archaic culture Helen is traditionally associated with love and beauty.[92] This characteristic makes the scene of the *Phaedrus* all the more fitting and significant. Stesichorus' initial incapacity to recognize Helen could hint at the temporary inability to identify real beauty. Generally speaking, one could conclude that, in the *Phaedrus*, authentic blindness is precisely mistaking a phantom of beauty for beauty itself. Stesichorus' sudden healing and Socrates' unveiling – as sudden shifts from darkness to light – symbolize this regained sharpness of vision.

Such risk is explicitly evoked in the scene that immediately follows the description of the manifestation of beauty. Meaningfully, it is expressed once again through the lexicon of mysteries: "he who is not newly initiated" (ὁ μὲν οὖν μὴ νεοτελής) or has been "corrupted" (διεφθαρμένος) may relapse in a state of blindness, for he cannot "quickly rise from this world to that other world and absolute beauty when he sees its namesake here, and so he does not revere (σέβεται) it when he looks upon it, but gives himself up to pleasure."[93] Real blindness does not lie in approaching sensible manifestations of beauty but in not venerating them as one does with representations of gods, which means negating the metaphysical horizon behind them. Only those who have a proper philosophical gaze grasp the ideal beauty behind beautiful bodies and, therefore, worship them as sacred *agálmata* – divine images – *of* something beyond them and other than them.[94]

Once again, the *Phaedrus* blames those who are not able to keep together sensible and intellectual *theōreîn*. In this case, the problem lies in not catching the sign-value of sensible appearances. The actual visually impaired souls are those who cannot see veritable beauty behind beautiful sensible appearances. Beautiful bodies are not perceived in their ulterior meaningfulness: they are not seen as traces of beauty or, to resume the metaphor of the mysteries, as torches in the night promising a brighter light.

In summary, one could say that blindness does not lie in not immediately seeing ideal beauty, but in not knowing how to move towards it. Coherently,

---

92 For an analysis of the development of this trope in Homer, see Edmunds (2019: 81–113).

93 *Phdr.* 250e.

94 See *Phdr.* 251a. Quoting Vernant and his well-known article on the *kolossós* (Vernant [1991: 153]), Nightingale underlines that *agálmata* were considered to have the ambiguous power of making the god present while highlighting its distance and its radical otherness (see Nightingale 2004, 164). The role of the *agálmata* in the *Phaedrus* is thoroughly discussed by Astorino, who underlines that, "as an *ágalma*" sensible images "represent the intelligible" and serve as "a means of connection and relation" between the empirical dimension and the intelligible one (Astorino, in this volume). On the difference between *agálmata* and *eídōla*, see also Vasiliu (2010: 143–187).

the same idea is reiterated in the last part of the dialogue, in which the dialectical method is at stake: those who do not possess a meaningful way of procedure (μέθοδος) resemble a blind man (τυφλός) groping in the dark.[95] On an epistemological level, blindness represents impairments in distinguishing ontological levels and the play of references among them and, more importantly, in clearly sharpening the authentic and ultimate object of the philosophical quest.

5 Conclusion

Blindness and vision, as well as the intermediate states between them, represent two central tropes in the *Phaedrus*. This importance is owing to the fact that one of the philosophical themes at stake is the analysis of *theōreîn*, from its sensible value to its intellectual and cognitive one. This particular focus may have led Plato to choose the mysteries of Eleusis as a metaphorical tool, because of the centrality of light and vision that characterized the progression of their rites.

The main phases of the mysteries – *epopteía* and *mýēsis* – are repeatedly evoked in the dialogue and shaped in order to express Plato's main philosophical thesis that sensible vision does not contradict and is not opposed to cognitive and intellectual one. On the contrary, there is a fundamental continuity between them. However, this continuity is ambivalent. It could represent an aid in philosophical research and recollection if sensible appearances are correctly envisaged as images of higher realities; alternatively, it could represent an obstacle to the pursuit of the research, if those appearances are not perceived as signifiers of ulterior entities. Seeing appears to be examined in light of all its advantages and risks, and its different modalities mark all the stages of the soul's itinerary back and forth from the *hyperouránion*. Thus, the voyage of the soul becomes also the voyage of sight.

Bibliography

Adkins, Arthur WH. 1970. "Clouds, Mysteries, Socrates and Plato." *Antichthon* 4: 13–24.

Arabatzis, Giorgos. 2003. "Εγκαλυψάμενος. La piété socratique dans le *Phèdre* de Platon." *Kernos* 16: 119–123.

95 *Phdr.* 270d–e.

Aronadio, Francesco. 2020a. "Cecità e sordità in Platone. Controprova di un dispositivo metaforico." *Antiquorum Philosophia* 14: 69–83.

Aronadio, Francesco. 2020b. "Epopteuo." In *Le forme del vedere. Studi lessicologici sui* verba videndi *nel* corpus platonicum, edited by Lorenzo Giovannetti, 119–126. Naples: Bibliopolis.

Baltzly, Dirk, Share, Michael, transl. 2022. *Hermias: On Plato's Phaedrus 245e–257c*. London: Bloomsbury.

Bénatouil, Thomas, Bonazzi, Mauro, eds. 2012. *Theoria, Praxis, and the Contemplative Life after Plato and Aristotle*. Leiden/Boston: Brill.

Betegh, Gábor. 2022. "Plato on Philosophy and the Mysteries." In *The Cambridge Companion to Plato*, edited by David Ebrey and Richard Kraut, 233–267. Cambridge: Cambridge University Press.

Beylache, Nicole *et. al.* eds. 2021. *Les mystères au II siècle de notre ère: un tournant.* Turnhout: Brepols.

Bonazzi, Mauro. 2021. "Plutarch and the Mysteries of Philosophy." In *Les mystères au II siècle de notre ère: un tournant*, edited by Nicole Beylache *et al.*, 267–278. Turnhout: Brepols.

Boyancé, Pierre. 1962, "Sur les mystères d'Éleusis." *Revue des Études Grecques* 75: 460–482.

Burkert, Walter. 1987. *Ancient Mysteries Cult.* Cambridge MA/London: Harvard University Press.

Cairns, Douglas L. 2001. "Anger and the Veil in Ancient Greek Culture." *Greece & Rome* 48: 18–32.

Cairns, Douglas L. 2009. "Weeping and Veiling: Grief, Display and Concealment in Ancient Greek Culture." In *Tears in Graeco-Roman World*, edited by Thorsten Fögen: 37–54. Berlin: De Gruyter.

Capra, Andrea. 2020. "'Total Reception': Stesichorus as Revenant in Plato's *Phaedrus* (with a New Stesichorean Fragment?)." In *The Reception of Greek Lyric Poetry in the Ancient World: Transmission, Canonization and Paratext*, edited by Bruno Currie and Ian Rutherford: 239–356. Leiden/Boston: Brill.

Chantraine, Pierre. 1968. *Dictionnaire étymologique de la langue grecque*. Paris: Klincksieck.

Christopoulos, Menelaos *et al.* eds. 2010. *Light and Darkness in Ancient Greek Myth and Religion*. Lanham/Oxford: Rowman & Littlefield.

Cingano, Ettore. 1982. "Quante testimonianze sulle palinodie di Stesicoro?." *Quaderni Urbinati di Cultura Classica*" 12: 21–33.

Clinton, Kevin. 2003. "Stages of Initiation in the Eleusinian and Samothracian Mysteries." In *Greek Mysteries. The Archaeology and Ritual of Ancient Greek Secret Cults*, edited by Michael B. Cosmopoulos: 50–78. London/New York: Routledge.

Cosmopoulos, Michael B. ed. 2003. *Greek Mysteries. The Archaeology and Ritual of Ancient Greek Secret Cults*. London/New York: Routledge.

Currie, Bruno, Rutherford, Ian eds. 2020. *The Reception of Greek Lyric Poetry in the Ancient World: Transmission, Canonization and Paratext.* Leiden/Boston: Brill.

D'Alfonso, Francesca. 1994. "Stesicoro versus Omero nel *Fedro* platonico." *Rivista di cultura classica e medievale* 36: 167–175.

Demos, Marian. 1997. "Stesichorus' Palinode in the *Phaedrus*." *The Classical World* 90: 235–249.

Denham, Alison E. ed. 2012. *Plato on Art and Beauty.* London: Palgrave-Macmillan.

Desclos, Marie-Laurence. 2000. "Idoles, icônes et phantasmes dans les dialogues de Platon." *Revue de Métaphysique et de Morale* 3: 301–322.

Des Places, Édouard. 1981. *Études platoniciennes, 1929–1979.* Leiden/Boston: Brill.

Dinkelaar, Bianca M. 2020. "Plato and the Language of Mysteries: Orphic/Pythagorean and Eleusinian Motifs and Registers in Ten Dialogues." *Mnemosyne* 73(1): 36–62.

Dowden, Ken. 1980. "Grades in the Eleusinian Mysteries." *Revue de l'histoire des religions* 197: 409–427.

Ebrey, David, Kraut, Richard eds. 2022. *The Cambridge Companion to Plato.* Cambridge: Cambridge University Press.

Edmunds, Lowell. 2019. *Towards the Characterization of Helen in Homer.* Berlin/Boston: De Gruyter.

Evelyn-White, Hugh G. transl. 1914. *The Homeric Hymns and Homerica with an English Translation by Hugh G. Evelyn-White.* Cambridge MA: Harvard University Press.

Fögen, Thorsten ed. 2009. *Tears in Graeco-Roman World.* Berlin: De Gruyter.

Fowlers, Harold N. transl. 1914. *Plato: Eutiphro. Apology. Crito. Phaedo. Phaedrus.* Cambridge MA: Harvard University Press.

Giovannetti, Lorenzo ed. 2020. *Le forme del vedere. Studi lessicologici sui* verba videndi *nel* corpus platonicum. Naples: Bibliopolis.

Gatti, Marialuisa. 2012. "'Soma-sema' l'Ade e l'iniziato. 'Giochi di parole' concernenti l'uomo, il filosofo e il retore-politico nella risposta di Socrate a Callicle nel *Gorgia* di Platone." *Rivista di Filosofia Neo-Scolastica* 104: 261–288.

Kerényi, Károly. 1962. *The Religion of the Greeks and the Romans.* Translated by Christopher Holme. New York: Dutton.

Mylonas, George. E. 1961. *Eleusis and the Eleusinian Mysteries.* Princeton: Princeton University Press.

Motte, André. 1995. "Le pèlerinage initiatique de la parole. Une lecture du *Phèdre* de Platon." *Methéxis* 8: 33–48.

Napolitano, Linda. 2007. *Platone e le ragioni dell'immagine. Percorsi filosofici tra metafore e miti.* Milan: Vita e Pensiero.

Nightingale, Andrea W. 2004. *Spectacles of Truth in Classical Greek Philosophy.* New York: Cambridge University Press.

Palumbo, Lidia. 2009. *Mimesis. Rappresentazione, teatro e mondo nei dialoghi di Platone e nella Poetica di Aristotele.* Naples: Loffredo.

Patera, Ioanna. 2010. "Light and Lightning Equipment in the Eleusinian Mysteries: Symbolism and Ritual Use." In *Light and Darkness in Ancient Greek Myth and Religion*, edited by Menelaos Christopoulos *et al.*: 261–275. Lanham/Oxford: Rowman & Littlefield.

Reeve, Charles DC. 2012. "Plato on Begetting in Beauty." In *Plato on Art and Beauty*, edited by Alison E. Denham: 142–172. London: Palgrave-Macmillan.

Riedweg, Christoph. 1987. *Mysterienterminologie bei Platon, Philon und Klemens von Alexandrien*. Berlin: De Gruyter.

Rutherford, Ian. 2013. *State Pilgrims and Sacred Observers in Ancient Greece. A Study of Theōria and Theōroi*. New York: Cambridge University Press.

Ustinova, Yulia. 2009. *Caves and the Ancient Greek Mind. Descending Underground in the Search for Ultimate Truth*. Oxford: Oxford University Press.

Vasiliu, Anca. 2010. *Eikôn. L'image dans le discours des trois Cappadociens*. Paris: PUF.

Velardi, Roberto. 2006. *Platone: Fedro*. Milan: BUR.

Vernant, Jean-Pierre. 1990. *Figures, idoles, masques*. Paris: Julliard.

Vernant, Jean-Pierre. 1991. *Mortals and Immortals*. Edited by Froma I. Zeitlin. Princeton: Princeton University Press.

Ward, Julia K. 2022. *Searching for the Divine in Plato and Aristotle. Philosophical Theoria and Traditional Practice*. New York: Cambridge University Press.

Woodbury, Leonard. 1967. "Helen and the Palinode." *Phoenix* 21: 157–176.

PART 4

The Lived Experience of Theôria *as Higher Cognition*

CHAPTER 11

Plato on Euripides and *Baccheia* as a Metaphor of Higher Cognition

Eva Anagnostou-Laoutides

1 Introduction: Wine and States of Altered Consciousness

Research in the past twenty years has proven that since the Neolithic period, when viniculture and wine production was developed in the upland regions of the Taurus and Zagros mountains, humans experimented with the psychoactive effects of wine by adding herbs, seeds, and other soluble substances to it.[1] The Greeks adapted this practice from an early time.[2] Apart from wine, they also used music, dance, hallucinogens, and strong emotions,[3] all of which were incorporated in religious mysteries and divination rites,[4] to stimulate cognition and provoke states of altered consciousness during which our perception is heightened or intensified beyond the norm.[5] As Ustinova noted, "[t]hese mental states, which today would be referred to as 'altered states of consciousness', were *enthousiasmos* ('engoddedness') or *mania* for the Greeks".[6]

Such episodes, typically affecting poets,[7] prophets,[8] and mystics,[9] were regularly described with vocabulary pertaining to *baccheia*,[10] an orgiastic rite

1 McGovern (2003: 305–306 and 308–312); Rinella (2010: 3–10). The ideas presented in this chapter are discussed extensively in Anagnostou-Laoutides 2025, 3–112.

2 McGovern (2003: 239–278); Valamoti et al. (2007). Cf. *Phdr.* 244b8–9.

3 Ustinova (2017: 265–293).

4 Ustinova (2017: 83, 117–129, 130, 134–136, 174–179, 182, 186, 193–195 and 332).

5 Druckman and Bjork (1994: 207–208) with Kihlstrom (1984: 207). Geertz (2014: 32–33) described the process as "mentalizing"; also, Ustinova (2017: 19–29, 69–70, 130–132 and 332–339).

6 Ustinova (2017: 57).

7 Ustinova (2017: 265–293); cf. Pl. *Io.* 533e7–535a1.

8 Ustinova (2017: 55–112 and 245–264).

9 Ustinova (2017: 113–168).

10 Hsch. s.v. Βακχεῖα (as in Ar. *Ran.* 357)· ἑορτὴ Διονύσου ("celebration of Dionysus"); s.v. βακχεύει (as in Aesch. fr. 57)· μαίνεται. τραγῳδεῖ ("to be manic. To represent in a tragedy"; Schmidt [1867: 287.9–10]); s.v. ὀρχεῖται· (Schmidt [1867: 1148.76]); s.v. χορεύει· (Schmidt [1867: 1561.3–4]); and s.v. ληνεύουσι (Schmidt [1867: 981]). Also, Ustinova (2017: 169–198).

 HTTPS://DOI.ORG/10.1163/9789004747128-013

celebrated by the Bacchants/Maenads[11] in honour of Dionysus, the god of religious enthusiasm and drunkenness,[12] and/or the Korybants, the frenzied dancers of Cybele.[13] Their symptoms, detailed in Euripides' *Bacchae*, exerted a normative effect on other claims to transcendent experiences. Furthermore, as a social activity, in line with Dionysus' civic profile,[14] inducing states of altered consciousness allowed Greek communities to forge certain sociocultural self-projections[15] and self-awareness among their members.[16]

Steeped in the literary traditions of his time,[17] Plato was able to tap into the prevalent cultural norms about wine drinking to defend the value of Socrates' well-known eccentricities,[18] often the target of fifth-century BCE Athenian anti-intellectualism.[19] In the *Phaedrus*, Plato famously identifies four modes of altered consciousness as the most beneficial types of *mania*:[20] next to the prophetic (associated with Apollo), telestic (associated with Dionysus), and

---

11 Suda s.v. βακχευούσας σὺν τῶι μέλει τῶι βακχείωι τε καὶ ἐνθέωι ("Maidens who celebrate with Bacchic singing inspired by the god"; Adler [1967: 1.449]). The Maenads derive their name from *mania*, a cognate of μάω/μά(ι)ομαι ("to desire/seek eagerly"), traced back to the IE root *men-. Anagnostou-Laoutides (2022a: 32) with relevant bibliography.

12 On Dionysian enthusiasm and drunkenness, see Ar. *Ran.* 218–219 referring to Dionysus' drunk-crazed mob (κραιπαλόκωμος … ὄχλος; *Ran.* 218–219); cf. schol. Ar. *Ran.* 216 s.v. Διὸς Διόνυσον ἐν Λίμναις (Dübner [1841: 281a.47–50]) and Suda s.v. κραιπαλόκωμος (Adler [1967: 3.188]); also, schol. Ar. *Nub.* 606 s.v. κωμαστὴς· ὅτι καὶ μεθύοντες βακχεύονται καὶ ὥσπερ ἐκμαίνονται ("Who participates in a festival: because those who are drunk experience deliria as if they were beside themselves"; Dübner [1841: 110b.37–38]). Cf. Suda s.v. Ἐνθουcιώcαc· […] τῇ πυρώcει τῆc μέθηc ἐπεξανῆπτον. ἀντὶ τοῦ ἐξανεγειρούcαc, μαινομέναc ("Possessed. […] by the burning of drunkenness. Instead of stirred up, maddened"; Adler [1967: 2.286]); cf. n51 below.

13 Eur. *Bacch.* 120–134; Graf and Johnston (2007: 146–147). In Orphic traditions the Korybants were said to guard infant Dionysus; Strabo 10.3.7 and 10.3.11; Edmonds (2006: 352–358); cf. Diod. Sic. 5.49. On Plato's references to the Corybantic rites in his dialogues, see Wasmuth (2015); Lynch (2018: 725–727).

14 Goldhill (1987); Seaford (1996: 34–52); Seaford (2021); Isler-Kerényi (2021).

15 Hobden (2013: 66ff). Cf. Papakonstantinou (2009 and 2012); Murray (1982 and 1983a/b) on drinking etiquette and early socio-political debates as echoed in Homeric and archaic poetry.

16 On the role of culture in fostering "metacognitive awareness", see Geertz (2014: 18 and 23), with Donald (2001: 285–287); cf. Dodds (1960: xx).

17 E.g. Belfiore and Pender in the same 2011 volume; cf. Hunter (2012: 38–108).

18 Socrates meditated for hours in public view (Pl. *Symp.* 175a1–b4 and 220c3–d5), talking to himself (*Tht.* 173e–174a and 189e–190a; *Phdr.* 249d1–4; cf. Frede [1989: 28–31]) or his *daimonion* (*Apol.* 31c–d; 40a–c; 41c–d; *Phdr.* 242b–d; *Rep.* 496c; *Tht.* 150e–151a; *Euth.* 3b; *Alc.* I 105e–106a; *Euthyd.* 272e–273a; *Thg.* 128d–129e; cf. Xen. *Mem.* 1.1.1–5; *Apol.* 12–13; cf. Ustinova [2017: 318–321]).

19 Whitehorn (2002: 33–34); Green (1979: 15–16); cf. Bromberg (2017: 32–35).

20 *Phdr.* 244a–245a; Ustinova (2017: 294–343).

poetic *mania* (inspired by the Muses), Plato added erotic frenzy (inspired by Aphrodite and Erōs) which can inspire exceptional lovers to devote themselves to the pursuit of wisdom – to become "lovers of wisdom", that is, philosophers. In this context, Plato also endorsed a fifth kind of altered consciousness – Socratic/philosophical inebriation – as the most beneficial form of *mania* of all.

However, like Socrates, Plato most likely had personal experience of self-transcendence;[21] thus, in the first part of the chapter, I relay my theoretical premises to explain why Plato had to resort to metaphors to relate the mental and ethical optimization achieved through philosophy. Then, I discuss Euripides' *Bacchae* in relation to Plato's metaphor of *baccheia,* palpable in the *Symposium*, the *Phaedrus*, and the *Phaedo*. In the final part of the chapter, I turn my attention to the late *Laws*, where Dionysian *choreia* is presented as underpinning the educational program of Magnesia.[22] Here, Plato prescribes regular wine consumption (which still operates at metaphorical level)[23] as a means for average citizens to rehearse, to the ability of each, the educational tenets of the Dionysian Choir and the exceptional perception of the philosophically inclined Nocturnal Council.[24] Thus, in the *Laws* Plato transforms *baccheia* into a metaphor about sharing philosophical and ethical perspectives across the citizenry.

2 Plato and the Role of Metaphors in Cognition

My approach draws on cognitive phenomenology, as anticipated in Merleau-Ponty's account of perception as "a system of meanings which makes the concrete essence of the object immediately recognizable".[25] By mapping drunkenness onto philosophical reverie, Plato employs the framework of – what Lakoff and Johnson defined as – "conceptual or structural metaphor", where "one concept is metaphorically structured in terms of another".[26] Hence, in

21 Ustinova (2017: 315–328). On the differences between trance and ecstasy, see Rouget (1990) with Ustinova (2017: 20–21).

22 Pl. *Law.* 665b; 666a2–c3; Belfiore (1986); Prauscello (2014: 105–181); Folch (2015: 215–219); Yu (2020). Cf. Belfiore (2006) on Corybantic *choreia* in the *Phaedrus*, and Périllié (2019) on Socratic midwifery as Corybantic initiation.

23 Anagnostou-Laoutides (2022b : 11–16).

24 Anagnostou-Laoutides (2022b: 17–21).

25 Merleau-Ponty (1962: 151); cf. Crowther (2003: 88) on metaphor as the "articulation of perception".

26 Lakoff and Johnson (1980: 14–15 and 327). On the way culture informs conceptual metaphors in Plato, also see Picciafuochi in this volume.

the manner of Lakoff and Johnson, Plato's metaphor might be formalized as *philosophy is inebriating*. Besides providing humans with cognitive frameworks in which "to create new meanings and new understandings of ourselves",[27] metaphors also allow us to mould "values, beliefs, attitudes, and behavior into cohesive political identities".[28] While Plato did not formulate a theory of metaphor, Aristotle paid sufficient attention to the ubiquitous presence of metaphors in speech,[29] and his views informed Ricoeur's appreciation of metaphor as a figure of speech which "presents in an *open* fashion, by means of a conflict *between* identity and difference, the process that, in a *covert* manner, generates semantic grids by fusion of differences *into* identity."[30]

According to Aristotle, "metaphor" refers to a range of linguistic phenomena, all of which involve the transference of a word associated with one domain to another.[31] Importantly, "the right use of metaphors is to perceive similarities"[32] for the sake of clarity.[33] Thus, metaphors also allow writers to specify things that do not have proper names.[34] Indeed, as Meno admits in the homonymous dialogue, articulating Socrates' philosophical genius and its impact on his interlocutors is a very demanding task which has left him speechless.[35] Likewise in the *Phaedrus* (255d3–5), the beloved, inspired by his attraction to the lover to pursue philosophy, is unable to express his mental and psychological distress. The most prominent type of metaphor is analogy:[36] it functions as an extended simile,[37] in which the points of similarity between two

---

27 Bougher (2014: 218).
28 Lakoff (2004: esp. 19–21), with Bougher (2014: 252).
29 Arist. *Rhet.* 1404b6: πάντες γὰρ μεταφοραῖς διαλέγονται ("we all engage in conversation using metaphors"); cf. *Rhet.* 1405a8 and *Poet.* 1459a5.
30 Ricoeur (1975: 252 = 1977: 234).
31 Arist. *Poet.* 1457b7: μεταφορὰ δέ ἐστιν ὀνόματος ἀλλοτρίου ἐπιφορά; O'Rourke (2006: 156–157); cf. Patsioti-Tsacpounidi (2006: esp. 180) drawing attention to Ricoeur (1996). Cf. Ricoeur (1975: 13–34 = 1977: 8–26).
32 Arist. *Poet.* 1459a8: τὸ γὰρ εὖ μεταφέρειν τὸ ὅμοιον θεωρεῖν ἐστιν; cf. Ricoeur (1978: 146–148).
33 Arist. *Poet.* 1458b1: εἰς τὸ σαφὲς τῆς λέξεως.
34 Arist. *Rhet.* 1405a12: ἐκ τῶν συγγενῶν καὶ τῶν ὁμοειδῶν μεταφέρειν ἐπὶ τὰ ἀνώνυμα. Kirby (1997: 542 and 547).
35 Pl. *Men.* 79e7–80b2; Kirby (1997: 529); Hunter (2004: 101–102).
36 Arist. *Poet.* 1457b9; *Rhet.* 1405a9 and esp. 1411a7: Τῶν δὲ μεταφορῶν τεττάρων οὐσῶν· εὐδοκιμοῦσι μάλιστα αἱ κατ' ἀναλογίαν. Johnson (1981: 25); cf. Black (1981: 68–71); Silk (2003: 127–128). For analogy as a way of "surpassing the limits of experience", see Grenet (1948: 39–68) with Anderson (1950: 125).
37 Ricoeur (1975: 34–40 = 1977: 26–30); cf. Henle (1981: 86–91); also, Mac Cormac (1985: esp. 34–36).

things are refined through the use of negation;[38] as Ricoeur put it, "the 'similar' is not the 'same.' To see the similar … is to apprehend the 'same' within and in spite of 'difference'."[39] Importantly, as we shall see, Socrates' "drunkenness" is *similar but not identical* to common intoxication since it is *not* caused by excessive alcohol consumption.[40]

Although Aristotle offers "the classic view of how metaphors work",[41] his theory has been mainly understood "as a matter of words, not concepts".[42] Thus, we must turn to Plato for a demonstration of the influence of images on cognition;[43] adhering to the Socratic credo that an expert should be able to give an account of his expertise,[44] Plato employs drunkenness as a preliminary or second-best way for explaining philosophy's ability to prompt cognitive transformation.[45] As Trabattoni argued, for Plato, truth is manifested in persuasion: if it can convince the soul to pursue and acquire certain knowledge, then speech is "sufficient".[46] Plato's application of philosophical inebriation is compatible with cognitive theories that accept similes and metaphors as integral means of devising and negotiating cognitive patterns, determined additionally by our physical experiences[47] and the cultural context in which they occur.[48] Thus, Merleau-Ponty defends man's use of language "to establish a living relation with himself or with his fellows", a need which transforms it from a mere instrument into "a manifestation, a revelation of intimate being and of the psychic link which unites us to the world and our fellow men."[49]

38 Arist. *Poet.* 1457b30: προσαγορεύσαντα τὸ ἀλλότριον ἀποφῆσαι τῶν οἰκείων τι ("by calling something by a name belonging to another to deny one of its characteristics"). Cf. Arist. *Rhet.* 1411a1–2.

39 Ricoeur (1975: 376 = 1977: 349–350).

40 Anderson (1950: 113–118) views Plato's use of analogy as "likeness mingled with unlikeness."

41 Lakoff and Johnson (1980: 234); cf. Johnson (1981: 7–8).

42 Lakoff and Johnson (1980: 295); cf. Johnson (1981: 35–42); Innes (2003: 12–13); Patsioti-Tsacpounidi (2006: 181–183). Still, Ricoeur (1975: 43–51 = 1977: 33–39) noted that, despite failing to situate metaphor in discourse, Aristotle did stress its cognitive value. Cf. Silk (2003: 141–146).

43 Pender (2003: 63–72).

44 Pl. *Phdr.* 268b3–5; cf. *Prot.* 319c–d; *Men.* 90b–e; *Gorg.* 453d–454a, 465a.

45 Cf. Pender (2003: 61); also, Szlezák (1999: 23, 34, 48–50 and 82–84).

46 Trabattoni (1993: 245–246).

47 Merleau-Ponty (1962: 28).

48 Lakoff and Johnson (1980); Mac Cormac (1985: 149 and 185); Camp (2006); Wentura (2019).

49 Merleau-Ponty (1962: 249–250; cf. 467).

In the next section, I explain Plato's adaptation of Euripides' iconic representation of Bacchic frenzy to convey the revelatory nature of Socratic philosophy.

3 Euripides' *Bacchae* and the Cultural Norms of Socratic Inebriation

In the *Bacchae* Dionysus is identified with his divine gift, as a "drink to be poured for the enjoyment of the other gods",[50] but is also hailed as the divine originator of wine, invented for the benefit of humanity.[51] Both views were widespread,[52] though Plato echoes the latter in the *Laws*.[53] While Plato's penchant for Euripides is well-established in scholarship,[54] in my view, Euripides' distinction between the genuine essence of Bacchic rites and their popular yet misleading association with drunken revelry proved invaluable for Plato's description of Socrates' cognitive superiority.[55] Throughout the play, Euripides employs the verbs *mainomai* and *oistreō* (and their respective nouns, *mania* and *oistros*)[56] to refer to Dionysus' effect on the Bacchants.[57] His description anticipates Plato's account in the *Phaedrus* of the symptoms of the philosophically inclined lover whose soul is raging with pain and smarting throughout[58] as he strives to grasp the realities that exist beyond the world of the senses. Plato also relates the impasse that the soul experiences[59] using vocabu-

50 *Bacch.* 284: οὗτος θεοῖσι σπένδεται θεὸς γεγώς.

51 *Bacch.* 279–283: Dionysus introduced wine to mortals, as relief from grief and hardships (ὃ παύει τοὺς […] βροτοὺς λύπης […] λήθην τῶν καθ' ἡμέραν κακῶν δίδωσιν), a medicine for their troubles (οὐδ' ἔστ' ἄλλο φάρμακον πόνων).

52 Jiménez San Cristóbal (2020: 387) on Diod. Sic. 3.62.2–3 and 3.63.1.

53 *Law.* 666b6–c1: Dionysus bestowed wine on people as medicine (φάρμακον) to forget their worries (δυσθυμίας λήθη).

54 Sansone (1996); Roochnik (2020: 45–74 and 79–94); Anagnostou-Laoutides (2022b: 4–11).

55 Seaford (2006: 15–22); Ustinova (2017: 169–171 and 175–176); see n12 above.

56 Meaning "to sting or goad to madness, typical said of a gadfly"; cf. Pl. *Apol.* 30e for Socrates as a gadfly for the citizens.

57 Eur. *Bacch.* 32–33; 119; 664–665; 979; 999; 1093–1094; 1229; 1295.

58 Pl. *Phdr.* 251d7–8: πᾶσα κεντουμένη κύκλῳ ἡ ψυχὴ οἰστρᾷ καὶ ὀδυνᾶται. Cf. Eur. *Bacch.* 794–795 on Dionysus' "stings" (κέντρα) which he unleashes upon anyone resisting him; cf. nn108–109 below.

59 Pl. *Phdr.* 251e1–3: … ἀδημονεῖ τε τῇ ἀτοπίᾳ τοῦ πάθους καὶ ἀποροῦσα λυττᾷ, καὶ ἐμμανὴς οὖσα …

lary that corresponds to Dionysus' ability to resolve such deadlocks in the *Bacchae*.[60]

Through his Bacchic energy, the god expands the human mind, endowing it with extraordinary skills, such as the ability to divine the future:

> But this god is a prophet – for Bacchic revelry and madness have in them much prophetic skill. For when the god enters a body in full force, he makes the frantic to predict the future.[61]

Furthermore, Dionysus' worshippers, we are told, become proficient in purification rites which they perform high in the mountains.[62] Plato engages with similar imagery early in the *Phaedrus* where Socrates links *mania* both with prophecy and purification rites,[63] while in the *Phaedo* he claims that the philosopher is dedicated to the purificatory rites of virtue.[64]

Euripides revisits the topic of Dionysus' power over human perception when the old-king Kadmos questions Agave, leader of the Maenads, about her Bacchic trance. Here is their exchange:[65]

> AGAVE: ...; why do you tell me to look at it? (i.e. the sky)
>
> KADMOS: Is it still the same, or does it appear to have changed?
>
> AGAVE: It is brighter than before and more translucent.

---

60 Pl. *Bacch.* 1102: ἀπορίᾳ λελημμένος; 1391: τῶν δ' ἀδοκήτων πόρον ηὗρε θεός. Cf. Wilson-Nightingale (2004: 113–115), on the soul feeling *aporia* and *atopia* after witnessing the Forms (in the *Republic*).

61 Eur. *Bacch.* 298–302: μάντις δ' ὁ δαίμων ὅδε: τὸ γὰρ βακχεύσιμον/ καὶ τὸ μανιῶδες μαντικὴν πολλὴν ἔχει:/ ὅταν γὰρ ὁ θεὸς ἐς τὸ σῶμ' ἔλθῃ πολύς,/ λέγειν τὸ μέλλον τοὺς μεμηνότας ποιεῖ./ μέγαν τ' ἀν' Ἑλλάδα.

62 Eur. *Bacch.* 73–82: the blessed (μάκαρ) and fortunate (εὐδαίμων) followers of the god, skilled in divine rituals (τελετὰς θεῶν εἰδὼς), lead a pious life (βιοτὰν ἁγιστεύει), "revelling in Bacchic frenzy at the holy purifications on the mountains" (ἐν ὄρεσσι βακχεύων ὁσίοις καθαρμοῖσιν).

63 Pl. *Phdr.* 244d9–e4; also, 243b2–5 for Socrates' first speech as a palinode delivered to achieve purification (ἐμοὶ ..., καθήρασθαι ἀνάγκη: ἔστιν δὲ τοῖς ἁμαρτάνουσι περὶ μυθολογίαν καθαρμὸς ἀρχαῖος, ...); cf. Hsch. s.v. κορυβαντισμός: κάθαρσις μανίας (Schmidt [1867: 906]); also, Ustinova (2017: 119–120).

64 Pl. *Phdo.* 69b11–69c3 naming several virtues as a kind of purification (κάθαρσίς τις ... καθαρμός τις ᾖ).

65 Eur. *Bacch.* 1265–1267: Ἀγαύη: ἰδού: τί μοι τόνδ' ἐξυπεῖπας εἰσορᾶν;/ Κάδμος: ἔθ' αὑτὸς ἤ σοι μεταβολὰς ἔχειν δοκεῖ;/ Ἀγαύη: λαμπρότερος ἢ πρὶν καὶ διειπετέστερος. Ustinova (2017: 189–192).

Agave's enhanced sight, notably distinct from Pentheus' distorted vision when disguised as a Bacchant,[66] agrees with Plato's claim in the *Phaedrus* that sight is the "sharpest of the physical senses" that stimulates our memory of the intelligible world of Ideas.[67] Just as the Socratic lover undergoes a profound mental change, when reminded of Beauty through his beloved's good looks and earthly charm,[68] so the Bacchants experience a change of consciousness when beholding the god "eye-to-eye"[69] – a process described in both texts as *metabolē*.[70] In the *Phaedrus* Socrates explicitly compares philosophically inclined lovers to the Bacchants:[71] "compelled to keep their eyes fixed upon the god", lovers are able to "grasp him by memory" and "once they become enthusiastic", they can "partake in his divine nature as far as this is humanly possible".[72] Overall, Socrates presents philosophical engagement as initiation in mystery rites,[73] to which he refers by the verb ὠργιάζομεν;[74] thus, he evokes Euripides' typical description of Dionysian rites as *orgia*[75] and clearly links mystic *mania* with Dionysian *teletai*.[76] Like Agave, philosophy initiates gain extraordinary sight,[77] and come to "witness" otherworldly truths: "they are initiated and behold in pure light perfect and simple and calm and blissful apparitions."[78]

66 Eur. *Bacch.* 918–919: hallucinating, Pentheus sees a double Thebes and two suns; Csapo (1997: 279–287).

67 Pl. *Phdr.* 250d6–8: ὄψις … ὀξυτάτη τῶν … αἰσθήσεων.

68 Pl. *Phdr.* 251a9–10: ἰδόντα δ' αὐτὸν οἷον ἐκ τῆς φρίκης μεταβολή.

69 Eur. *Bacch.* 470: ὁρῶν ὁρῶντα, καὶ δίδωσιν ὄργια; cf. Csapo (1997: 256–260).

70 Eur. *Bacch.* 1266: μεταβολή; cf. *Bacch.* 1270: γίγνομαι δέ πως ἔννους, μετασταθεῖσα τῶν πάρος φρενῶν ("I have somehow come to my senses, having changed from my previous state of mind").

71 Pl. *Phdr.* 253a8: ὥσπερ αἱ βάκχαι.

72 Pl. *Phdr.* 253a3–6: διὰ τὸ συντόνως ἠναγκάσθαι πρὸς τὸν θεὸν βλέπειν, καὶ ἐφαπτόμενοι αὐτοῦ τῇ μνήμῃ ἐνθουσιῶντες […], καθ' ὅσον δυνατὸν θεοῦ ἀνθρώπῳ μετασχεῖν.

73 For philosophers as initiates, see Pl. *Phdr.* 249c8–9; 250e1; 251a3; 253c3.

74 *Phdr.* 250c1–2: ὠργιάζομεν ὁλόκληροι μὲν αὐτοὶ ὄντες.

75 Eur. *Bacch.* 34, 78–79 (in connection with Cybele's cult), 262, 415–416, 470 and 471, 482, 998, 1080. On the connection of *mania* and truth in the *Phaedrus*, cf. Serranito (2020: 105–109).

76 Pl. *Phdr.* 265b4: Διονύσου δὲ τελεστικήν … μανίαν.

77 Pl. *Phdr.* 250b7–c1: μακαρίαν ὄψιν τε καὶ θέαν … εἶδόν τε καὶ ἐτελοῦντο τῶν τελετῶν … μακαριωτάτην ("The blessed sight and vision … they witnessed and performed … the most blessed of rites"); *Phdr.* 250d4–e2: … τὸ κάλλος θεώμενος; for *theōria* in the *Symposium* (i.e. 210d4–6: τοῦ καλοῦ … θεωρῶν), see Evans (2006: 11–13); Wilson-Nightingale (2004: 116); Edmonds (2017).

78 Pl. *Phdr.* 250c3–5: ὁλόκληρα δὲ καὶ ἁπλᾶ καὶ ἀτρεμῆ καὶ εὐδαίμονα φάσματα μυούμενοί τε καὶ ἐποπτεύοντες ἐν αὐγῇ καθαρᾷ. For *phasmata* in Dionysian cult, see Petridou (2015: 71); for

The wording draws on the rite of the *epopteia/ theōria*, enacted as part of the Bacchic-Orphic and Eleusinian mysteries.[79] The experience causes the initiates "to be struck with amazement and beside themselves" (ἐκπλήττονται καὶ οὐκέτ' ἐν αὑτῶν γίγνονται) since the physical senses only afford them an imperfect semblance of the Ideas.[80] Importantly, Plato insists, no earthly poet has ever been able to describe adequately "the colourless, formless, and intangible truly existing essence" of this otherworldly space, "visible only to the mind" of every soul[81] which must spend time "gazing upon truth" so to be nourished and made happy.[82]

However, attaining an altered state of consciousness has little to do with wine abuse or debauchery. Euripides rejects the suspicions of king Pentheus that the Bacchic rites are a pretext for indulging in drunken wantonness,[83] in line with evidence suggesting that the role of wine in Dionysian and related rites was limited and likely symbolic.[84] In the play, a royal messenger arrives from the mountains having witnessed the conduct of the Bacchants; he explicitly reports that the Bacchants behaved[85]

> *modestly*, not as you say drunk with the goblet and the sound of the flute, hunting out Aphrodite through the woods in solitude.

---

phasmata in the Eleusinian mysteries, see Burkert (1983: 286–288), and Burkert (1987: 92–93).

79 Mylonas (1961: 274–278); Kerényi (1967: 45–47 and 113); Burkert (1983: 275 and 282–283); Wilson-Nightingale (2004: 173–180); Bremmer (2014: 11–16); Ustinova (2017: 126–130 and 142–143); Dinkelaar (2020); cf. Csapo (1997: 281 and 284) for Dionysus as *pompos theōrias* ("parade leader of the festival").

80 Pl. *Phdr.* 250a7–b5; cf. nn67–69 above; note the wording on visual similarities: ὁμοίωμα ἴδωσιν, ... οὐκ ἔνεστι φέγγος οὐδὲν ἐν τοῖς τῇδε ὁμοιώμασιν ... ὀλίγοι ἐπὶ τὰς εἰκόνας ἰόντες θεῶνται τὸ τοῦ εἰκασθέντος γένος.

81 On the eye of the mind, see Pl. *Tht.* 165c; cf. Frede (1999); Anagnostou-Laoutides (2021: 218–220).

82 Pl. *Phdr.* 247c4–e7: τὸν δὲ ὑπερουράνιον τόπον οὔτε τις ὕμνησέ πω τῶν τῇδε ποιητὴς οὔτε ποτὲ ὑμνήσει κατ' ἀξίαν. ... ἡ γὰρ ἀχρώματός τε καὶ ἀσχημάτιστος καὶ ἀναφὴς οὐσία ὄντως οὖσα, ψυχῆς κυβερνήτῃ μόνῳ θεατὴ νῷ, ... καὶ θεωροῦσα τἀληθῆ τρέφεται καὶ εὐπαθεῖ.

83 Eur. *Bacch.* 224–225 and 260–262.

84 Jiménez San Cristóbal (2009: 46–47 and 55); Ustinova (2017: 119, 124, 134 and 137); cf. Dodds (1960: xiii); Graf and Johnston (2007: 148–149). On the rite of *kraterismos*, in the Korybantic festival at Erythrai, see Graf (2010).

85 Eur. *Bacch.* 686–688: [...] **σωφρόνως**, οὐχ ὡς σὺ φῄς/ ᾠνωμένας κρατῆρι καὶ λωτοῦ ψόφῳ/ θηρᾶν καθ' ὕλην Κύπριν ἠρημωμένας.

Their behaviour accords with Dionysus' "wise mind and perception, kept hidden from presumptuous people".[86] Notably, the god does not compel the Bacchants to "act modestly" (*sōphronein*) but merely encourages them to exercise a feature inherent in their nature: "... remember: even in ecstatic worship a modest woman (*sōphrōn*) will not be corrupted."[87]

Euripides' Bacchants, then, are endorsed as genuine because they are not drunk, like Socrates who, as often repeated in the *Symposium*, is inebriated and inebriating *without* being intoxicated.[88]

In the *Symposium*, using repeatedly the verb *eoikenai* and the adjective *homoios*, Alcibiades compares Socrates to the Silens and Satyrs,[89] also included alongside the Maenads in Dionysus' entourage and often portrayed in ancient art as drunk.[90] He also likens him to Marsyas,[91] the Phrygian satyr famous for his musical skill and his *sōphrosynē*, exemplified in his discovery of the laws of flute-playing.[92] Marsyas "charmed"[93] his audiences to ecstasy with his soul-piercing aulos, inspiring those ready to perceive the gods and their *teletai* into a unique trance.[94] Likewise, those listening to Socrates' arguments are "struck with amazement and possessed".[95] Their symptoms include heart throbbing

86 Eur. *Bacch.* 427–428: σοφὰν δ' ἀπέχειν πραπίδα φρένα τε / περισσῶν παρὰ φωτῶν. On mysteries and secret knowledge, see Burkert (1987: 72); Segal (1997: 355).

87 Eur. *Bacch.* 314–318: οὐχ ὁ Διόνυσος σωφρονεῖν ἀναγκάσει/ γυναῖκας ..., ἀλλ' ἐν τῇ φύσει/ τὸ σωφρονεῖν ἔνεστιν ... / καὶ γὰρ ἐν βακχεύμασιν/ οὖσ' ἥ γε σώφρων οὐ διαφθαρήσεται; cf. *Bacch.* 940: ... σώφρονας βάκχας.

88 Pl. *Symp.* 176c4–6; 214a4–6; esp. 220a5–7: Σωκράτη μεθύοντα οὐδεὶς πώποτε ἑώρακεν ἀνθρώπων ("no one ever saw Socrates drunk"); Anagnostou-Laoutides and Payne (2021); cf. Roochnik (2020: 86–90).

89 Pl. *Symp.* 215b1–5: ὁμοιότατον αὐτὸν εἶναι τοῖς σιληνοῖς; 215b8: ὡς δὲ καὶ τἆλλα ἔοικας; 216c8–10: ὡς ὅμοιός τ' ἐστὶν οἷς ἐγὼ ᾔκασα αὐτὸν; 221d6–e1: εἰ οἷς ἐγὼ λέγω ἀπεικάζοι τις αὐτόν, [...], τοῖς δὲ σιληνοῖς καὶ σατύροις, αὐτὸν καὶ τοὺς λόγους. [...], ὅτι καὶ οἱ λόγοι αὐτοῦ ὁμοιότατοί εἰσι τοῖς σιληνοῖς τοῖς διοιγομένοις; cf. 216d5–7: ὡς τὸ σχῆμα αὐτοῦ τοῦτο οὐ σιληνῶδες; [...] ὥσπερ ὁ γεγλυμμένος σιληνός); also, Xen. *Symp.* 5.7 (τοὺς Σειληνοὺς ἐμοὶ ὁμοιοτέρους). On Socrates' representation as a satyr, see Catoni and Giuliani (2019: 696–713); cf. Belfiore (2006: 199–203). Cf. n91 below.

90 Eur. *Cycl.* 139–161; Paus. 6.24.8; also, Hedreen (1992: 161–165).

91 Pl. *Symp.* 215b5–6: ἐοικέναι αὐτὸν τῷ σατύρῳ τῷ Μαρσύᾳ; τό γε εἶδος ὅμοιος εἶ τούτοις.

92 Pl. *Minos*, 318b and *Law.* 677d5 with Dušanić (2006: 32); Lynch (2018: 724–728); Fehr (2008: esp. 140–143).

93 Pl. *Symp.* 215c2: ἐκήλει.

94 Pl. *Symp.* 215c6–8: μόνα κατέχεσθαι ποιεῖ καὶ δηλοῖ τοὺς τῶν θεῶν τε καὶ τελετῶν δεομένους [...]; cf. Prior (2006: 153).

95 Pl. *Symp.* 215d6: ἐκπεπληγμένοι ἐσμὲν καὶ κατεχόμεθα; 216d4: ἐκπέπληκται as in *Phdr.* 250a7 above.

and involuntary weeping,[96] typically observed during Korybantic trance.[97] Just like Teiresias and Kadmos seemed ridiculous to Pentheus in their Dionysian attire though they were perceptive of the god's alter-reality,[98] so these Bacchic figures look absurd but are wise; thus, the funny-looking Silenoi statues sold in the Athenian agora contained images of gods hidden inside them.[99] Similarly, despite appearing frenzied, Socrates is "full of *sōphrosynē*";[100] his speeches may appear ridiculous at first (φανεῖεν ἂν πάνυ γελοῖοι τὸ πρῶτον) and fitting of a mocking satyr (σατύρου δή τινα ὑβριστοῦ), yet

> … first, they are the only speeches that have any sense in them, then, they are most divine and contain countless "statues" [i.e. images/representations] of virtue and on … every subject fitting for study by anyone who is going to become a fine, decent man.[101]

Thus, the essence of Socratic *mania* and *baccheia*, familiar to all present at Agathon's dinner party[102] and vividly contrasted to Alcibiades' unmissable drunkenness,[103] is his admirable *sōphrosynē*,[104] defined as the ability to "exercise control over pleasures and desires".[105] Socrates' ability to control his drinking,[106] similar but different to convivial intoxication and the popular perception of mystery religions,[107] is a metaphor for his determined, even obsessive,

---

96 Pl. *Symp.* 215e2: ἥ τε καρδία πηδᾷ καὶ δάκρυα ἐκχεῖται; cf. *Phdr.* 251d5–6 where the soul, overwhelmed with yearning, pulsates (πηδῶσα); *Phdr.* 251b9: ἀνακηκίει; Eur. *Bacch.* 307–308 for a similar description of Dionysus as "leaping with torches over the saddle of a two-peaked mountain, brandishing and shaking the Bacchic branch": πηδῶντα σὺν πεύκαισι δικόρυφον πλάκα, πάλλοντα καὶ σείοντα βακχεῖον κλάδον. Cf. n58 above.

97 Pl. *Symp.* 215b4–5: οἳ διχάδε διοιχθέντες φαίνονται ἔνδοθεν ἀγάλματα ἔχοντες θεῶν

98 Eur. *Bacch.* 248–254; cf. *Bacch.* 186 where Kadmos praises Teiresias as *sophos*; cf. 195–196 where Kadmos asks: "are we the only ones of the city to dance for Bacchus? (μόνοι δὲ πόλεως Βακχίωι χορεύσομεν;) to which Teiresias responds: "we alone think correctly, the others incorrectly (μόνοι γὰρ εὖ φρονοῦμεν, οἱ δ' ἄλλοι κακῶς).

99 Pl. *Symp.* 215b4–5.

100 Pl. *Symp.* 216d7–9: γέμει σωφροσύνης; cf. n117 below.

101 Pl. *Symp.* 222a2–6: πρῶτον μὲν νοῦν ἔχοντας ἔνδον μόνους εὑρήσει τῶν λόγων, ἔπειτα θειοτάτους καὶ πλεῖστα ἀγάλματ' ἀρετῆς ἐν αὑτοῖς ἔχοντας καὶ … ἐπὶ πᾶν ὅσον προσήκει σκοπεῖν τῷ μέλλοντι καλῷ κἀγαθῷ ἔσεσθαι. cf. *Symp.* 216e7–217a2, on the "statues" that Socrates has in him as θεῖα καὶ χρυσᾶ … καὶ πάγκαλα καὶ θαυμαστά.

102 Pl. *Symp.* 218b3–4: πάντες γὰρ κεκοινωνήκατε τῆς φιλοσόφου μανίας τε καὶ βακχείας.

103 Pl. *Symp.* 212e3–4; 213a1; 214c7; 212d5; cf. 214a9–b2.

104 Pl. *Symp.* 219d4–8.

105 Pl. *Symp.* 196c4–5: τὸ κρατεῖν ἡδονῶν καὶ ἐπιθυμιῶν; cf. Pl. *Rep.* 439c5–d2.

106 See n88 above; also, Xen. *Mem.* 1.3.6–8.

107 Ar. *Ran.* 326, 328–330; Pl. *Rep.* 363d1.

focus on philosophical enquiry which allows him to exceed the realm of the senses, leading him to a philosophical altered state of consciousness.[108] During this state, the philosopher

> Becoming detached from human concerns and veering toward the divine, is rebuked by the many as mad, and although he is inspired, he eludes the many.[109]

The philosopher's bacchic state of mind is also related in the *Phaedo*, where drawing yet again on the language of the mysteries,[110] Socrates determines that the true Bacchoi are those who "practice philosophy correctly".[111] Equally, in the *Phaedrus* Plato rejects physical drunkenness as a distraction from philosophical pursuit and a symptom of the soul's yielding to *hubris*.[112] Bacchic philosophers dedicate their lives to minimizing the interference of the senses in their perception[113] and to solitary contemplation during which "the soul inquires alone by itself" departing "into the realm of the pure, the everlasting, the immortal and the changeless".[114] Socrates, then, appreciates that genuine Bacchic experience is not necessarily frenetic,[115] just as in the *Bacchae* Dionysus remains calm when confronted by Pentheus, repeatedly advising him to behave likewise.[116] Crucially, in the *Phaedo* too, Plato defends *sōphrosynē* as the quintessential virtue of the Bacchic philosopher which leads him to

---

108 Cf. Pl. *Phdr.* 265a10–12 on the two kinds of *mania*: one arising from human diseases (ὑπὸ νοσημάτων ἀνθρωπίνων), and one from a divine release from our customary habits (τὴν δὲ ὑπὸ θείας ἐξαλλαγῆς τῶν εἰωθότων νομίμων γιγνομένην).

109 Pl. *Phdr.* 249d1–4: ἐξιστάμενος δὲ τῶν ἀνθρωπίνων σπουδασμάτων καὶ πρὸς τῷ θείῳ γιγνόμενος, νουθετεῖται μὲν ὑπὸ τῶν πολλῶν ὡς παρακινῶν, ἐνθουσιάζων δὲ λέληθεν τοὺς πολλούς; cf. *Phdr.* 249d10–e1: τῶν κάτω δὲ ἀμελῶν, αἰτίαν ἔχει ὡς μανικῶς διακείμενος ("Neglecting the things below, he is accused of being mad"); also, *Tht.* 173c–e and n121 below.

110 Pl. *Phdo.* 69c9–10: "as they say in the mysteries (οἱ περί τάς τελετάς), 'the thyrsus-bearers are many (ναρθηκοφόροι μὲν πολλοί), but the mystics few (βάκχοι δέ τε παῦροι)". Anagnostou-Laoutides (2023: 29–30).

111 Pl. *Phdo.* 69d2–3: οἱ πεφιλοσοφηκότες ὀρθῶς.

112 Pl. *Phdo.* 256b20–c1; also, 238b3–5; cf. 240e.

113 Hence, the bodily senses are said to "deceive, disturb and confuse" the soul (πλανᾶται καὶ ταράττεται καὶ εἰλιγγιᾷ), "as though it were drunk" (ὥσπερ μεθύουσα; *Phdo.* 79c8–10).

114 Pl. *Phdo.* 79d1–2: ὅταν δέ γε αὐτὴ καθ' αὑτὴν σκοπῇ, ἐκεῖσε οἴχεται εἰς τὸ καθαρόν τε καὶ ἀεὶ ὂν καὶ ἀθάνατον καὶ ὡσαύτως ἔχον.

115 On Socrates' "mental withdrawal", see Ustinova (2017: 316–317).

116 Eur. *Bacch.* 636: ἥσυχος; 647: ἥσυχον πόδα; 789–790: ἡσυχάζειν.

impassibility[117] and self-fulfilment,[118] while in the *Phaedrus sōphrosynē* is defined as the mental state "when opinion leads *through reason* toward the best".[119] In the *Laws*, again, *sōphrosynē* is an integral part of divine virtue, elements of which still remain in humans, and which citizens must activate so that they and their city may become as godlike as possible.[120] Thus, despite the obvious risk that philosophers may be misunderstood as drunk or mad,[121] Plato is determined to employ metaphors[122] to reinstate the benefits of their Bacchic insights in the *polis* – a point he further elaborates in the *Laws*.

4 Wine and *Baccheia* at Civic Level: Sharing Philosophical Perspectives

According to Euripides, wineless *sōphrosynē*, the main attribute of Dionysus,[123] represents an ethical mindset suitable for sensible citizens and their leaders.[124] In contrast, Pentheus, the harsh, young king of Thebes is portrayed as "drunk" with the license of his newly acquired power and thus, unable to exercise his duties with the sobriety that befits his position. Teiresias accuses Pentheus thus:[125]

117 Pl. *Phdo.* 68c6–11 with Anagnostou-Laoutides (2022b: 7 and 23–24). *Sōphrosynē* is an integral part of Socratic *elenchus* (cross-/self-examination); Anagnostou-Laoutides (2022a: 33–38).

118 Pl. *Alc.* 1, 134s12–13 with Anagnostou-Laoutides (2022b: 20 and 52 with n74) on the purpose of *sōphrosynē* being happiness.

119 Pl. *Phdr.* 237e2–3: δόξης μὲν οὖν ἐπὶ τὸ ἄριστον λόγῳ ἀγούσης [...] σωφροσύνη ὄνομα; cf. *Tim.* 71e4–72a6; also, n160 below.

120 Pl. *Law.* 906a3–b4 with Anagnostou-Laoutides (2022b: 16); Carone (2005: 183–184); cf. *Tht.* 176b2–3 and n72 above.

121 See Pl. *Symp.* 216d9–e6, *Apol.* 29e7–30a2 and 28a8–10; and *Euth.* 3c with Ustinova (2017: 320).

122 On resorting to extended metaphors to explain why and how the true philosopher is systematically excluded from public life, see Pl. *Rep.* 487e4–488a2 and 489a4–b2. Cf. Ricoeur (1975: 318 = 1977: 299) who notes: "In brief, critical consciousness of the distinction between use and abuse leads not to disuse but to re-use of metaphors ... .".

123 Eur. *Bacch.* 329: τιμῶν τε Βρόμιον σωφρονεῖς ("by honouring Dionysus, you act in accordance with *sōphrosynē*"); *Bacch.* 1150–1151: τὸ σωφρονεῖν δὲ καὶ σέβειν τὰ τῶν θεῶν κάλλιστον ("the best thing is to act in accordance with *sōphrosynē* and worship the gods"); cf. *Bacch.* 504; also, *Athn. Gr.* 11.32.

124 See n98 above; also, Eur. *Bacch.* 39–40: δεῖ γὰρ πόλιν τήνδ' ἐκμαθεῖν ... τῶν ἐμῶν βακχευμάτων ("the city must learn ... my Bacchic revels").

125 Eur. *Bacch.* 268–271: σὺ δ' εὔτροχον μὲν γλῶσσαν ὡς φρονῶν ἔχεις,/ ἐν τοῖς λόγοισι δ' οὐκ ἔνεισί σοι φρένες./ θράσει δὲ δυνατὸς καὶ λέγειν οἷός τ' ἀνήρ/ κακὸς πολίτης γίγνεται νοῦν οὐκ ἔχων.

> You have a sharp tongue *as though you are sensible*, but there is no sense in your words. A bold and powerful man, one capable of speaking well, becomes *a bad citizen if he lacks sense.*

Plato further shares Euripides' preoccupation with pseudo-ecstasy and its civic aspects[126] in the *Laws*, where he describes at length the legal code of Magnesia, an imaginary polity, and its educational model which relies on music,[127] since *nomos* meant both law and song/musical genre.[128] Magnesia's citizens ought to practice *sōphrosynē*,[129] responding to the merciful invitation that gods extend to humans to become their "fellow dancers".[130] However, in this context, *baccheia* is rejected as a "dubious type of dance",[131] performed

> by those who *imitate in their drunken state* the so-called Nymphs and Pans and Silenoi and Satyrs, as they say, while performing purifications and certain rites.[132]

Baccheia is unsuitable for Plato's second ideal city (*Leg.* 815d2: *ouk … politikon*)[133] since it does not produce real *sōphrosynē*.[134] Like in the *Symposium*, the speakers in the *Laws* know that drinking sessions tend to become boisterous,[135] and that drunkenness, like all excesses, "renders people out of their minds by intoxicating them with pleasure".[136] Thus, a sober symposiarch ought

126 Eur. *Bacch.* 218: πλασταῖσι βακχείαισιν; on pseudo-*baccheia* as a feature of tyrants who instead revel in drunken license, see Pl. *Rep.* 561c–d; 571c; and 573c1–10 (also n136 below); also, Anagnostou-Laoutides (2020).

127 Pl. *Law.* 653b1–c4.

128 Anagnostou-Laoutides (2022b: 19 with n72).

129 See n123 above; Carone (2005: 177–178); cf. Pl. *Phdo.* 114d–115a, for *sōphrosynē* as an ornament of the soul.

130 Pl. *Law.* 665a5: τοὺς θεοὺς συγχορευτάς. Anagnostou-Laoutides (2023: 29–33).

131 Pl. *Law.* 815b8: ἀμφισβητουμένην ὄρχησιν; Anagnostou-Laoutides (2022b: 14–15); Ustinova (2017: 123); Folch (2015: 217–219).

132 Pl. *Law.* 815c1–4: βακχεία τ' ἐστὶ …, αἷς Νύμφας τε καὶ Πᾶνας καὶ Σειληνοὺς καὶ Σατύρους ἐπονομάζοντες, ὥς φασι, μιμοῦνται κατῳνωμένους, περικαθαρμούς τε καὶ τελετάς τινας ἀποτελούντων.

133 Cf. Pl. *Euthyd.* 277d6–e2 in Edmonds (2006: 349) on the verbal gyrations of the sophists compared to Corybantic rites; Belfiore (2006: 206).

134 Cf. Pl. *Phdo.* 68e2–69c9 (mistaking *sōphrosynē* for the exchange of some virtues for others leads to a σκιαγραφία/ "rough painting" of virtue).

135 Pl. *Law.* 671a3–b1; 640c1–2; *Symp.* 212d5–7; 223b6–8; *Prot.* 347c3–348a6; Anagnostou-Laoutides and Payne (2021: 63–67).

136 Pl. *Law.* 649d4–8: δι' ἡδονῆς αὖ μεθύσκοντα παράφρονας ποιεῖ; cf. 775c4–d3; cf. *Rep.* 573a7–b5: περὶ αὐτὸν βομβοῦσαι αἱ ἄλλαι ἐπιθυμίαι, … καὶ οἴνων καὶ τῶν ἐν ταῖς τοιαύταις συνουσίαις ἡδονῶν ἀνειμένων, …, τότε δὴ δορυφορεῖταί τε ὑπὸ μανίας καὶ οἰστρᾷ … note the

to preside over convivial gatherings,[137] while the city's law-keepers ought to be "sober and calm men to act as commanders over the un-sober".[138]

Nevertheless, education in Magnesia is entrusted above all to the leaders of the Dionysian Choir, citizens of advanced age distinguished for their virtue,[139] including their sympotic integrity.[140] Under the rubric of "right education",[141] they must instruct the citizens to discern between the right kind of *choreia* and pseudo-Bacchic dances. Thus, they pass laws forbidding all young people under eighteen from participating in wine drinking gatherings;[142] men under thirty, though explicitly deterred from wine abuse and drunkenness, are allowed moderate wine consumption; only after reaching the age of forty, by which time educational tenets tend to become solid,[143] citizens are invited to

> join in the convivial gatherings and invoke Dionysus, above all other gods, in the rites of the elders which is also their recreation.[144]

Furthermore, Plato, who imagines the State as a mixing-bowl of wine, where "the sober deity of water" must be stirred to balance out its madness,[145] introduces an important wine metaphor about sharing Bacchic *sōphrosynē* across all citizens. Thus, he recommends the so-called "spectation/*theōria* of Dionysus"[146] as a playful test[147] that allows citizens to ponder on their ethical

---

vocabulary of genuine *baccheia* (n56 above), here used to describe the drunken decadence of tyrants.

137 Pl. *Law.* 640d4–7; cf. *Symp.* 213e10. Belfiore (1986: 424); Sauvé-Meyer (2015: 326–327); also, Stalley (1983: 9, 14–15 and 17–18) claiming that Plato revises here relevant points from the *Republic.*

138 Pl. *Law.* 671d6–8: τοὺς ἀθορύβους καὶ νήφοντας τῶν μὴ νηφόντων στρατηγούς. Cf. Sauvé-Meyer (2015: 327 s.v. 671d5–7).

139 Pl. *Law.* 665d1–3: ποῦ δὴ τοῦθ' ἡμῖν τὸ ἄριστον τῆς πόλεως, ἡλικίαις τε καὶ ἅμα φρονήσεσιν πιθανώτατον ὂν τῶν ἐν τῇ πόλει. Prauscello (2014: 117–118); Murray (2018); Pfefferkorn (2021); Sauvé-Meyer (2015: 328–330); cf. n24 above.

140 Pl. *Law.* 671b8–9.

141 Pl. *Law.* 653a1–2: ὀρθὴν παιδείαν; 652a5: ἐν τῇ κατ' ὀρθὸν χρείᾳ τῆς ἐν οἴνῳ συνουσίας.

142 Pl. *Law.* 666a1–c7; common drunkenness is totally excluded from Magnesia: hence in 775c4–d3 Plato emphasizes the importance of sobriety for newlyweds.

143 Pl. *Law.* 653a9–11.

144 Pl. *Law.* 666a8–b5 at 666b4–6: καλεῖν τούς τε ἄλλους θεοὺς καὶ δὴ καὶ Διόνυσον παρακαλεῖνεἰς τὴν τῶν πρεσβυτέρων τελετὴν ἅμα καὶ παιδιάν.

145 Pl. *Law.* 773c8–d4; cf. *Phil.* 61c5–8 where Plato imagines humans as wine-pourers.

146 Pl. *Law.* 650a1: τῆς τοῦ Διονύσου θεωρίας.

147 Pl. *Law.* 649e1: τῆς ἐν οἴνῳ βασάνου καὶ παιδιᾶς.

disposition.[148] A cheap, readily available training substance,[149] wine inspires fear (of public humiliation) in the citizens.[150] Thus, it enhances their self-awareness, urging them to conduct themselves with *sōphrosynē*,[151] understood literally as sobriety and metaphorically as moderation. Given that in Magnesia, the city's leaders are tasked with introducing the citizens to the divine hierarchy of virtues (where *phronēsis* features first, followed by *sōphrosynē*),[152] the Test of the Wine enables the Magnesians to reiterate these core lessons of moral reasoning and develop a shared code about the civic values they endorse.[153] Most importantly, although so far Plato reserves Bacchic *sōphrosynē* for true philosophers, the Test encourages average citizens to engage with the philosophical perspectives of the lawgivers,[154] and the Nocturnal Council, the chief regulatory body of Magnesia responsible for examining the ethical principles of law and their implementation.[155] The Council, especially focused on the study of virtue and its various manifestations (*sōphrosynē*, courage, justice and *phronēsis*),[156] is also responsible for appointing *theōroi*,[157] men tasked with travelling across good and bad polities in search of wise

148 Pl. *Law.* 649d11: εἰς τὸ μελετᾶν; 650a4–5: ἦθος ψυχῆς θεάσασθαι.

149 Pl. *Law.* 649d8: εὐτελῆ τε καὶ ἀσινεστέραν; cf. 646d11: τὴν σωμασκίαν ("for bodily training"). Wine is characterized as a *pharmakon*, a medicine/drug for the soul's modesty; 649a4; 672d8–9 with Rinella (2010: 149–165) (on the medical discourse underpinning ecstatic intoxication).

150 Pl. *Law.* 647e1; 649a–b; Belfiore (1986: 433–434); Baima (2018); cf. Baima (2017: 67–73) on Plato's use of drunkenness in the *Republic* and the *Laws*; also, Rinella (2010: 49–63).

151 Pl. *Law.* 648e6–7: Ναί· σωφρονοῖ γὰρ ‹ἄν›, ὦ ξένε, καὶ ὁ τοιοῦτος οὕτω πράττων (by leaving a banquet before getting drunk). Cf. 673e3–6 where Plato concedes that the aim of the Test, provided it is applied in "a lawful and orderly manner" (μετὰ νόμων καὶ τάξεως), is to achieve *sōphrosynē* … (ὡς τοῦ σωφρονεῖν ἕνεκα)."

152 Pl. *Law.* 631c5–d7.

153 Cf. Folch (2015: 82).

154 According to Plato, lawgivers must legislate (*Law.* 657a10–11) in accordance with the divine reason which exists among the stars (967d10) and follows the principles of music (967e2). However, philosophers are skilled in "the greatest form of music" (*Phdo.* 61a4–5; cf. *Rep.* 591c1–d3; *Laches*, 188d3–9). Anagnostou-Laoutides (2022b: 19).

155 Klosko (1988); Lewis (1998); Verslinsky (2016); Anagnostou-Laoutides (2022b).

156 Pl. *Law.* 965c9–d3; cf. 964b3–6. The Nocturnal Council also debates theological matters, including the existence of gods and our ability to grasp their powers (966c1–d3), the divine origin of the soul, and the rational principle that brought about the order of the universe (966d10–e5).

157 Rutherford (2013: 142–148 and 324–338); Ker (2000: 308–309); Wilson-Nightingale (2004: 63–64) and Wilson-Nightingale (2007); Anagnostou-Laoutides (2022b: 17–18). Plato expands on the concept of *theōria* as developed by Solon (*Symp.* 209d7), the Athenian lawgiver and Plato's distant relative. Solon also used the sympotic etiquette to discuss civic *arētē* and advocate *sōphrosynē*.

people,[158] whose insights ensure that the Magnesians remain perceptive of alternative worldviews against which they frequently review their own perceptions.[159]

5 Conclusion

Using the vital distinction between wine-induced ecstasy and wine abuse (or pseudo-ecstasy), illustrated compellingly by Euripides in the *Bacchae*, Plato coins a metaphor about the Bacchic mindset (i.e. superior cognition) of inspired philosophers, especially Socrates. While conscious that his metaphor may be misunderstood, Plato offers a relatable account of the inebriating (i.e. mind-altering) effects of philosophy, a topic he focuses upon in the *Phaedrus*, the *Symposium*, and the *Phaedo*. In essence, he argues, philosophy amounts to unremitting dedication to *sōphrosynē* and its divine essence, which may appear frantic but is always distilled through reason, as the philosophically trained magistrates of Magnesia exemplify in the *Laws*.[160] Here, revising the usual slander against philosophers,[161] the only reliable witnesses of the eternal form of *sōphrosynē*,[162] Plato accommodates their cognitive insights in his alternative model of the ideal city.[163] Further, through the Test of the Wine, he invites average citizens to share the philosophical perspectives that underpin the city's legal and educational systems. Above all, by examining truth through images and concepts,[164] Plato aims to promote social cohesion among the Magnesians, while enriching their reality.

158 Pl. *Law*. 951a4–5; 951b5–8; Verlinsky (2016,:182 and 192–194).
159 Pl. *Law*. 951c3–5. For more on the theoretical activity of the *theôroi* and the Nocturnal Council in the *Laws*, see Bjelde's contribution in this volume.
160 Pl. *Law*. 967d3–e5.
161 Pl. *Law*. 967c8–d1.
162 Pl. *Rep*. 500d1–e2.
163 Cf. Pl. *Rep*. 592a5–b4 (the philosopher will only participate in the politics of the ideal city).
164 Pl. *Phdo*. 99d5–100a3.

Bibliography

Adler, A., ed. 1967. *Suidae Lexicon.* 5 vols. Stuttgart.

Anagnostou-Laoutides, E. 2020. "A Toast to Virtue: Drinking Competitions, Plato, and the Sicilian Tyrants." In *Conflict and Competition: Agon in Western Greece*, edited by H. Reid, J. Serrati, and T. Sorg, 123–138. Sioux City: 123–138.

Anagnostou-Laoutides, E. 2021. "The Eye of the Soul in Plato and Pseudo-Macarius: Alexandrian Theology and the Roots of Hesychasm." In *Hymns, Homilies and Hermeneutics in Byzantium*, edited by A. Mellas and S. Gador-Whyte, 216–238. Leiden: 216–238.

Anagnostou-Laoutides, E. 2022a. "Plato's Manic Sōphrosynē and the Meaning of Virtue in Heraclitus and Pythagoras." In *Aretē in Plato and Aristotle*, edited by J. Elliott and R. Brown, Siracusa, Sicily/South Dakota, SD: 25–44.

Anagnostou-Laoutides, E. 2022b. "Beyond Human Reason: *Baccheia* and *Theōria* in Plato and Clement of Alexandria." *Journal of Hellenic Religion* 14: 1–55.

Anagnostou-Laoutides, E. 2023. "Man before God: Silence and Altered States of Consciousness in the Phaedo and Clement of Alexandria." In *Later Platonists and their Heirs: Christian, Jewish, and Muslim*, edited by E. Anagnostou-Laoutides and K. Parry, Leiden: 25–60.

Anagnostou-Laoutides, E. 2025. *Wine and Ecstasy in Plato: A Metaphor of Sorts and its Early Reception.* New York.

Anagnostou-Laoutides, E., and A. Payne. 2021. "Drinking and Discourse in Plato." *Methexis* 33: 57–79.

Anderson, JF. 1950. "Analogy in Plato." *The Review of Metaphysics* 4 (1): 111–128.

Armstrong, JM. 2004. "After the Ascent: Plato on Becoming Like God." *Oxford Studies in Ancient Philosophy* 26: 171–183.

Baima, NR. 2017. "On the Value of Drunkenness in the Laws." *History of Philosophy and Logical Analysis* 20 (1): 65–81.

Baima, NR. 2018. "Playing with Intoxication: On the Cultivation of Shame and Virtue in Plato's Laws." *Apeiron* 51 (3): 345–370.

Belfiore, E. 1986. "Wine and Catharsis of the Emotions in Plato's Laws." *The Classical Quarterly* 36 (2): 421–437.

Belfiore, E. 2006. "Dancing with the Gods: The Myth of the Chariot in Plato's 'Phaedrus'." *The American Journal of Philology* 127 (2): 185–217.

Belfiore, E. 2011. "Poets at the Symposium." In *Plato and the Poets*, edited by P. Destrée and F-G. Herrmann, Leiden: 155–174.

Black, M. 1981. "Metaphor." In *Philosophical Perspectives on Metaphor*, edited by M. Johnson, Minneapolis, MN: 63–82.

Bougher, LD. 2014. "Cognitive Coherence in Politics: Unifying Metaphor and Narrative in Civic Cognition." In *Warring with Words: Narrative and Metaphor in Politics,* edited by M. Hanne, WD. Crano, and JS. Mio, New York, NY and East Sussex: 250–271.

Bremmer, JN. 2014. *Initiation into the Mysteries of the Ancient World.* Berlin/Boston.

Bromberg, JA. 2017. "A Sage on the Stage: Socrates and Athenian Old Comedy." In *Socrates and the Socratic Dialogue*, edited by A. Stavru and C. Moore, Leiden: 31–63.

Burkert, W. 1983. *Homo Necans: The Anthropology of Ancient Greek Sacrificial Ritual and Myth.* Translated by P. Bing. Berkeley/Los Angeles/London.

Burkert, W. 1987. *Ancient Mystery Cults.* Cambridge, MA/London.

Camp, E. 2006. "Metaphor in the Mind: The Cognition of Metaphor." *Philosophy Compass* 1 (2): 154–170.

Carone, GR. 2005. *Plato's Cosmology and Its Ethical Dimensions.* Cambridge.

Catoni, ML., and L. Giuliani. 2019. "Socrates Represented: Why Does He Look Like a Satyr?" *Critical Inquiry* 45 (3): 681–713.

Crowther, P. 2003. "Literary Metaphor and Philosophical Insight: The Significance of Archilochus." In *Metaphor, Allegory, and the Classical Tradition*, edited by GR. Boys-Stones, Oxford/New York: 83–100.

Csapo, E. 1997. "Riding the Phallus for Dionysus: Iconology, Ritual, and Gender-Role De/Construction." *Phoenix* 51 (3/4): 253–295.

Dinkelaar, BM. 2020. "Plato and the Language of Mysteries." *Mnemosyne* 73 (1): 36–62.

Dodds, ER. 1960. *Euripides: Bacchae.* Oxford.

Donald, M. 2001. *A Mind So Rare: The Evolution of Human Consciousness.* New York.

Druckman, D., and RA. Bjork, eds. 1994. *Learning, Remembering, Believing: Enhancing Human Performance.* Washington, DC.

Dübner, F., ed. 1841. *Scholia Graeca in Aristophanem.* Paris.

Dušanić, S. 2006. "The Noble Virtue of ΣΩΦΡΟΣΥΝΗ." *ŽAnt* 56: 25–39.

Edmonds, RG. 2006. "To Sit in Solemn Silence? 'Thronosis' in Ritual, Myth, and Iconography." *American Journal of Philology* 127 (3): 347–366.

Edmonds, RG., III. 2017. "Alcibiades the Profane: Images of the Mysteries in Plato's Symposium." In *Plato's Symposium: A Critical Guide*, edited by P. Destrée and Z. Giannopoulou, Cambridge: 194–215.

Evans, N. 2006. "Diotima and Demeter as Mystagogues in Plato's Symposium." *Hypatia* 21 (2): 1–27.

Fehr, B. 2008. "Bilderzählung und Handlungsmuster: Athena, Marsyas, der Aulos und die jungen Leute von Athen." In *Komplexe Bilder*, edited by M. Seifert, Berlin/Basel: 129–150.

Folch, M. 2015. *The City and the Stage: Performance, Genre, and Gender in Plato's Laws.* Oxford.

Frede, D. 1989. "The Soul's Silent Dialogue: A Non-Aporetic Reading of the Theaetetus." *Proceedings of the Cambridge Philological Society* 215: 20–49.

Frede, D. 1999. "Plato on What the Body's Eye Tells the Mind's Eye." *Proceedings of the Aristotelian Society* 99: 191–209.

Geertz, AW. 2014. "Whence Religion? How the Brain Constructs the World and What This Might Tell Us about the Origins of Religion, Cognition and Culture." In *Origins of Religion, Cognition and Culture*, edited by id., Abingdon, Oxon/New York: 17–70.

Goldhill, S. 1987. "The Great Dionysia and Civic Ideology." *Journal of Hellenic Studies* 107: 58–76.

Graf, F. 2010. "The Kyrbantes of Erythrai." In *Studies in Greek Epigraphy and History in Honor of Stephen V. Tracy*, edited by G. Reger, FX. Ryan, and TF. Winters, Pessac/ Paris: 301–309.

Graf, F., and SI. Johnston. 2007. *Ritual Texts for the Afterlife: Orpheus and the Bacchic Gold Tablets*. London/New York.

Green, P. 1979. "Strepsiades, Socrates and the Abuses of Intellectualism." *Greek, Roman, and Byzantine Studies* 20: 15–25.

Grenet, P. 1948. *Les origins de l'analogie philosophique dans les dialogues de Platon*. Paris.

Hedreen, GM. 1992. *Silens in Attic Black-figure Vase-painting: Myth and Performance*. Ann Arbor, MI.

Henle, P. 1981. "Metaphor." In *Philosophical Perspectives on Metaphor*, edited by M. Johnson, Minneapolis, MN: 83–104.

Hobden, F. 2013. *The Symposion in Ancient Greek Society and Thought*. Cambridge.

Hunter, R. 2004. *Plato's Symposium*. Cambridge.

Hunter, R. 2012. *Plato and the Traditions of Ancient Literature: The Silent Stream*. Cambridge/New York.

Innes, D. 2003. "Metaphor, Simile, and Allegory as Ornaments of Style." In *Metaphor, Allegory and the Classical Tradition*, edited by GR. Boys-Stones, Oxford: 7–27.

Isler-Kerényi, C. 2021. "Dionysos, the Polis and Power." In *Dionysus and Politics. Constructing Authority in the Graeco-Roman World*, edited by F. Doroszewski and D. Karłowicz, Abingdon, Oxon/New York: 9–17.

Jiménez San Cristóbal, AI. 2009. "The Meaning of βάκχος and βακχεύειν in Orphism." In *Mystic Cults in Magna Graecia*, edited by PA. Johnston and G. Casadio, Austin, TX: 46–60.

Jiménez San Cristóbal, A. 2020. "At the Origins of Dionysus and Wine: Myths, Miracles, and Festivals." *Archiv für Religionsgeschichte* 21–22: 387–409.

Johnson, M. 1981. "Introduction: Metaphor in the Philosophical Tradition." In *Philosophical Perspectives on Metaphor*, edited by id., Minneapolis, MN: 3–47.

Ker, J. 2000. "Solon's '*Theôria*' and the End of the City." *Classical Antiquity* 19 (2): 304–329.

Kerényi, K. 1967. *Eleusis: Archetypal Image of Mother and Daughter*. London.

Kihlstrom, JF. 1984. "Conscious, Subconscious, Unconscious: A Cognitive Perspective." In *The Unconscious Reconsidered*, edited by KS. Bowers and D. Meichenbaum, New York, NY: 149–211.

Kirby, JT. 1997. "Aristotle on Metaphor." *The American Journal of Philology* 118 (4): 517–554.

Klosko, G. 1988. "The Nocturnal Council in Plato's Laws." *Political Studies* 36: 74–88.

Lakoff, G. 2004. *Don't Think of an Elephant: Know Your Values and Frame the Debate.* White River Junction, VT.

Lakoff, G., and M. Johnson. 1980. *The Metaphors We Live By.* Chicago.

Lewis, VB. 1998. "The Nocturnal Council and the Platonic Political Philosophy." *History of Political Thought* 19 (1): 1–20.

Lynch, T. 2018. "The Seductive Voice of the Aulos in Plato's Symposium: From the Dismissal of the Auletris to Alcibiades' Praise of Socrates-auletes." In *Music Cultures in Sounds, Words and Images*, edited by A.Baldassarre and T. Markovic, Vienna: 709–723.

Mac Cormac, ER. 1985. *A Cognitive Theory of Metaphor.* Cambridge, MA/London.

McGovern, PE. 2003. *Ancient Wine. The Search for the Origins of Viniculture.* Princeton/Oxford.

Merleau-Ponty, M. 1962. *Phenomenology of Perception.* Translated by C. Smith. London.

Murray, O. 1982. "Symposion and Männerbund." In *Concilium Eirene: Proceedings of the 16th International Eirene Conference (Prague 31.8 – 4.9. 1982)* 1, edited by P. Oliva and A. Frolikova, Prague: 94–112.

Murray, O. 1983a. "The Symposion in History." In *Tria Corda: Scritti in Onore di Arnaldo Momigliano*, edited by E. Gabba, Como: 257–272.

Murray, O. 1983b. "The Symposion as Social Organisation." In *The Greek Renaissance of the 8th C B.C.: Tradition and Innovation*, edited by R. Hägg, Stockholm: 195–199.

Murray, O. 2018. "The Chorus of Dionysus: Alcohol and Old Age in the Laws." In *The Symposion: Drinking Greek Style*, edited by V. Cazzato, Oxford: 251–260 [= 2013, A-E. Peponi (ed.), *Performance and Culture in Plato's Laws*, Cambridge, 109–122].

Mylonas, GE. 1961. *Eleusis and the Eleusinian Mysteries.* Princeton, NJ.

O'Rourke, F., with commentary by I. Patsioti-Tsacpounidi. 2006. "Aristotle and the Metaphysics of Metaphor." *Proceedings of the Boston Area Colloquium of Ancient Philosophy* 21 (1): 155–190.

Papakonstantinou, Z. 2009. "Wine and Wine Drinking in the Homeric World." *L'Antiquité Classique* 78: 1–24.

Papakonstantinou, Z. 2012. "A Delight and a Burden (HES., Sc. 400): Wine and Wine-Drinking in Archaic Greece." *Ancient Society* 42: 1–32.

Patsioti-Tsacpounidi, I. 2006. See under O'Rourke.

Pender, EE. 2003. "Plato on Metaphors and Models." In *Metaphor, Allegory and the Classical Tradition*, edited by GR. Boys-Stones, Oxford: 55–81.

Pender, EE. 2011. "A Transfer of Energy: Lyric Eros in Phaedrus." In *Plato and the Poets*, edited by P. Destrée and F.-G. Herrmann, Leiden: 327–348.

Périllié, J-L. 2019. "Corybantic Rituals of Socratic Mysteries and the Doctrine of Principles." *Revue de métaphysique et de morale* 103 (3): 267–285.

Petridou, G. 2015. *Divine Epiphany in Greek Literature and Culture*. Oxford.

Pfefferkorn, J. 2021. "The Three Choruses of Plato's Laws and their Function in the Dialogue: A New Approach." *Phronesis* 66 (4): 335–365.

Power, T. 2010. *The Culture of Kitharôidia*. Washington, DC.

Prauscello, L. 2014. *Performing Citizenship in Plato's Laws*. Cambridge.

Prior, W. 2006. "The Portrait of Socrates in Plato's Symposium." *Oxford Studies in Ancient Philosophy* 31: 137–166.

Ricoeur, P. 1975. *La métaphore vive*. Paris.

Ricoeur, P. 1977. *The Rule of Metaphor: The Creation of Meaning in Language*. Translated by R. Czerny, K. McLaughlin, and J. Costello. Toronto.

Ricoeur, P. 1978. "The Metaphorical Process as Cognition, Imagination, and Feeling." *Critical Inquiry* 5 (1): 143–159.

Ricoeur, P. 1996. "Between Rhetoric and Poetics." In *Essays on Aristotle's Rhetoric*, edited by A. Oksenberg Rorty, Berkeley, CA: 324–384.

Rinella, MA. 2010. *Pharmakon: Plato, Drug Culture, and Identity in Ancient Athens*. Lanham/Boulder/New York/Toronto/Plymouth, UK.

Roochnik, D. 2009. "What is Theoria? Nicomachean Ethics Book 10.7–8." *Classical Philology* 104 (1): 69–82.

Roochnik, D. 2020. *Eat, Drink, Think: What Ancient Greece Can Tell Us about Food and Wine*. London.

Rouget, G. 1990. *La musique et la transe*. Paris.

Rutherford, I. 2013. *State Pilgrims and Sacred Observers in Ancient Greece: A Study of Theoria and Theoroi*. Cambridge.

Sansone, D. 1996. "Plato and Euripides." *Illinois Classical Studies* 21: 35–67.

Sauvé-Meyer, S. 2015. *Plato Laws 1 and 2: Translation, Introduction, Commentary*. Oxford.

Schmidt, M. 1867. *Hesychii Alexandrini Lexicon*. Jena.

Seaford, R. 2006. *Dionysos*. London / New York.

Seaford, R. 1996. *Euripides: Bacchae*. Warminster.

Seaford, R. 2021. "The Politics of Euripides' *Bacchae* and the Preconception of Irresolveable Contradiction." In *Dionysus and Politics: Constructing Authority in the Graeco-Roman World*, edited by F. Doroszewski and D. Karłowicz, Abingdon, Oxon/New York: 18–31.

Segal, C. 1997. *Dionysiac Poetics and Euripides' Bacchae*. Princeton.

Serranito, F. 2020. "ΜΑΝΙΑ and ΑΛΗΘΕΙΑ in Plato's Phaedrus." *The Classical Quarterly* 70 (1): 101–118.

Silk, M. 2003. "Metaphor and Metonymy: Aristotle, Jakobson, Ricoeur, and Others." In *Metaphor, Allegory, and the Classical Tradition*, edited by GR. Boys-Stones, 115–148. Oxford/New York.

Stalley, R. 1983. *An Introduction to Plato's Laws*. Indianapolis, IN.

Szlezák, TA. 1999. *Reading Plato*. Translated by G. Zanker. London/New York.

Trabattoni, F. 1993. *Scrivere nell'anima: Verità, Dialettica e Persuasione in Platone*. Firenze.

Ustinova, Y. 2017. *Divine Mania: Alteration of Consciousness in Ancient Greece*. London/New York.

Valamoti, SM., Mangafa, M., Koukouli-Chrysanthaki, Ch., and Malamidou, D. 2007. "Grape-pressings from Northern Greece: The Earliest Wine in the Aegean?" *Antiquity* 81: 54–61.

Verlinsky, A. 2016. "The Nocturnal Council in Plato's Laws." *Philologia Classica* 11: 180–222.

Wasmuth, E. 2015. "ΩΣΠΕΡ ΟΙ ΚΟΡΥΒΑΝΤΙΩΝΤΕΣ: The Corybantic Rites in Plato's Dialogues." *The Classical Quarterly* 65 (1): 69–84.

Wentura, D. 2019. "Cognition and Emotion: On Paradigms and Metaphors." *Cognition and Emotion* 33: 85–93.

Whitehorn, J. 2002. "Aristophanes' Representations of 'Intellectuals'." *Hermes* 130 (1): 28–35.

Wilson-Nightingale, A. 2004. *Spectacles of Truth in Classical Greek Philosophy: Theoria in Its Cultural Context*. Cambridge.

Wilson-Nightingale, A. 2007. "The Philosopher at the Festival: Plato's Transformation of Traditional Theōria." In *Pilgrimage in Greco-Roman and Early Christian Antiquity: Seeing the Gods*, edited by J. Elsner and I. Rutherford, Cambridge: 151–180.

Yu, KW. 2020. "The Politics of Dance: Eunomia and the Exception of Dionysus in Plato's Laws." *The Classical Quarterly* 70 (2): 605–619.

CHAPTER 12

Theôria in *Laws* 12

Joseph Bjelde

1 Introduction

When Nightingale discusses *theôria* in *Laws* 12, she does not take it to be a special philosophical kind of *theôria*.[1] That is, she does not take it to be the "transformed" or "appropriated" philosophical kind of *theôria* which, on her account, is characterized by detachment from the perceptible world. Instead, the role of the *theôros* in *Laws* 12 is just to gather information abroad, and then hand that information over to the nocturnal council to discuss. So, on her account, theorizing in *Laws* 12 does not involve philosophizing. At most, it would involve enabling the nocturnal council to philosophize. And the other recent treatment of the passage essentially agrees with Nightingale on this. For Long (2013: esp. 153–5, though his focus is not on the activities of the *theôroi*) also explains their job as essentially the gathering and reporting of information: the *theôroi* "inspect foreign customs" and "report back" to the council. It is then the council which reflects on their reports.[2]

It might come as a surprise that this is *not* supposed to be a kind of philosophical *theôria*. For Nightingale takes as a paradigm of this kind of *theôria* the very same person, Solon, whom Herodotus famously has Croesus characterize as philosophizing and travelling for the sake of the *theôria* (*Histories* 1.30). Nightingale's idea, of course, is that Plato "appropriates" and "transforms" earlier notions of "wisdom" (or being "sophos") and "philosophy," so that, while Solon's *theôria* and the *theôria* in *Laws* 12 are philosophical in this older sense, they aren't philosophical activities in *Plato's* sense.[3] I am inclined to doubt this account of "wisdom" and "philosophy" in general. But in this paper, I focus on *theôria* in the *Laws* and explain why we should take even Plato to regard this *theôria* as a philosophical activity.

My case is built on two nontrivial but eminently defensible assumptions. First, I'll assume that *Laws* 12 is Plato's work, including the nocturnal council and some of the striking claims about it. Second, I'm just going to assume that

1 Nightingale (2004: 65–66).
2 Long (2013: 154).
3 Cf. Nightingale (2004: 64).

HTTPS://DOI.ORG/10.1163/9789004747128-014

the nocturnal council's activity meets Plato's standard for philosophizing. Naturally, both of those assumptions are disputable – the first is famously disputed, the second somewhat less famously.[4] But my contention here will build on those assumptions, by arguing that *theôria* is relevantly similar to the activity of the nocturnal council. So, given that the nocturnal council's activity is philosophical, *theôria* will be philosophical too. Three considerations, taken together, strongly support that contention. The *theôroi* in *Laws* 12

1. share a dual purpose with nocturnal council, namely perfection by preservation
2. take the same means to that purpose: understanding *via* conversation with the wise
3. become members of the nocturnal council (if they pass a test)

2 Shared Purpose

The purpose of the relevant kind of *theôroi* is made clear when they are introduced. Just before, the Athenian has introduced another kind of *theôria*, less interesting for present purposes, in the context of a broader discussion of who will be able to leave Magnesia. This kind will involve going to festivals and teaching the youth that the *nomima* in other cities are worse than in Magnesia. Then the Athenian adds that

> Some other *theôroi* will have to be sent out, with the law-guardians' permission. If some of the citizens want to observe the affairs of other humans at somewhat greater leisure, let no law stop them. For no city without experience of human bads and goods can ever, so long as it has no contact with others, be sufficiently civilized or perfect. Moreover, it will never be able to preserve its laws without grasping them by insight rather than by habit alone. For among the masses of humanity there are always some divine people – not many – worth everything to converse with, and no more of them are born in cities with good laws than are born in those without. Those who make their home cities with good laws and are themselves incorruptible must cross land and sea to seek the traces of these people, both to confirm those of the customs which are well-laid, and correcting others, if any are deficient. For without this *theôria* and

4 Pro: Laks (2022: 30); contra: Bartels (2017: 197).

> inquiry, no city will ever remain perfect – nor will it if they do the *theôria* badly.[5]
>
> 951a4–c4

Why must (χρεὼν) another kind of *theôroi* be sent out? For two reasons (951b1: γὰρ). First, an isolated city cannot be sufficient or complete. Second, an isolated city cannot adequately guard its laws without having insight (951b3: ἄνευ τοῦ γνώμῃ λαβεῖν) as opposed to habit. These reasons are reprised in combination at 951c3–4: "without this sort of *theôria* and inquiry, no city will ever remain complete." In short: these *theôroi* contribute to the preservation of the city, and to its perfection.

That purpose is a marked one in *Laws* 12. It is the same purpose for which the auditors (Εὐθύνων) were introduced earlier in Book 12 (945b–8b). And it is the purpose for which the nocturnal council is re-introduced at 960b:

> … so that our lawmaking is nearly perfected (τέλος). But nothing is ever perfected by just doing something, or by possessing and settling something. Rather, figuring out how to preserve (σωτηρίαν) forever what has been produced makes it perfect (τελέως).
>
> 960b4–c1

> … what one needs to furnish the city and constitution is not only health and preservation of the body, but also good order in the soul, or rather preservation (σωτηρίαν) of the laws. And it seems to me that this is still lacking for our laws, a way to make the unalterable a part of their natural power.
>
> 960d1–6

The Athenian's idea is apparently that, as a matter of general principle about perdurant things, perfection requires preservation. A city which was otherwise ideal at a time, but which would not persist through time, would not be perfect. There is, naturally, a hard question about *how* the nocturnal council will accomplish this preservation of the laws in the citizen's souls – and that is a question I will return to very shortly. But the mere fact that the *theôroi* share this purpose already shows something about their activity, in light of something the Athenian had said about preservation when introducing the auditors:

5 Translations are my own, from Burnet's text.

> There are many opportunities for a *politeia* to dissolve, just as for a ship or an animal, which is braced by sinews and ropes and tendons, which because they are scattered are often called by many different names, although they are of one nature (μίαν οὖσαν φύσιν).
>
> 945c3–6

This suggests, though of course it does not entail, that the Magnesian analogues of the sinews and ropes and tendons, though they are of many names – auditors, *theôroi*, and nocturnal council members – similarly share a single nature, defined by their preservative/perfective purpose. That suggestion is reinforced by the fact that the auditors and nocturnal council members are also described as ruling the other office-holders in Magnesia, however those other office-holders are described (as rulers at 945c1 or as assistants at 968a3). That is to say that, in addition to sharing a function, these two groups share a characteristic activity – even if it is not immediately clear what that activity comes to.[6]

Naturally, a philosopher and their research assistants might in some sense have the same goal, although their activities may be so discontinuous that the research assistants did not properly speaking philosophize. In the same way, one might think that the *theôroi* are like research assistants, whose travel is necessary for their contribution to the goal, even though that contribution is not philosophical. But I think it is worth noting that other important offices in Magnesia are *not* introduced for the purpose of preservation/perfection – including other offices which one might think contribute centrally to that purpose, viz. the guardians of the laws introduced at 6.752d, and the superintendent of education introduced at 6.765d. That is of course not to say that the theme is entirely new.[7] But if what the *theôroi* do is discontinuous with what the nocturnal council does – if they are like research assistants – then one might wonder why they, and not the law-guardians, are explicitly cited as contributing to the preservative/perfective purpose.

3 Similar Means

In fact, however, there is a considerable overlap in how the *theôroi* and council both contribute to that purpose: both contribute by learning things *via* con-

6 For a dissenting view on which the 968a claim is not about an activity, see Lane (2023: 113–4). For a response, see chapter 6 of Bjelde forthcoming.

7 Cf. *Laws* 4.714a on a ruler who breaks the laws.

versations with the wise, and both then turn around and make use of these epistemic achievements in conversation. There are two parts to this story: a bit about epistemic achievements, and a bit about conversations. I will take these in turn.

First, it is clear that the nocturnal council is supposed to contribute to the preservation of the law in the citizens' souls in virtue of their own epistemic achievements – whatever name those achievements deserve. And it is also clear that this is part of how the *theôroi* contribute to preservation. For one thing, they contribute to the preservation of the city because they contribute to the city as a whole "grasping the laws by insight rather than by habit alone" (951b3: ἄνευ τοῦ γνώμῃ λαβεῖν). For another, the *theôroi* are supposed to report not only divine things they heard from others, but also things they themselves apprehended (952b9: αὐτὸς νενοηκὼς). Moreover, in the middle of the description of the nocturnal council that will test returning *theôroi*, they are explicitly said to *learn* while abroad:

> Their conversation and accounts are always to concern laws and their home city, and also anything significant they learn while abroad about these things, especially concerning anything they learned which seems potentially beneficial in their inquiry, whether because it makes more lawful those who learn it, or because to those who don't learn it, matters about laws appear darker and indistinct.
>
> 951e5–2a6

The fact that it is the *theôroi* who are here said to learn is obscured by translators who make the council members rather than *theôroi* the subject of πυνθάνωνταί at 952a2, although (with an important caveat, discussed below) it is the *theôroi* rather than the council members who have been abroad (ἄλλοθι), and so might have learned something there. Finally, in a closely following passage about how to receive "the counterparts of our *theôroi*" (953c4: τῶν παρ' ἡμῖν θεωρῶν ἀντίστροφος), those counterparts are not only described as wise (953d2: σοφῶν), they are also said to both learn from and teach the wise (953d6: τὸ μὲν διδάξας, τὸ δὲ μαθὼν).

> [...if a] counterpart of our *theôroi* [arrives] from another land, thinking they can see something exceeding in beauty anything in other cities or can show anything of which that is true to another city, let this sort go unbidden to the doors of the rich and wise, since they are themselves another of that sort. For let them go to the household of the superintendent of education ... and conversing with them, sometimes teaching,

> sometimes learning, let them go away, honored as a friend among friends with gifts and fitting honors.
>
> 953c4–d7

So the *theôroi* are not regarded as mere research assistants, but as really making epistemic achievements of their own.

Second, although the Athenian famously has very little to say about what the nocturnal council does, as opposed to their epistemic achievements, what he does say focuses on conversation. Even when they are initially mentioned in Book 10, their job is to persuade recalcitrant but otherwise virtuous atheists by conversing with them (10.909a3–5: νουθετήσει … ὁμιλοῦντες). The main novelty about their activities in Book 12 is that they also converse with each other and test *theôroi*. But the *theôroi* and their counterparts also converse with the wise.[8] So it is not simply that *theôroi* pursue the same goal as the nocturnal council. They also do so in largely the same way. The major difference is that the *theôroi* converse with the wise in other cities, while the nocturnal council (again with an important caveat, discussed below) does not leave Magnesia. The nocturnal council may in that way be more like Socrates: but I see no reason why they should not for that reason be more philosophical.

The auditors present an interestingly different case. They will be members of the nocturnal council, but have a surprising number of other jobs in addition. Their main other job is to investigate other office holders, but they will have other side hustles too, as ambassadors and priests. But the job of investigating the other office holders seems to make a very different contribution to preserving the laws in the souls of the citizens than does, for instance, persuading the atheists. So there seems to be important variations in the way different jobs contribute to the preservative/perfective purpose. My point here is thus not that the *theôroi* contribute in exactly the same way as the nocturnal council, but that, like the auditors, they make a substantially similar kind of contribution to that goal.

4 Council Membership

Some additional corroboration of this is that some *theôroi* are taken up into the nocturnal council itself. This information comes hard on the heels of the re-introduction of the nocturnal council, at 961a4–8:

---

8 As Long (2013) correctly emphasizes.

> [The ten eldest guardians of the laws are to meet together with prize winners,] and also those who went abroad in search of anything they could hear which might guard the laws, and, having been tested (961a7: διαβασανισθέντας) by these [eldest guardians and prize winners], seem to them to be worthy of the council.

While the Athenian here leaves entirely open how liberal the criteria for inclusion will be, he makes clear that the *theôroi* are regarded as specially qualified for membership. They are, after all, on a par with the eldest guardians of the laws, the superintendent of education, and the auditors, all of whom are supposed to be exceptional with respect to virtue even for Magnesia. But why should the *theôroi* be specially qualified, in this way? It is hard for me to see what about the *theôroi* would qualify them other than their activity which so closely resembles doing what the council does – which, I am still assuming, is to philosophize.

The inclusion of some *theôroi* in the council also clarifies a point from earlier. I objected that it was the *theôroi rather than* the council who had been abroad, and so the *theôroi* rather than the nocturnal council members who might have learned things abroad for the council to discuss. But this now turns out to have been an illusory tension. For it may be that the *theôroi*-members will contribute the things they have learned abroad to the council's discussions (it is also possible that the auditors learn something while abroad on embassies), while other members bring different things to the table. But that again is wholly compatible with these *theôroi* philosophizing: all actual philosophers bring different things to the table.

5 Three Objections and Replies

None of this is likely to persuade those already convinced of a view like Nightingale's. Sure, they may say, the *theôroi* are introduced for the same purpose, and they genuinely learn stuff abroad, and they do so by conversing with the wise. Maybe that shows that they are not like research assistants. But still, all of that could be true of journalists who went abroad with that assignment. And those journalists might be then tested upon their return for inclusion in the council. Of course, thereafter they might philosophize. But before that inclusion, they do journalism rather than philosophy – or so a strong version of the objection might run. A weaker version might admit that what the *theôroi* are doing is suitable training for philosophy, but deny that that training itself constitutes philosophizing.[9] I do not think there is conclusive reason or a proof text in the *Laws* which shows that these objections cannot be right. But the following broader picture of *Laws* 12 tells against it, though the picture itself is contentious, and not something I can argue for here.[10]

The nocturnal council's main activity is to persuade, not as rhetors do, but as philosophers do, and in ways that are tailored to specific problems, groups, and individuals. That is what is required to make the citizens well-persuaded of the laws, and so to preserve the laws in the souls of the citizens. That persuasive work is what counts as real ruling, so that the nocturnal council counts as ruling Magnesia because they do that work – and not because they have the power to change the laws willy-nilly, for instance, or to disregard them. So, both the strong version of the objection is wrongheaded because that persuasive work is substantially similar to what the *theôroi* are said to do, and that similarity is what explains why the *theôroi* are even marked out as candidate members of the council. It is, for instance, easy to imagine a *theôros* excelling at counselling atheists, if they spent their 10 years discussing divine substance with foreign wise people. But it is difficult to imagine a journalist, even one who reported back about theology, succeeding at persuading the kind of philosophically sophisticated atheists who end up in the prudentiary. The weaker version of the objection similarly underestimates the activity of the *theôroi*, by putting their qualifications on a par with the junior members of the nocturnal council, who receive training in ways that suit them to become senior members. But that transition is not automatic – those trained as junior members must distinguish themselves in other ways, in order to join as senior members. By contrast, *theôroi* join immediately after returning. Moreover, their foreign counterparts who come to Magnesia are explicitly said to be themselves of the same sort as the wise (952d2–3), rather than (for instance) to have the same nature.

But by emphasizing the parallels between the *theôroi* and others who are inducted as senior members into the council, this response brings a subtler objection into focus. For, renewing a question from earlier, one might ask, why *are* the *theôroi* supposed to be able to make a distinctive philosophical contribution, and what could it be? When we select graduate students or hire junior faculty, we try to assess the applicants' ability to make a distinctive philosophical contribution. And – the subtle objection goes – it does not make a difference to that ability where they have been. Being well-travelled is (arguably) not a job qualification for philosophers, even if they have been doing philoso-

9 I am grateful to Mehmet Erginel for this objection.

10 I argue for this picture in chapter 6 of Bjelde forthcoming.

phy while travelling. So, what does *travelling*, in specific, contribute? After all, if (as I am assuming) there are wise people at home, why not converse with them? If that was good enough for Socrates, the objector might go on, why is it not good enough for the nocturnal council?

My suggestion is that this would also not be good enough for Socrates, for a reason that only becomes clear if we pay careful attention to (or perhaps: put undue pressure on) a passage from the *Apology* where Socrates considers and rejects a hypothetical offer of acquittal conditional on no longer philosophizing:

> [...] but I will obey the god rather than y'all, and so long as I draw breath and am able, I will not stop philosophizing and exhorting y'all and always showing whoever I run into, by saying the sorts of things I am accustomed to say...
>
> *Apology* 29c7–d7

The way Socrates rejects the hypothetical offer is remarkable. He says that he will not stop conversing with "whoever [he] run[s] into." Now, that *modus operandi* is quite different from how we philosophize. And it is quite different to a standard picture of how Plato thinks we or the nocturnal council philosophizes. For us, as for that standard picture of philosophizing for Plato, there may really be no reason not to restrict oneself to philosophizing within one's circle. So why does Socrates characterize the proposal as ceasing to philosophize, rather than, for instance, ceasing to bother people or conversing with the young – and then accept the proposal, stay home, and philosophize with other old people? I strongly suspect that the reason is that, on Socrates' view, this would not be to philosophize at all. Instead, on Socrates' conception, philosophizing requires a remarkable *breadth* of interlocutors. That being said, there remain important questions about how to interpret Socrates' obedience to the god, which must be thoroughly examined to fully substantiate this claim regarding his conception of philosophy. Into that calculation would have to go the idea at *Phaedo* 78a[11] that those present will have to work together to scour the world – both Greece and beyond – for someone to replace him as an interlocutor when he is gone.

Whatever the truth is about Socrates in the *Apology*, what I want to suggest is that Plato's Athenian in the *Laws* requires something like this of philosophizing. Real philosophizing requires a broad group of interlocutors. If that is right, then what is distinctive about the contribution of the *theôroi* in *Laws* 12

11 As Schöpsdau (1994: 554) points out, in connection with the *theôroi* of *Laws* 12.

will be that they put the nocturnal council in touch with a suitably broad group of interlocutors. They do this not by reporting what those interlocutors say, but by representing their views in philosophical discussion as only a philosopher can. Think for instance of how interlocutors in Plato take over theses from others. Protarchus in the *Philebus* is in some sense representing Philebus' views. Glaucon and Adeimantus are in some sense representing Thrasymachus' views. Protagoras is not present in the *Theaetetus*, but his views certainly are. In sum: *theôroi* in *Laws* 12 perform the indispensable job – indispensable, that is, *for philosophy* – of representing a diversity of philosophical views in the discussions of the nocturnal council, and they are enabled to do this not by looking inwards, but by looking outwards.

Bibliography

Bartels, Myrthe. 2017. *Plato's Pragmatic Project: A Reading of Plato's Laws*. Stuttgart: Franz Steiner Verlag.

Bjelde, Joseph. Forthcoming. *Plato on the Practical Value of Episteme*. Cambridge: Cambridge University Press.

Lane, Melissa. 2023. *Of Rule and Office: Plato's Ideas of the Political*. Princeton: Princeton University Press.

Laks, André. 2022. *Plato's Second Republic: An Essay on the Laws*. Princeton: Princeton University Press.

Long, Alex. 2013. *Conversation and Self-Sufficiency in Plato*. Oxford: Oxford University Press.

Nightingale, Andrea. 2004. *Spectacles of Truth In Classical Greek Philosophy: Theoria in its Cultural Context*. Cambridge: Cambridge University Press.

Schöpsdau, Klaus. 1994. *Nomoi VIII–XII*. Göttingen: Vandenhoeck & Ruprecht.

CHAPTER 13

Understanding the Epistemic Journey of Plato's Philosophical *Theôros* as Uprootedness

Shane Fair

1 Introduction

The word "theôria" in Greek (θεωρία) has a very interesting history. It has now come to have an almost exclusively epistemological meaning in its philosophical use, but its origins are found in cultural, religious practices which are much more akin to religious pilgrimages and the participation in rites and mysteries. Though the topic of *theôria* has been explored at some depth both in its traditional, cultural context and as a concept which is core to Greek philosophy in their own right, it seems that study of the transition of *theôria* from the former, religious context to the latter, philosophical usage has only recently become a topic of interest among scholars of Greek philosophy – particularly as it relates to interpreting the texts of Plato and Aristotle.

Given it would be impossible to provide a comprehensive account of the parallels between traditional and philosophical *theôria* as it appears in Plato's dialogues, I have opted to frame the present discussion through the lens of "uprootedness." I hope to show that this concept is quite essential to both expressions of *theôria*. To this end, I will begin by providing a brief account of uprootedness in traditional *theôria*. I will then look at the conditions for uprooting oneself in the context of philosophical *theôria* in Plato's dialogues. Having given an overview of these conditions, I will give a very brief account of the vision of the forms as the goal of uprooting oneself. Finally, I will speak to some degree about the effects of attempting to uproot the forms for the purposes of rendering an account of the experience of the philosophical *theôros*.

2 Overview of the Theoric Journey

Let us begin, then, with a quick overview of traditional civic *theôria*. In her book, *Spectacles of Truth in Classical Greek Philosophy*, Andrea Wilson Nightingale outlines three stages of traditional *theôria*, namely, (1) the departure from one's city and journeying abroad, (2) the so-called 'liminal' stage where

 HTTPS://DOI.ORG/10.1163/9789004747128-015

one witnesses spectacles and participates in rituals, and (3) the return home of the *theôros*.[1] In the case of civic *theôria*, where the "*theôros*" (the person(s) participating in the *theôria*) was an official delegate of the city, the duties of the *theôros* included participation in the festival and events abroad and giving an account or report of the festival upon their return to the city.[2] In terms of the aim of the theoric journeys themselves, in general, they were of a religious nature and had at their centre the participation in or the sight of some religious mystery, normally within the context of some festival.[3] Because of this religious aspect, the aim of the theoric journey was to take in or experience some divine truth, some spectacle[4] related to the divine.

While Nightingale's assessment is fairly straightforward, one thing that bears emphasis is the idea that the experience and the sights that the *theôros* travels abroad to witness are not available to them in their city of origin.[5] In order to participate, the *theôros* needs to leave their[6] familiar surroundings to have this new, divine, and true experience; they must uproot themselves, as it were, from their community for a time, otherwise the experience remains closed off to them. As Nightingale says,

1 See Andrea Wilson Nightingale (2004: 42). Although this tripartite division may work for traditional *theôria*, it is not so clear that the same division obtains for philosophical *theôria*. Including both the journey to the festival and the viewing of spectacles as part of the liminal stage of the journey suggests a more seamless transition between these two moments of the *theôria* than is suggested by Plato's dialogues. Though it is not possible to delve into this issue here, I would propose dividing the journey of the philosophical *theôros* into five sections, namely: (1) the initial state of the *theôros*; (2) the *periagoge* and journey to the forms; (3) the spectacle or vision of the forms itself, the principal goal of the journey; (4) the return journey; and (5) the rendering of an account.

2 See Pierre Voelke (2014: 136): "L'action qui incombe au θεωρός n'est donc pas la seule consultation de l'oracle, mais lui revient la mission de ramener le plus fidèlement possible la réponse transmise ... le terme θεωρός implique donc un mouvement d'aller et retour."

3 Julie K. Ward (2022: 23). Ward points out that the official *theôros* was not just a delegate of the city but was often expected to play an active role in the events of the festivals.

4 That is, some "θέα" or "θέαμα" in Greek, which share the same etymological "ΘΕΑ" root as "theôria."

5 For more on the topic of traditional *theôria*, see, Ian Rutherford (2013). Rutherford is one of the scholars with perhaps the most publications on the topic of traditional *theôria*. The work cited provides a detailed overview of the different elements of *theôria* as a cultural practice in ancient Greece, including a short section on the appropriation of the term "theôria" by philosophers.

6 Throughout this paper I try to employ as consistently as possible "they/them" pronouns as well as the English impersonal, third person singular pronoun "one" rather than singling out a particular gender for the philosophical *theôros*.

> In short, theoric spectating could only take place at a distance from the pilgrim's hometown or city: geographical distance was a precondition for the special kind of viewing and apprehensions that characterized *theôria*.[7]

It is not difficult to see how this could map onto certain of Plato's dialogues which deal with the philosopher's interactions with the forms. If, for example, one never leaves the cave, then one never has the experience of seeing the form of the Good; one is consigned to living with the shadows on the wall.

Another aspect to consider in order to reinforce the connection between traditional and philosophical *theôria* is that just as traditional *theôria* is aimed at some divine experience, in Plato too, the aim of the philosophical *theôros* is also to experience something divine, namely a vision of the forms, which are in almost every instance described as divine (θεῖος) in the dialogues.

3 The Two Regions

This preliminary discussion on *theôria*, and the notion that the *theôros* must be in some sense uprooted from his community, leads us in a somewhat obvious-sounding direction, but one which should be made explicit regardless – namely, a brief discussion of geography. In both traditional *theôria* and philosophical *theôria* in Plato's dialogues, we see a distinction between two geographical locations (whether actual or metaphorical). These regions are separated by some distance which must be traversed. In both kinds of *theôria*, the destination is reached only after one has crossed over from one territory to the other. Otherwise, there is no experience of the sights, there is no report tendered to the city – there is no *theôria*.[8]

It is worth noting that in the actual spectating of a *theôros*, the events often took place in a sacred space, frequently requiring some measure of personal preparation or purification before being allowed to engage in the event. In

7 Nightingale (2004: 42). See also James Ker (2000: 308): "… *theôria* would not normally take place within one's own city."

8 Cf. Voelke (2014 : 143) : "la participation à ces fêtes citadines nécessite toujours un déplacement … et c'est la θεωρία, le spectacle, qui apparaît comme la composante pouvant motiver le déplacement." We might compare this to the accounts of the *Symposium* and the *Phaedrus*, where the sight of the beautiful beloved triggers the memory of the Beautiful itself.

other words, the *theôros* would be aware that they will be engaging in something other, divine, and extra-ordinary.[9]

In Plato's accounts of the forms, he uses metaphor and analogy to refer to the intelligible realm (the realm of the forms). The forms are not physical objects, and hence they don't occupy physical space, yet, because Plato insists that the forms are what truly are (τὸ ὄντος ὄν), he must therefore make use of numerous analogies to demonstrate both the ontological distinctness of the forms, and the distinctness of the intelligible and visible realms in order to speak about them at all. Moreover, Plato consistently speaks of these realms as being "geographically distinct," and that they are, in some sense, mutually exclusive. For example, there is the cave entrance (*Republic*), the air-aether barrier (*Phaedo*), and the ridgeline separating off the hyperouranian realm (*Phaedrus*). In each of these examples, there is a firm demarcation between one region and the other, a threshold that must be traversed before there can be any interaction with the forms. Moreover, as we will discuss below, in the *Republic* there seems to be a period of transition whereby one must adjust one's focus away from one realm in order to truly see the other.

In the dialogues which more explicitly investigate the forms (notably, *Phaedo*, *Phaedrus*, *Symposium*, *Republic*), we see the philosopher described in various literal and metaphorical ways making their way from one region to another, but how exactly does one make this epistemic journey? What are the conditions that must be met to make the transition from one region into the other – i.e., how does one go about uprooting oneself from the visible realm? In the majority of cases, despite the facility with which Plato describes this ascent, the language employed implies a thoroughly unpleasant and arduous process.[10] This difficulty, I believe, can be understood by means of (at least) two factors: (1) the nature of the regions themselves and (2) the psychological conditions necessary to make the journey.

---

9 Cf. Andrea Wilson Nightingale (2021: 69). See also Miguel Herrero de Jáuregui (2015: especially 599, 604, and 608).

10 In the *Republic*, for example, the route of the philosopher is described more as being dragged, bleeding and bruised, rather than led out of the cave. The vocabulary of necessity (ἀναγκάζοι) and dragging/laceration (ἑλκόμενον) are used throughout the passage in Book 7 (circa 515e–516a) describing the ascent from the cave. See also *Phdo* 109e2–110a1, "If anyone got to this upper limit, if anyone came to it or reached it on wings and his head rose above it, then just as a fish on rising from the sea sees things in our region, he would see things there and, *if his nature could endure to contemplate them*, he would know that there is the true heaven, the true light, and the true earth" (emphasis added). One need not be an ichthyologist to recognise that removing a fish from water is not likely to end well for the fish.

Starting with the latter, what are these psychological conditions? How does one get from one realm to another? One finds different answers depending on which dialogue one reads. In the *Republic*, once one has been set free of the chains that bind them to their place of origin, one must then come to a greater understanding of things by degrees, until finally reaching the realm of the forms and seeing the things themselves in the light of the Good. This could be considered an epistemic purification.[11] Although it is not explicit in the Allegory of the Cave how exactly this purification comes about, we can deduce, based on the surrounding text,[12] that it is through some kind of dialectical exercise, such as the elenchus, the maieutic method, or the method of *diairesis* and *sunagoge*, which we see employed in other dialogues. This epistemic purification seems to be more easily identifiable as an "uprooting," insofar as there is explicit mention of breaking chains, and the very clear outcome of the nascent philosopher having their false beliefs pulled out of them. Before this *periagoge* or "turning around" moment, the prisoner in the cave does not make any attempt at the upward journey, because they are by all accounts ignorant of such a possibility.[13] The uprooting of the prisoner from their received beliefs (opinions about the shadows on the wall of the cave) is, therefore, a necessary condition to make this journey. On this model, the initial state (one's region of origin) is that of merely existing in the world, accepting the received opinions and commonly held beliefs (*endoxa*) of the *polis*, and this is more or less sufficient to get by. It takes an external agent to question these beliefs (some elenchic cross-examiner or spiritual midwife).[14] In the early dialogues, this

11 On the topic of purification rituals in the religious practices of ancient Greece, see Herrero de Jáuregui (2015: 603): "The process of personal purification that in temple 'there' religion is physically expressed as an external movement, e.g., 'entering the pure space,' in itinerant cults is completely internalized as 'becoming pure.' This internal process is … frequently envisaged in spatial categories." Moreover, he states (2015: 604) that that which is incompatible with the sacred space (anything impure of body or mind) must be left behind before entering the ritual space.

12 E.g., the language in the Divided Line passage which immediately precedes the Allegory of the Cave, and which speaks about using hypotheses to make one's way to the unhypothesised.

13 The breaking of the chains lends further credence to the more dialogical character of the epistemic purification, as the nascent philosopher is not freed by their own hand, but by another. This guide both frees the prisoner and turns them around toward the harsh, blinding light of the fire.

14 In the passage following the Allegory of the Cave (at *Rep.* 523aff.), Plato speaks of subjects which naturally "pull" (ἑλκτικῷ, whose "ἑλκῶ" root suggests some measure of violence) or "summon" (παρακαλοῦντα) the mind beyond mere sense data toward understanding (πρὸς τὴν νόησιν ἀγόντων). Whereas the senses can only perceive objects as mixed (and indeed, as far as the senses are concerned, as *unproblematically* mixed) agglomerations

examination frequently results in *aporia* (perplexity) arising from contradictions among these beliefs.[15] It is by realising the flaws, falsehood, or incompleteness of one's beliefs that one is free to pursue truth elsewhere. In this way, the budding philosophic *theôros* begins to examine and refine their understanding of a given definition, topic, or idea by exploring it from all angles, continually ridding oneself of the false beliefs which prevent the upward journey of the soul or *noûs* – a kind of doxological or epistemic uprooting.[16] Having undergone this epistemic purification, or uprooting of false beliefs, the philosopher is able to see the forms. As Plato says in his *Seventh Letter*,

> Only when all of these things – names, definitions, and visual and other perceptions – have been rubbed against each other and tested, pupil and teacher asking and answering questions in good will and without envy – only then, when reason and knowledge are at the very extremity of human effort, can they illuminate the nature of any object [i.e., see the forms].
>
> *Epis.* 344b3–c1

---

of qualities, reason is able to perceive in this sense data when a quality and its opposite are present in the same thing. Such occurrences induce perplexity and demand that the philosophical soul investigate the matter and distinguish each quality and understand them as separate and unmixed (αὐτὸ καθ' αὑθὸ ὁρᾶται). Subjects which bring about this summoning are mandatory for philosophers as they "must rise up out of becoming and grasp being if they are ever to become rational (διὰ τὸ τῆς οὐσίας ἁπτέον εἶναι γενέσεως ἐξαναδύντι)" (*Rep.* 525b4–6).

15 See Nightingale (2004: 106): "To become a philosopher – to escape from the bondage of opinion and set out on a quest toward the unknown truths – one must be exiled from one's original 'home' and experience a profound sense of *aporia* ([*Rep.*] 515e). The fact that the perplexed soul wishes to flee back to its original dwelling is truly poignant: *aporia* is depicted as (among other things) a state of homelessness. The soul has, both cognitively and existentially, left home, and thus suffers deracination and disorientation."

16 One could also argue that there is a conative dimension to this uprooting. Because one does not desire that which is in one's possession, one does not seek it out. In the case of false opinions, because one mistakenly believes oneself to be in possession of truth, one does not see the need to look elsewhere. However, once one comes to know that one does not know, then one can (and indeed must) search for truth. This is most clearly seen in the *Meno* (81dff.), which recounts the serial removal of false beliefs about the double of the square from the slave boy, through a cross-examination by Socrates. After two attempts, the slave boy becomes "*aporon*" and his mind is free of the confidently held false beliefs, Socrates asks Meno, "Do you think that before he would have tried to find out that which he thought he knew though he did not, before he fell into perplexity and realized he did not know and longed to know?" (*Meno* 84c4–6). This new outlook makes the slave boy an ideal candidate for recollection.

This epistemic evolution is also mirrored in the language used in the *Republic*, which describes the interactions of the inhabitants of the visible realm in terms of "opining" and "competition" (e.g., *Rep.* 516d2–3). Since there is an indefinite plurality of ways an object can be experienced in this realm, the opinions formed on the basis of these experiences compete and must be examined and justified. The objects in the intelligible realm, however, are not "opined" about, but rather are "discerned and contemplated" (κατιδεῖν καὶ θεάσασθαι, *Rep.* 516b6). On the epistemic level, we see that as one moves from the visible realm to the intelligible realm, the opinion which characterises the visible realm is replaced by knowledge; the plurality of the objects of opinion is replaced with the unity of the form.[17]

This epistemic purification, however, does not seem to be sufficient in itself to get to the realm of the forms. In the text of the Allegory of the Cave, there are numerous passages which show that the philosophical *theôros* tends to prefer the shadows to the light, given that they are more accustomed to these shadows.[18] Indeed, the sudden exposure to light comes as an uncomfortable and deeply distressing experience. As a result, the preference for the shadows indicates a dynamic tendency toward comfort.

We see that there is a need for a further kind of purification – one that is not mentioned in the *Republic*, but is examined in the *Phaedrus*, and which we will call, for lack of a better term, a kind of "moral" purification. In the section of the dialogue which speaks of the procession of souls, using the analogy of the charioteer and the two horses (*Phdr.* 246aff.), the necessary condition for making the ascent and reaching the hyperouranian plane of the forms is not necessarily related to knowledge or any epistemic conditions, but rather a kind of education of the soul,[19] an alignment of the aims of the tripartite soul, most notably between the desirous part of the soul (*epithumia*) and the rational part of the soul (*noûs*).[20]

---

17 *Rep.* 507b5–10, "And beauty itself and good itself and all the things that we thereby set down as many, reversing ourselves, we set down according to a single form of each, believing that there is but one, and call it 'the being' of each… And we say that the many beautiful things and the rest are visible but not intelligible, while the forms are intelligible but not visible."

18 This is true even when they have reached the outside of the cave, preferring first to contemplate the reflections in water, etc. There is, in some sense, a natural inclination downward, back toward the cave (*Rep.* 515e1–4; 516a5–b2).

19 Adopting and adapting the term from Olivier Renaut (2014).

20 Cf. Ward (2022: 78): "Second, the epistemic confusion arises from a deeper, moral defect: having the wrong desires, the lover of spectacles loves and pursues perceptual goods, pleasant sights and sounds, instead of what is truly good." On the topic of purifying the desires of the soul and aligning the aims of the tripartite soul, see Mauro Bonazzi's paper,

Without this alignment, the journey is not possible to complete – and, indeed, even with the alignment of the soul and its desires toward intelligible objects, the philosophical soul still struggles to focus on the forms, preoccupied as they are with ensuring the course of their chariot. Whereas the gods in the *Phaedrus* account have perfect control over their chariots, and are hence able to complete the entire revolution in procession around the hyperouranian plane, seeing the entire spectacle of the forms from all angles,[21] the human soul – even the one capable of making the upward journey – will, at best, catch the briefest of glimpses as they struggle to keep pace with the gods, and keep their heads above the clouds. Again, here we have Plato stating that, without training the soul to desire the objects which satisfy the *noûs*, there can be no successful upward journey ("the horse of evil nature weighs the chariot down," *Phdr*. 247b2–5) – and even the successful soul will just barely see reality (καὶ μόγις καθορῶσα τὰ ὄντα, *Phdr*. 248a4–5).[22]

The union of these two kinds of purification, which are both seemingly necessary to make this theoric ascent, appears to find an expression in the *Symposium*. In this dialogue, in the "Ladder of Love" section (*Symp*. 210aff.), we see the lover of beauty both deepening their understanding of beauty and also (it is implied) loving and desiring these newly discovered expressions of beauty, until they eventually see Beauty itself, a beauty which always is, itself by itself, monoeidetic and neither becoming nor ceasing to be. In short, it is an object of cognition unlike anything encountered on the previous rungs of the Ladder. This new understanding becomes the object of the philosopher's erotic love and informs and purifies their love of particulars.[23]

---

in this volume. In his paper, M. Bonazzi provides a rather comprehensive overview of competing interpretations of the tripartition of the soul with a special focus on the *Republic*. His argument, which is not strictly speaking incompatible with my own, challenges the notion that the rational soul by its very nature seeks to rule, but is rather uniquely motivated by the desire to know or contemplate the truth. This, however, presupposes some measure of education or purification of the aims of the other two parts of the soul, the *epithumetikon* and the *thumoeides*.

21 We won't go into detail here; however, it bears noting that one practice common to theoric festivals was the procession around religious idols or *agalmata*, which we will look at briefly below. *Theôroi* would make their way around these *agalmata* to see them from all angles (*cf*. Ward (2022: 62, 72, 84)). It is interesting to note that, for Plato, only the gods are able to accomplish this feat; the human souls "see some things and fail to see others" (τὰ μὲν εἶδεν, τὰ δ' οὔ, *Phdr*. 248a6) on their processional revolution of the hyperouranian plane.

22 This limited view almost certainly plays a role in the difficulties we will explore later, when I talk about rendering an account of the forms.

23 Nightingale (2021: 64ff, 97) has a lot to say on the topic of Beauty both as playing a key role in philosophic *theôria* and what she calls "metaphysical desire" in Plato.

4 The Nature of the Vision

In addition to the travelling aspect of *theôria* (which requires the epistemic and moral uprooting), there is also a visual aspect – a *theôreô*. The journey abroad culminates in some sight or spectacle of the divine.[24] We have already begun to look at some characteristics of this vision as it translates to philosophical *theôria*, namely, that it is short-lived and achieved with great difficulty.[25] In addition to these characteristics, we might add those which follow from the "noetic" nature of contemplation.[26] To understand this point, it should be enough to simply look at the vocabulary that is associated with these philosophical acts: sudden, immediate vision of a perfect, eternal form. We have in this vision a knowledge act that is fundamentally different from the normal process of human cognition (that is, discursive reasoning) and which has an object that is fundamentally different from the regular objects of human experience (i.e., perfect, true, and unchanging objects). In the act of contemplating the forms, the philosophic *theôros* is, as it were, completely uprooted from their normal situation.[27]

It is important to note that the vocabulary which Plato employs to describe the interaction between the philosopher and the forms is sufficiently different to see that we are confronted with a different epistemological situation. Plato uses a number of words to denote the philosophic practices of cross examin-

---

24 See Voelke (2014 : 142) : "Alors que les simples particuliers voyagent et se rendent ... pour voir ce qui mérite le plus d'être vu ..."

25 For reasons of scope, we will have to forego a possible discussion about the short-lived nature of the vision of the forms due to the experience being painful and/or unsustainable. If one looks at the imagery invoked in the *Phaedo* and the *Republic*, one can see that this activity cannot be sustained by the human soul. In the analogy of a fish in the *Phaedo* noted above, though the fish is able to peek its head out of the water to glimpse the land above, it cannot stay there, as it cannot breathe the air, and must therefore return below, lest it suffocate. Similarly, in the *Republic* (516b4–7) we have the philosophical soul looking directly at the sun. Based on even the most cursory understandings, we can see this being a short-lived activity, given that no eye could endure to look directly at the sun for long. On this, see Ward (2022: 73–74) and Nightingale (2004: 98).

26 "Noetic" in the sense outlined in the analogy of the divided line, where Plato draws a distinction between *dianoia* and *noesis* (*Rep.* 511d6–e2), where dianoia is associated with discursive reasoning and noesis tends to be associated with more intuitive, non-discursive thought.

27 Cf. Nightingale (2021: 89). Quoting Jás Elsner (2000), she says, "The ritualized mode of seeing 'denies the appropriateness ... of interpreting images through the rules and desires of everyday life. It constructs a ritual barrier to the identifications and objectifications of the screen of [social] discourse and posits a sacred possibility for vision ....' This 'ritualized visuality' effectively screened out secular modes of viewing and allowed the celebrant to experience the presence of a god." See Jás Elsner (2000: 61).

ation which carry the meaning of discourse, dialectic, plurality of interlocutors and plurality of ideas (opinion, hypotheses, premises, conclusions). These tend to convey a sense of continuous dialogue, whereas the vocabulary associated with the noetic contemplation of the forms is sight-based. No attempt is made by the philosopher to engage dialogically with the forms in the moment of contemplation (*Sym.* 211e3–212a2 uses the terms "sight," "contemplation," and "communion" (κατιδεῖν, θεωμένου, συνόντος)). This cognitive shift is also represented more subtly on a grammatical level, in that the verb tense used to describe noetic contemplation is almost exclusively in the aorist, which generally carries an instantaneous, moment-in-time aspect. In other words, we have a vocabulary suggesting an immediacy between subject and object, an experience which is not experienced temporally in the same way that dialectical reasoning is experienced.[28]

There is one additional point to further emphasise the notion of uprooting oneself and ties the vision of the forms back to the theme of *theôria*. This is the notion of "initiation."[29] There are at least two dialogues which explicitly mention "initiation" in relation to the forms, namely, the *Symposium* and the *Phaedrus*. In the *Phaedrus*, the uninitiated (ἀτελεῖς, *Phdr.* 248b5) are those who are unable to make the upward journey to the hyperouranian plane and turn back, nourished only by their opinions (those that are incapable of the moral uprooting necessary to ascend); whereas the initiates (μυούμενοί) are permitted to see the perfect, simple, calm, and happy apparitions of the forms (*Phdr.* 250c2–4). In the *Symposium*, Diotima also uses the word *muêtheiês* (*Sym.* 210a1, μυηθείης).

28 This is further amplified by the frequent use of the term "ἐξαίφνης" ("suddenly," "immediately") in describing the nature of the vision of the forms in the dialogues. In the *Parmenides* (155e4–157b5), the term ἐξαίφνης is used to speak about the moment between motion and rest, where a thing simply is. See Mary Louise Gill (2012: 65). See also the paper of Abida Malik in this volume, which discusses the different interpretations of "understanding"/knowledge in Plato's dialogues and proposes a synthesis of two of the main interpretations: understanding as intuition (Understanding_1), and understanding as being able to render an account (Understanding_2). That is, true understanding (on this reading) would contain (or at a minimum recognize the importance of) elements of both intuitive understanding and propositional understanding.

29 For reasons of scope, I am not able to delve too deeply into this topic here. For those interested in a more comprehensive account of "initiation," I would direct the reader to the paper of my co-conferencier Guilia Cervato in this volume. In her paper, she discusses the topics of the Eleusinian mysteries, blindness, and μυησις in much greater detail. Of particular interest for my paper is her argument, rooted in a reading of the *Phaedrus* and of Hermias of Alexandria's commentary of the *Phaedrus*, noting that the closing of the eyes of the body allows the eyes of the soul (i.e., reason) to see more clearly.

This term is closely related to the practice of the Eleusinian mysteries.[30] The word itself comes from the "ΜΥΩ" root, meaning (predominantly) "to close one's eyes."[31] The implication being that one must close one's eyes to the visible realm as a necessary condition for participating in the mysteries, decidedly severing one's connection to and reliance on this visible realm. In the context of an initiation/*muêtheiês* for the philosophical *theôros*, one is effectively abandoning the cognitive activity which characterizes the vast majority of one's lived experience.[32] In order to open one's eyes to the realm of the forms, one must close one's eyes to the visible realm.[33] Initiation in traditional *theôria* is, moreover, frequently tied to purification rituals, which aligns will with the practices of epistemic and moral uprooting outlined above as necessary conditions for completing the philosophical theoric journey.

5 The Return and the Rendering of a Report

Having participated in the theoric spectacle, the civic *theôros* then returns to their city of origin and renders their report or account of the experience, to bring back some measure of their experience to the *polis*.[34] We see a similar trajectory in the *Republic*. Plato's philosopher is not permitted to stay in the realm of the forms, but must return to the cave and legislate for their community based on their vision of the form of the good. Just as with traditional civic *theôria*, Plato describes this return to the cave for the purpose of legislating as an official duty of the philosophical *theôros* (*Rep.* 519c8–520a4; *Sym.* 209a3–7). Rendering an account comes with its own difficulties, however. The philosophic *theôros* must find some way to translate the noetic into the discursive, to describe in detail one's vision, to uproot, in some sense, something of the realm of the forms and transplant it in the visible realm.

30 Unfortunately, we cannot explore the topic of the Eleusinian mysteries in depth; however, there is a rather large amount of research on the topic. Much of the literature on the philosophical appropriation of traditional *theôria* touches to some degree on the Eleusinian mysteries – especially in Plato's dialogues. See, e.g., Nightingale (2021: 88–95).

31 See Pierre Chantraine (1999: 728) on "μύω."

32 See Sylvana Chrysakopolou (2012: 110); see also Nightingale (2004: 84).

33 I think that it is also important to note that, although the imagery used to describe the act of contemplation is generally sight-based, Plato also uses touch vocabulary in the *Republic* (see for example *Rep.* 490a8–b7 (ἐφάπτεσθαι), and 511b3–4 (ἅπτεται)). This tactile expression of contemplation seems to be absent from the literature related to philosophical *theôria*, but bears further exploration.

34 See Rutherford (2013: 190).

It does not take long to see how this rendering an account becomes problematic. As we briefly discussed, it is the nature of the forms to be perfect, simple, self-contained, analytically self-referential, and, importantly, ontologically independent of any understanding of them. How does one give an account of something which is simple without complicating it?[35] One has only to look at the *Parmenides* to see how quickly this exercise becomes problematic. This, again, goes back to the difference between the cognitive acts of noesis and dianoia. Aporia has no place in noetic contemplation, because there is no need to hypothesize about an object that is seen, itself by itself, in its own place. Dianoesis and acts which seek to set down definitions, however, introduce multiplicity by their very nature. For example, to say "the Beautiful itself is good," (i.e., "x is y") is effectively to say that Beauty is more than Beauty (x is greater than x). This is tantamount to an invitation for aporia – which would be incongruent with the nature of the forms; if a form is simple by its nature (and therefore analytically true, in the sense of x = x), then it cannot admit of any contradictions.[36]

Nightingale attributes this difficulty in rendering an account (in part) to the fact that the theoric experience, the "theoric gaze" is not panoptic. As she states,

> Plato's theoretical philosopher, in sum has blind spots: the vision of the theorist is not panoptic, since he sees differently – and different objects – in each realm. When contemplating the Forms, he does not see the world, which now appears dark; and when he pursues practical and political activities back in the cave, he sees the Forms less clearly.[37]

This goes back to what I raised briefly earlier about the need in the *Republic* for a transition period to fully remove oneself from one realm to be present and

35 Cf. Nightingale (2021: 48): "In short, we want to understand the essence of this individual Form. In addition, we want to know that the Form we conceive of actually exists, and that the definition of the essence we come up with matches the Form's essence. The philosopher who grasps a Form must provide a definition of its essence that is incontrovertibly true."

36 As stated in Francis Cornford (1932: 188–189): "A defining formula is expressed in words; it can be written in a book and intellectually understood, like the definition of any other Idea. ... But the intellectual understanding of formulae expressed in words is not the same as the intuitive vision of the reality which the formulae profess to describe." A form, in its authentic simplicity, must be ineffable.

37 Nightingale (2004: 103–104).

interact in the other.[38] Though I agree with Nightingale's statement, I think that we can also look at this difficulty in light of the nature of the two realms and the objects that reside in them. Just as in Plato's account of the Divided Line in the *Republic*, the philosopher hypothesises until they make it to the unhypothesised (that is, an object whose truth value is independent of any hypothesis about it). In the realm of the forms we have objects which are ontologically distinct and the truth about these forms is independent of the philosophical *theôros*' perception (or definition) of them. In the visible realm, by contrast, the objects are dependent upon the forms for both their being and their meaning.[39] More straightforwardly, one might ask whether the converse of the image of the fish in the *Phaedo* does not obtain in the case of uprooting the forms: just as the fish will suffocate in the air, the human will drown if submerged in the water. There are inherent limitations to how the objects of one realm interact with the other; the objects of one realm cannot fully inhabit the other.

We see a bit of how Plato may have envisioned the interplay between these two realms and the relationship between the forms and particulars in the *Timaeus* (37c6–d7), where he speaks of time originating from the demiurge's attempt to emulate the perfection of the forms in generated and corruptible things – i.e., in the visible realm.

> Now when the Father who had begotten the universe observed it set in motion and alive, a thing that had come to be as a shrine for the everlasting gods, he was well pleased, and in his delight he thought of making it more like its model still [i.e., the forms]. So, as the model was itself an everlasting Living Thing, he set himself to bringing this universe to completion in such a way that it, too, would have the character to the extent that was possible. Now it was the Living Thing's nature to be eternal, but it isn't possible to bestow eternity fully upon anything that is begotten. And so he began to think of making a moving image of eternity: at the same time as he brought order to the universe, he would make an eternal image, moving according to number, of eternity remaining in unity. This number, of course, is what we now call 'time.'

38 See *Rep*. 518a1–3, "But anyone with understanding would remember that the eyes can be confused in two ways and from two causes, namely, when they have come from the light into the darkness and when they have come from the darkness into the light."

39 Cf. Nightingale (2021: 48).

In this passage, we see on the one hand that the forms cannot be translated fully into the visible realm, and on the other hand, we begin to appreciate how the act of rendering an account is a kind of demiurgic activity on the part of the philosophical *theôros*. The philosopher, in seeking to uproot the forms by bringing their limited, imperfect vision back with them to the visible realm, creates a kind of "moving image of the eternal." Which is to say that the truth value of any account of the forms in the visible realm will not be absolute, but will vary based on how adequately it describes the forms.

There is a sort of homologue to this in the context of traditional *theôria*, namely, the *agalma*. *Agalmata* were carvings, statues, or casts of divinities and were often one of the main attractions at theoric festivals. In other words, they were objects which represented or pointed to some divinity, while being distinct, corruptible, lower-order objects; objects which could more or less closely resemble the object intended for emulation, something which can be changed and improved upon. In other words, they reminded the *theôros* of the divinities which they represented; despite being material, they evoked the divine in the minds of the *theôros*.[40] We see an example of the power of *agalmata* in the *Phaedrus*,

> Now Beauty, as I said, shone out among the other objects, and now that we have come down here we grasp it sparkling through the clearest of our senses ... A recent initiate (ἀρτιτελής), one who has seen much in heaven (πολυθεάμων), when he sees a *godlike face or bodily form that has captured Beauty well* ... he gazes at him with the reverence due a god ... and if he were not afraid people would think him completely mad, he would even sacrifice to his boy as if he were the image of a god (θύοι ἂν ὡς ἀγάλματι καὶ θεῷ)... When [the soul] looks upon the beauty of the boy and takes in the stream of particles flowing into it from his beauty ... then all its pain subsides and is replaced by joy.[41]
>
> emphasis added, *Phdr*. 250c8 ff.

---

40 On the subject of *agalmata*, we might point here to *Phdr*. 246a3–6, which draws a distinction between describing what the nature of the soul *is*, versus what it *seems to be* (οἷον μέν ἐστι ... ᾧ δὲ ἔοικεν). The former is a divine task, the latter is within the abilities of a human. See Yancy Hughes Dominick (2013: 563): "Images necessarily point beyond themselves ... toward the originals of which they are images. To recognize an image is to recognize something that falls short, something that calls its model to mind, though the image itself is not the model."

41 It is worth noting that this passage is full of text alluding to the participation in mysteries and rites, which I have skipped over for brevity's sake.

We see here that when a material body well represents the form in which it participates, it transports the well-ordered soul to the realm of the forms.

The inability of the philosopher to render an adequate account of the forms means that there is a need to constantly revisit the realm of the forms, to renew the vision and perfect their account. This is taken by some scholars to be the reason why the philosopher persists in the liminal state of the theoric journey – permanently uprooted from their community, and constantly seeking to perfect their understanding.[42] Moreover, although the philosopher does not stay in the intelligible realm of the forms, they do not necessarily re-root themselves in the visible realm either. The detachment which began the journey persists.[43]

Thus, we see that, not only is uprooting oneself necessary for the philosopher to make this upward journey – this *theôria* to the forms – but it also appears to account to some degree for the difficulties of the philosopher in providing an account of their journey upon their return to the visible realm. Indeed, even their journey abroad is not simple; their uprooting requires a kind of epistemic and moral purification. An additional effect of this purification (more than being the conditions of making the philosophical journey) is that the restructuring and reprioritization of the desires toward the good of the rational part of the soul seems to create a restlessness in the soul that cannot be satisfied by the (apparent) goods of the visible realm. This further induces a constant desire or drive to revisit the realm of the forms. Per the *Republic* and *Symposium*, this desire is somewhat hampered by a civic duty to rule. This period of ruling is, however, made less burdensome by the accounts they render of their experience of the forms, the "verbal *agalmata,*" which serve both as the basis for the laws they will maintain, and to remind them of the forms they hope to revisit. The philosopher must, therefore be contented with the reproductions until such time as they are able to return to their life of philosophical *theôria*.

42 See Nightingale (2021: 44): "Indeed, in Plato's texts, we confront an oscillation between the effable and the ineffable: the philosopher knows and gives an account of the essence of a Form, but he cannot put his encounter with divinity into language."

43 See Nightingale (2004: 106–107). We have almost entirely skirted the issues of whether the philosophical *theôros* will be welcomed back upon their return. The *Republic* (517a) suggests that the philosopher, having returned to the cave could be put to death should they try to convince others to make the ascent. See Rutherford (2013: 189–191, 338) and Ker (2000: 314).

Bibliography

Chantraine, Pierre. 1999. *Dictionnaire étymologique de la langue Grecque: Histoire des mots*. Paris: Klinsieck.

Chrysakopolou, Sylvana. 2012. "Wonder and the Beginning of Philosophy in Plato." In *Practices of Wonder: Cross-Disciplinary Perspectives*, edited by Sophia Vasalou. Eugene, OR: Pickwick Publications.

Cornford, Francis. 1932. "Mathematics and Dialectic in the *Republic* VI–VII." *Mind* 41 (162): 173–190.

Dominick, Yancy Hughes. 2013. "Images for the Sake of the Truth in Plato's 'Symposium'." *The Classical Quarterly* 63 (2): 558–566.

Elsner, Jás. 2000. "Between Mimesis and Divine Power: Visuality in the Greco-Roman World." In *Visuality Before and Beyond the Renaissance: Seeing as Others Saw*, edited by R. Nelson, 45–69. Cambridge: Cambridge University Press.

Gill, Mary Louise. 2012. *Philosophos: Plato's Missing Dialogue*. Oxford: Oxford University Press.

Herrero de Jáuregui, Miguel. 2015. "The Construction of Inner Religious Space in Wandering Religion of Classical Greece." *Numen* 62 (5/6): 596–626.

Ker, James. 2000. "Solon's 'Theoria' and the End of the City." *Classical Antiquity* 19 (2): 304–329.

Nightingale, Andrea Wilson. 2004. *Spectacles of Truth in Classical Greek Philosophy: Theoria in its Cultural Context*. Cambridge: Cambridge University Press.

Nightingale, Andrea Wilson. 2021. *Philosophy and Religion in Plato's Dialogues*. Cambridge: Cambridge University Press.

Plato. 1900–1910. *Platonis Opera*, edited by John Burnet. 5 vols. Oxford: Oxford University Press.

Plato. 1989. *Plato: The Collected Dialogues*, edited by Edith Hamilton and Huntington Cairns. Princeton: Princeton University Press.

Plato. 1997. *Plato Complete Works*, edited by John M. Cooper. Indianapolis: Hackett Publishing.

Renaut, Olivier. 2014. *Platon: La médiation des émotions: L'education du thymos dans les dialogues*. Paris: Librairie J. Vrin.

Rutherford, Ian. 2013. *State Pilgrims and Sacred Observers in Ancient Greece: A Study of* Theôriâ *and* Theôroi. Cambridge: Cambridge University Press.

Voelke, Pierre. 2014. "Entre communauté civique et communauté hellénique: Theoria l'hétaire dans la 'Paix' d'Aristophane." *Quaderni Urbinati di Cultura Classica* 106 (1): 135–152.

Ward, Julie K. 2022. *Searching for the Divine in Plato and Aristotle: Philosophical Theoria and Traditional Practice*. Cambridge: Cambridge University Press.

CHAPTER 14

Theôria in Islamic Platonism

Victoria Rowe Holbrook

1 Introduction

This chapter began as a paper read to a meeting in Ankara on "*Theôria* as Cognition in Plato," attended by scholars working in Greek philosophy. It is addressed to those with knowledge of Plato's work, but for whom things Islamic are new. I argue that a historical path for *theôria* as Plato understood it may be traced in the early 7th-century Quran, and set forth a context for recognizing its flowering in later literature and intellectual life expressed in Arabic, Persian and Turkish – the earliest languages of Islamicate[1] literatures, although the later are no less important – through the 18th century. While scholarly attention has often focused on the early Arabic, one can see how early concepts developed by looking back over those centuries from the perspective of trilingual Turkish intellectual history.

The first monograph to be written on the Islamic reception of Plato is a 2017 Yale dissertation by Geoffrey Mosely. 2017 is an astoundingly late date for such a first. To be sure, articles and prefaces to translations had been published, but no book. The reasons are many; the reception is fragmentary and complex, and the disciplinary ground filled with rabbit holes crisscrossing university career tracks. Mosely begins by saying that Franz Rosenthal's pioneering article, "On the Knowledge of Plato's Philosophy in the Islamic World," published in 1940, remains unsurpassed in its analysis although additional texts have been found since.[2] Despite the remarkable efforts of Rosenthal, Raymond Klibansky, Richard Walzer, Francesco Gabrieli, and Paul Kraus, frustrated by conditions in Germany during the second World War, Islamic Platonism has come to be explained away by generalizations of Neoplatonist influence,

1 "Islamicate" is a term coined by Marshall GS. Hodgson on the model of "Italianate" to refer to the broad range of activities carried out in the poly-ethnic, poly-lingual, poly-confessional Muslim-majority polities that flourished subsequent to the early community of Arabia. It can seem thick-headed to refer to all artefacts of polities ruled by Muslims as "Islamic." If pornography is written in Arabic, are we to call it Islamic? Furthermore, the major contributions of Jewish and Christian philosophers to intellectual life are obscured.

2 Moseley (2017: 9). The work includes a review of previous scholarship and collects almost all known extant fragments of Plato translated into Arabic.

 | HTTPS://DOI.ORG/10.1163/9789004747128-016

and obscured under rubrics of "mysticism," "perennial wisdom," and "Sufism," which found a home in sometimes ahistorical corners of Oriental, Near/ Middle East, or Religious Studies university departments.

The habit of referring to Islamic philosophy as Aristotelian derives from European Enlightenment grand narratives treating medieval Latin translators' selection of an early slice of philosophical texts in Arabic, concerned largely with the works of Aristotle,[3] as the whole of Islamic philosophy. If we want to understand the Platonism of Islamicate cultures, we will broaden our view to include earlier and later thought expressed in a variety of texts, first of all the holy Quran, which according to Muslim faith, was revealed by God through the angel Gabriel to the illiterate[4] caravan leader, Muhammed, from 610 to 632 C.E.

When tracing Plato's Islamic reception, it is customary to begin with what Dimitri Gutas named "the Graeco-Arabic Translation Movement," which took off in the mid-800s CE, two centuries after the Quran. The translation work centered in Baghdad, not only from Greek but Syriac, Persian, and Sanskrit, was patronized under Abbasid rule (750–950), which is to say in an increasingly Perso-Turkic cultural environment. The written past of Arabic was brief, while Persian has a history of writing as old as c. 600 BCE and provided models for the Arabic prose which flowered after 750 CE to become a literary language for peoples of varied mother tongue, ethnicity and religion.[5] As the great "gunpowder" empires developed, Persian and Turkish ("Turkic") largely took its place.

D. Gutas opened his 1998 study by saying that "almost all non-literary and non-historical secular Greek books [not lyric, not Homer, not Herodotus] that were available throughout the Eastern Byzantine Empire and the Near East were translated into Arabic." His student Moseley's dissertation begins there. But although Moseley collected many, and remarkably accurate, quotations in Arabic from Plato's works, as far as is known, no translation of a work of Plato

3 Although these are customarily referred to as translations, many are original works building upon Classical through Late Antique philosophy in Greek.

4 Muhammed is traditionally called "illiterate" (*ummī*) not as a disparagement but rather as pointing to one proof of the Quran's divine origin – because he was "illiterate," he could not have composed the Quran himself. The term *ummī* does not mean he could not read or write to the extent needed in his business, but that he was not a reader of literature.

5 During early Muslim conquests and administrations, Persian, Turkish, and other languages of the conquered territories were flooded with Arabic. A competition between cultural elements resulted in a certain Persian ascendency that took political form in 750 with the revolution of the Abbasids, who founded the city of Baghdad.

was done in those centuries or before, into Arabic or any other language.[6] In other words, the sources of the translated quotations are mostly unknown, and seem too exact, too true to Plato's prose as we know it today, to be paraphrase or quotation from epitome. It is a philological puzzle.

I find telling a remark Peter Adamson made in a 2019 article, where he observed that al-Kindi (c. 801–873), an early translator and the first person to be called an Islamic philosopher, writes of God as if He "functions like a Platonic form."[7] Many terminological habits of the Movement seem to spring forth fully armed from its start. Such habits do not arrive without a past; they develop through long preparation. Syriac played a role as a bridge translation language, but the Arabic translation terms had their own heritage. I am making the novel claim that the concept and practice of *theôria* as Plato used the term are evident in the Quran and Arabian cultural life prior to any written translation from Greek, and prepared the way for the distinctively Islamicate Platonism we see in Arabic, Persian and Turkish up to the modern era.

I said prior to any *written* translation. Although Arabic verse had a sophisticated oral heritage as early as the mid-400s CE, the Quran, revealed aurally to Muhammed, is the first Arabic book to be written down. When I say we can see preparation for Islamic Platonism in the Quran, I am not suggesting transmission by "influence," an 18th-century notion organizing anxieties characteristic of modern individuality in the context of an expanding print culture.[8] This was not the situation in Late Antique Arabia. I am not pointing to "sources" of the Quran, but to structures of thought that move through language by way of (oral) translation. I propose a Late Antique transmission of traces and tracks of structures of thought – themes, images, idioms, quotations, metaphors and other figures of speech – through language, which in ways we do not yet understand, carried some of Plato's philosophy into Arabic as spoken in Muhammed's lifetime. These constitute a Platonic inheritance, however mediated.[9]

---

6 With the possible exception of Armenian (I. Tinti [2022]), but not Syriac: "No literary or non-literary data permit the assumption that full translations of the genuine works of Plato ever existed in Syriac" (Arzhanov [2019: 6]).

7 P. Adamson (2020).

8 Clayton & Rothstein (1991).

9 For Arabian oral legends of Socrates, see I. Alon (1991). Much attention has been given to relationships between the Quran and Jewish and Christian scripture, but no connection with philosophy has been suggested. It is as if Islam could be related only to other religions. Philological reconstruction of paths of mediation for the Platonic inheritance will require collaboration between scholars specializing in at least Attic and koine Greek, Hebrew, Aramaic, Armenian and Syriac as well as the various Arabian languages (not to mention

Let me make clear what I mean by *theôria*. Most briefly, it is the spectating of higher realities by the faculty of intellect, and is often rendered as "contemplation." Andrea Nightingale has traced the development by which the ancient civic institution of *theôria* was appropriated in 4th-century BCE Athens for a new understanding of philosophy. Ancient civic *theôria* ambassadors traveled to oracles and religious festivals, and then returned to their cities to recount what they had witnessed and heard, transmitting what the god revealed. This was practiced by private individuals in a non-official way also. But in the 4th-century, *theôria* emerged as an intellectual activity by which the philosopher gazes "with the eye of the soul upon divine and eternal verities." For Plato, philosophers are intellectual ambassadors who view things as they really are, in the intelligible Forms, and return to teach "divine *theôria*" to others so that they may live out virtue in the civic realm.[10]

T. Bénatouïl and M. Bonazzi point out that scholarly emphasis on *theôria* in the works of Plato and Aristotle, and dismissal of Hellenistic "metaphysical dogmas about the *cosmos* and skepticism about knowledge of nature," have left the impression that the contemplative life has "no history outside the *Republic* and the last chapters of the *Nicomachean Ethics*."[11] My effort may be seen as an extension of their project to identify *theôria* were it is found. They recapitulate Nightingale's narrative as a list of four crucial elements for identification:

> First, an ethical justification of the superiority of knowledge over other human activities, and the ensuing advocacy of a life chiefly (but not necessarily exclusively) devoted to it. Second, a psychological and epistemological elucidation of contemplation, assigning it to a separable and immortal, faculty, νοῦς, and distinguishing it from other lower cognitive activities such as sensation, opinion, experience, practical reason, etc. Third, an ontological and cosmological definition of the entities contemplated as superior divine beings, which are both objects of knowledge and models to be imitated. Fourth, an implicit or explicit analogy between the intellectual activity defined by the first three aspects and the witnessing of a religious and cultural spectacle (θεωρία) or festival, which justifies the name θεωρία given to the philosopher's activity.[12]

Sanskrit); ancient through Late Antique Greek and pre-Islamic ("old" and "middle") Persian philosophy as well as religion.

10 This is an extremely brief summary of Nightingale (2001).

11 Bénatouïl & Bonazzi (2012: 1).

12 Bénatouïl & Bonazzi (2012: 4).

I take this brief summary as ground for my argument that Plato's understanding of *theôria* is expressed in Quranic Arabic and later Islamicate literature by the term *sayr*. *Sayr* is a verbal noun that literally means "traveling," but has the figurative meaning of spectating/contemplating higher realities. It becomes incorporated into the phrase *al-sayr al-sulūk*, which names a practical discipline and is the subject of a large and widely-known body of literature I am here defining as a literary genre. *Sulūk* means travel along a path with designated stops, as on a caravan route or progress in a profession. In the phrase *al-sayr al-sulūk*, *sulūk* is the adjective of *sayr*, thus *al-sayr al-sulūk* is the progressive spectating of higher realities.[13] The phrase is customarily rendered in English as "spiritual wayfaring," which apparently makes the meaning of Platonic *theôria* inherent in it difficult to detect.[14]

I define the *al-sayr al-sulūk* genre as a diverse body of Islamicate works that describe progressive, transformative cognition of higher realities, and refer to their own history as such works. By the 11th century, if not before, intellect is often identified with spirit. But it is not only that discussions of spirit may be discussions of intellect, and if we don't understand that, we miss their philosophical import.

In the Islamicate understanding of "existence" – more accurately, "being" (*wujūd*) – everything derives its being from God, and in that sense is nothing but God's being, although no single thing is being in itself.[15] Again, when I say "Islamicate," I am not referring to the small, early Muslim community and its debates, but to larger-scale cultural practices in much larger polities over the longest span of time. Being is "present" to us in countless degrees of intelligibility and perceptibility. The scientific achievements of medieval Islamicate

13 The phrase *al-sayr al-sulūk* underwent alterations as a loan-phrase in Persian and Turkish. In Arabic, usage of the definite article *al-* tells us whether the relationship between two words is adjectival or genitive. Persian and Turkish have no definite articles, and often drop the *al-* in loanwords and loan-phrases. The relationship has been interpreted as copulative in Persian *sayr va sulūk*, and in Turkish as *seyr u sülûk* (*va*/*u* meaning "and"). Or, the adjectival/genitive *izāfet* vowel is retained, but the distinction between the two becomes unclear: taken adjectivally, Perian *sayr-e sulūk* and Turkish *seyr-i sülûk* mean "the traveling spectating" or genitively, "the traveling of the spectating." Even in Arabic, it can be tricky to distinguish the adjectival from the genitive. Usage shows that *'ishq al-ilāhī*, "divine love," has been understood both as man's love of God and God's love of man.

14 If the noun *sayr* is taken literally rather than figuratively, and the adjective *sulūk* means "traveling a path," then the phrase *al-sayr al-sulūk* must mean, rather redundantly, "traveling the path traveling." The translation "spiritual wayfaring" conflates the two terms *sayr* and *sulūk* and adds the adjective "spiritual."

15 The "oneness of being" has been called "pantheism," but this is a philosophically naive choice.

societies may be understood in this context: an analysis of any single thing, not only "divine and eternal verities," but a chair or a tree, is an analysis of a thing that derives its being from God. In the contemplation of intelligibles, one has followed being back almost to its source.

"Almost" because intellect, for all its wonders, is ultimately limited. God's essence (*dhat*) is not an intelligible; it is not an "entity" (*'ayn*). *Logos* deals in language, and language is representation, but God in Himself cannot be represented. *Al-sayr al-sulūk* literature treats of the "journey" through intelligibles, and the highest degree of intelligibility in the cosmos is God's knowledge of Himself. His knowledge is expressed in His Names and Attributes, which correspond to Plato's Forms.[16] These are objects of *theôria*. Intellect goes only thus far; there must be a leap beyond.

Al-sayr al-sulūk literature is customarily dated post-Translation Movement, and al-Qushayri's (d. 1072) *Treatise*[17] cited as the first. But "firsts" of a genre are a matter of how one defines the genre. I prefer to leave that question open. Extant works include brief treatises (as opposed to Qushayri's some 400 pages) or chapters in longer texts, while some are short or very long verse narratives, and some are great works of art comparable to Dante's *Divina Commedia* or Milton's *Paradise Lost*. Although I leave overview of the literature to another time, I can mention a late 18th-century masterpiece in translation, the philosophical romance *Beauty and Love* by Galip.[18]

But let us begin with festivals.

2 Festivals

Frequent travel and pilgrimage were features of Muhammed's social geography, and the revelation of the Quran occurred against the background of a profound and ancient history of oracle in the Near East. Muhammed was a caravan leader, and the temple of the Kaaba in his native city, Mecca, displaying the gods of the various Arabian peoples, urban and nomad, was the major pilgrimage site in Arabia.[19] In the Quran, God speaks directly, and also tells Muhammad to "Say:.." The analogy with oracle was inescapable, and the

16 Chittick (1989: 84).

17 al-Qushayri (2007).

18 By far, most works in the *al-sayr al-sulūk* genre are not titled *al-Sayr al-Sulūk*. If they had been, we would have hundreds of works with the same title. But Galip uses the term *seyr* (Turkish spelling) in the sense of spectating higher realities dozens of times in his work. Galip (2005b: passim).

19 TH. Weir (2012).

Quran emphatically distinguishes Muhammed from oracles precisely because people thought he was an oracle… or a poet: "So remind [them]; you are not, by the grace of your Lord, an oracle"(52:29–30). "I swear by what you see and what you do not see, it [the Quran] is truly the speech of a noble messenger and not the speech of a poet. …And it is not the speech of an oracle. …It is a revelation from the Lord of the worlds" (69:38–43).

The oracle sites now most widely-known were to the north and south – the temples at Baalbek, Awām – while the topography and climate of central Arabia, where Mecca is located, are hard on built evidence, and extant written record of pre-Islamic oracle sites there consists of epigraphy and grafitti. When we look for comparison between civic *theôria* and Arabian religious practices, we find "religious festivals" in ancient Greece, and "fairs" in Late Antique Arabia. "Fair" translates "souq" in its familiar meaning of a city bazaar, but temporary souqs of the past, like the pan-Arabian 'Ukāz – set up near Mecca just before the annual pilgrimage there and to Arafat – were sites of religious ceremonies and oracle as well as trade and negotiation of agreements and treaties. Ukaz, which played a crucial role for the new religion of Islam, has been compared to "the national festival of Classical Greece, the πανήγυρις."[20]

Arabian cities and tribes had their own male and female *kāhin* oracles[21] *and* poets.[22] Ukaz was famed for its poetry competitions. Like Socrates, Muhammad worked in an environment of revered oral literature in verse. As in Greece, poets were thought to be divinely inspired. "Eloquence" (*faṣāhat*) had a powerful aura of incontrovertible truth. The economy of Arabian tribes included raiding,[23] and one of the poet's several duties was to engage in insult and curse duels in verse, considered effective, with a poet who represented the enemy. The Quran criticizes poets and poetry severely: "…poets, followed by those who go wrong, wander from valley to valley and say what they do not do" (26:226). The commentator al-Razi noted that poets were known for saying one thing one day and the opposite the next, to please a different audience,[24] like the Sophists Plato's Socrates condemned.

---

20 I. Shahid (2012).

21 Scholars differ in interpretation of distinctions and overlap between Greek oracles and Arabian *kāhins* (A. Fischer [2012], T. Fahd [2012]); the relevant parallel here is that in both cases, a new kind of *theôria* emerges from analogy with the old.

22 A. Fischer (2012).

23 Confined to specific months of the year; Ukaz was held during one of the "sacred" (warfare forbidden) months.

24 Quoted in Nasr et al (2015.). Fakhr al-Din al-Razi (1150–1210) wrote a renowned 32-volume commentary on the Quran, *Tafsīr al-Kabīr*.

The Quran challenges poets to compose something like it: "If you are in doubt about what We have revealed to Our servant, then bring forward a *sūra* [chapter] like it" (2:23). "This Quran is not a thing someone other than God could invent ... Do they say, 'he made it up'? Say: 'Bring a *sūra* like it'" (10:37–38). Muhammed recited the Quran at Ukaz,[25] and poets who came from all over to compete were unable to do so. Their failure to rival the eloquence of the Quran is taken as proof of its divine origin (the doctrine of *i'jāz*[26]).

The point to be taken here is the parallel between Ancient Greek civic *theôria* and these practices. In both cases, a new kind of *theôria* emerged from analogy with the old. Muhammed's recitation of the Quran at Ukaz broke with the old customs of oracle and inspiration of poets to inaugurate a new mode of *theôria*. It had been his habit to go outside his city to meditate in caves, and it was during one such meditation that the angel Gabriel appeared to him as a messenger conveying the speech of God. Gabriel ordered Muhammed to "Recite!" (96:1),[27] thus turning the mission of God's ambassador over to him, not as oracle but as builder of a virtuous society.

3 *Theôria* as *Sayr* in the Quran

I said that Plato's notion of *theôria* is expressed in Quranic Arabic by the term *sayr*. For definition of the term, I look first to E. W. Lane, who collated Classical Arabic dictionaries from the earliest in the 8th century through the 19th. According to these, and the many Quran commentaries Lane also consulted for his eight-volume lexicon, *sayr* means "moving along, traveling," and by extension, "conduct." But I also said that *sayr* has the figurative meaning of spectating/contemplating; it absorbs the meanings of another term, *naẓar*, which itself develops along a different track. In the Quran, we most often find *sayr* coupled with *naẓar*.[28]

---

25 I. Shahid (2012).

26 GE. Grunebaum (2012).

27 The order of the chapters in the Quran is by length; one should attribute no other significance to the order.

28 Almost all Arabic words have triliteral roots, which are thought of as "producing" verbal and corresponding nominal forms. When I say *sayr* or *naẓar*, I refer to the verb-noun group of their respective roots. One can use the online *Quranic Arabic Corpus* (K. Dukes, [2009–2017]) to find instances of roots in the Quran; forms of *sayr* occur 27 times. Twelve are verbs coupled with *naẓar*. Exhortative: "travel through the land and view..." (3:137.6, 6:11.2, 16.36.21, 27:69.2, 29:20.2, 30.42.2). Narrative: "they traveled through the land and viewed..." (12:109.13, 30:9.2, 35:44.2, 40:21.2, 40:82.2, 47:10.2). In four instances, *sayr* occurs

The Quran typically exhorts us to "*travel* [*sayr*] through the land and *view* [*naẓar*]…"[29] The travel is unspecified and the things we are exhorted to view are not sensible objects. The kind of spectating demanded involves contemplating the meanings of what we see, often theorizing from cause to effect: to give a simplest example, wrongdoing ends badly. Or, from effect to cause: there is such order in the cosmos that there must have been a Creator. *Naẓar* in this sense is an active kind of looking that includes spectating, contemplating, and theorizing; the abstract noun, *naẓariya*, which does not appear in the Quran, is used to mean "theory." But although *sayr* is most often coupled with *naẓar* in the Quran, we also find the coupling dropped and meanings of *naẓar* incorporated figuratively into *sayr*. Here is an instance very often cited, in which the heart is referred to as possessing reason, hearing, and "sight" (22:46):

> Have they not traveled [*sayr*] through the land, having hearts to reason and ears to hear? Indeed it is not the eyes that are blind, but blind are the hearts in the breasts.

Naẓar does not occur here, yet in order for the passage to make sense, spectating and reasoning must be implicit in *sayr*; if not, those who travel would not be chided for their "blindness." The trope of sight for cognition or knowing for seeing occurs hundreds of times in the Quran. The connection between heart and reason is compatible with Muhammed's Arabian Jewish and Christian milieu; heart and *nous* are linked in the Torah, Pentateuch and the New Testament.

But why do I choose sayr, rather than *naẓar*, as the key term for *theôria*? *Naẓar* also has the meanings "to wait," or "give respite," and occurs in the Quran contrasted with the verb-noun group "sight" (*ruʾyā*) in such phrases as "to see" (*ruʾyā*) while "looking on" (*naẓar*). Seven instances of *ruʾyā* are translated as "sights," "visions" or dreams: two are seen by Muhammed (17:60 and 48:27), and the rest by Joseph, the ruler of Egypt, and Abraham. There is the verb-noun group *baṣira*, used for insight: "they see [*ru'ya*], but they do not see [*baṣira*]" (7:198). The noun used for eyes in the passage above is a noun of *baṣira*. There are several terms one might choose out of context to render

as a verb without *naẓar* (22:46.2, 28:29.5, 34.18.13, 52.10.1). Six instances are nouns, words for travelers, caravans, journeys (5:96.8, 12:10.13, 12:19.2, 34:18.12), the stage or phase a thing is in (20:21.6), or movement (52.10.3). Five instances are causative verbs narrating how God controls people or inanimate things such as mountains, making them move (10:22.3; 12:109.13, 13.31.4; 18:47.2; 78:20:.1; 81:3.3).

29 Translations from the Quran are mine, with help from several others, especially AY Ali (1937).

theôria (*mushahada, tafakkur*), but despite all apparent indication to the contrary, including the fact that *naẓar*, not *sayr*, was among a handful of words used to render *theôria* two hundred years later during the Translation Movement,[30] it is the very large *al-sayr al-sulūk* genre that details transformative progression in cognition, and authors based their usages on the Quran's.

Naẓar acquired the meaning of an outward-directed gaze that affects the thing viewed rather than the viewer, such as the evil eye, or the powerful gaze of an enlightened person that can transform those it falls upon (or in rare cases, cripple or kill them). Persian usage and dictionary definitions of *sayr* as a loanword include "spectacle," "to go to see anything" (Steingass), and Turkish, "a looking on, seeing any spectacle or action" (Redhouse). Lane did not report any figurative meaning of "spectating" or "contemplating" among the definitions of *sayr* given by the works he consulted, but we observe in the above passage that it exists in Quranic usage.[31] In uncounted Arabic, Persian, and Turkish works written over more than a thousand years, it is *sayr* that refers to the contemplation of higher realities,[32] and we find many phrases such as *al-sayr fi'l-'asmâ*, "the *sayr* in the Divine Names,"[33] which as I said, correspond to Plato's Forms.

The above passage also mentions hearing, crucial to the aural revelation of the Quran. One of the accounts of Moses' experience of God as a burning bush or tree on Mount Sinai does also (28.29–30):

> …he was traveling [*sayr*] with his family when he sensed fire on the mountainside. He said to his family, "Stay here. I sense fire…" And when

30 In the online corpus *Glossarium Graeco-Arabicum*, one can enter a Greek term and see a list of the Arabic words used to render it in each work translated during the Translation Movement, and vice-versa. For *theôria*, see https://glossga.bbaw.de/results.php?gr_lexeme=θεωρία, retrieved 22.04.2023.

31 I observe that Arabic dictionaries can be distinctively conservative with regard to figurative usage, and Persian and Turkish dictionary definitions of Arabic loanwords can, in some cases, be better guides to the Arabic usage. The first Arabic language dictionary, *Kitab al-'Ayn*, dated to the 700s, one of the first dictionaries in any language, was not based on usage. Instead, the author theorized all possible words that could be produced from Arabic roots, and extracted the language presented as actual from the possible language produced in this way.

32 For example, in the influential works of Abu Hamid al-Ghazali (d. 1111; see Alexander Trieger [2012: 54]), in narratives such as Ruzbihan Baqli's (1128–1209) Persian *Sayr al-Arwah* ("The Spectating of the Spirits/Intellects", see K. Murata [2018]), the polymath Nasir al-Din Tusi's 1246 Persian autobiography, *Sayr va Sulūk* published in English translation as *Contemplation and Action* by SJ. Badakhchani (1999) and Mahmud Hüdayi's (1541–1628) Turkish *Seyr u Sülûk*.

33 See M. H. Ali (2022: 68).

> he came to it... a voice called from the tree, "O Moses! Truly I am God, Lord of the worlds!"

The verb for "he sensed," *'anisa,*[34] used rarely in the Quran, has the meaning of being familiar with, recognizing, and shares a root with the noun *insān*, "human being." Moses senses a fire nearby, and goes to find it. When he comes face-to-face with it, a burning tree speaks to him. He sees, but he does not see what he sees – he sees a burning tree, but what he is really seeing is God self-disclosing in the form of fire. In another account, Moses is also away from home and God is speaking to him. Moses says, "Show me, that I may look upon You." God replies: "You cannot see Me, but look to the mountain; if it remains in place, you will see Me." God's self-disclosure to the mountain reduces it to dust (7.143).

These passages are some of the most-interpreted in the Quran, and obviously have precedents, not only in scripture. Here, what is relevant is that reality must be "seen" without eyes. The notion that God has no sensible form is familiar to us now, but it was inconceivable to many in the past. Augustine wrote that he had great difficulty conceiving of God as incorporeal until he read "the Platonists."[35]

4 Soul and Spirit

The heart is possessed of reason, but the soul (*nafs*) also thinks and knows: "Do they not reflect (*yatafakkaru*) in their souls? (30:8); "No soul knows (*ta'lamu*) what delights are kept in store for it" (32.17); "Each soul will know (*'alimat*) what it has prepared" for itself when it arrives in the Hereafter (81:14, 82:5). The psychology we can assemble from the Quran mentions the soul in connection with perception, desire, emotion, feeling, patience, faith, thought, knowledge, and judgment. Sight is not mentioned specifically, but as I mentioned, sight/knowledge is a common trope.

As for Plato in *Republic* and *Phaedrus*, the soul is tripartite. It is worth stressing that the tripartite soul, from Plato to the Quran to Freud, is distinctive; there is no tripartite soul in Jewish, Christian, Vedic, Buddhist, or Taoist

34 In another rendition of the incident, Moses "saw" fire (20:10).

35 Augustine (1984: 133): "Although I did not imagine you in the shape of a human body, I could not free myself from the thought that you were some kind of bodily substance extended in space..." and 154: "By reading these books of the Platonists I had been prompted to look for truth as something incorporeal, and *I caught sight of your invisible nature...*" (referring to NT, Romans 1:20, incorrectly cited in this edition).

thought. As in Plato, the Quranic soul struggles within itself to progress from desire for sensible to intelligible loves. Two centuries later during the Translation Movement, *nafs* became the most common rendering for *psuché*.[36] *Nafs* then acquires an Aristotelian terminology in Islamicate texts.[37]

As for Plato, the distinction between the visible and the invisible organizes both epistemology and cosmology. The Quranic cosmos has two major "worlds": the world present to the senses, and the "Unseen" (*al-ghayb*), which is the object of *theôria*. The Unseen is ontologically prior to the sensible; all things "descend" from the Unseen through ontologically defined realms. The soul has often been characterized as having one foot in the Unseen and the other in the sensible realm. It is a bridge between the two; as later authors emphasized, the soul is a *barzakh*. A *barzakh* is, literally, an "isthmus," a thin strip of land connecting two land masses. In the Quran the term occurs once as a location after death that we cannot cross until Resurrection Day (23:100). The term occurs twice more as an invisible partition between two bodies of water, one salty, one sweet (25:53, 55:20). The customary interpretation is that the salty sea represents the sensible realm, and the sweet, the Unseen. Because our souls are *barzakhs* between the two, we access both by way of our souls. In later Islamicate literature, the "Imaginal Realm," populated by incorporeal images that are objects of *theôria*, is called a barzakh.

In some Quranic instances one may understand *nafs* as "a life," in the sense Achilles means when he says that he risks his *psuché* in battle (*Iliad* 9.322),[38] or as in the old-fashioned English usage referring to souls lost on a ship that sank. *Nafs* is also the reflexive pronoun, "self." Often, one can distinguish between noun and pronoun only by interpretation. If the action mentioned pertains to perception, thought, and the rest, or occurs after death, I take *nafs* to mean soul. However one defines "self," the self of self-deception is in a class different from the reflexive "He dressed himself." I take *nafs* to mean soul in the many instances customarily translated as "they do wrong to themselves" or "do good to themselves"; for in these cases, it is always at least implied if not outright asserted that what one does to the *nafs* leads either to Paradise or hellfire.

Although the three parts of the soul are specifically named only once each in the Quran, they have been the subject of extensive commentary and elaboration. In the Quran, they are referred to, not introduced or explained, and

---

36 *Glossarium Graeco-Arabicum*, https://glossga.bbaw.de/results.php?gr_lexeme=ψυχη&ar_lexeme=&ar_root_1=&ar_root_2=&ar_root_3=&ar_root_4=&ar_root_5=&submit-button=, retrieved 08/01/2024.

37 For an influential example, see Naṣīr al-Dīn Ṭūsī's Persian manual of ethics, trans. GM. Wickens (1964).

38 H. Lorenz (2009).

whether or not one or more of the three is intended in the very many unnamed references to soul is a matter of interpretation. We remember that the Quran is not something written at a desk, but revealed aurally to a man who did not write. It was revealed over more than two decades, in pieces that often respond to events during years when the early Muslim community was subject to ostracization in a society where affiliation with tribe was a matter of life or death. Muhammed delivered it as a verbal address, often to multiple people in dire circumstances, and it no doubt assumes knowledge common at the time, but not known to us. While there are narrative, and "poetic" passages – in prose often rhymed, not poetry in established forms of verse, but highly compressed – many mentions of soul refer to persecution, betrayal, war, and the needs of a newly founded society governed by new customs and laws detailed in the Quran's more systematic passages.

The three parts are, with their customary translations: 1. "the commanding soul" (*al-nafs al-ammara*, 12:53:6); 2. "the blaming soul" (*al-nafs al-lawwama*, 75:2); and 3. "the tranquil soul" (*al-nafs al-muṭma'inna*, 89:27:3).[39] Only in the case of the commanding soul does context provide much basis for interpretation. From the story of Joseph in the Quran, we understand that the commanding soul is what moves us to do things that seem to happen of themselves, as when powerful attraction moves us to have sex with a person we should not, and we say, "It just happened." Joseph repels the attempt of Potiphar's wife to seduce him, but says that he does not absolve his soul, for it "commands to evil." In other words, Joseph admits that his soul moved him to desire her, but he did not gratify his desire.

As we can all attest, most people live their lives batted back and forth by desire and regret; the commanding soul carries us away, and later we wish we had not obeyed it. The blaming soul's single mention is in connection with Resurrection Day, severe as a stern superego. The tranquil soul is welcomed into Paradise: "O tranquil soul! Return to your Lord, pleased and pleasing."

As when reading Plato's *Republic*, we may wonder if there are three different souls, or one soul with three parts or states. But we recognize Plato's appetitive soul (*epithumetikon*) in the commanding soul, led around by its desires. As "the ally of reason" (*Rep.* 440b), the blaming soul may correspond to the Platonic part translated as "spirited," the *thumoeides*. But is the tranquil soul the *logistikon*?

39 Later in Islamicate intellectual history, further levels to the soul are named. But these designate states of the soul according a schema of advanced development, rather than describing a given psychological structure.

The translation choice "spirited" for *thumoeides* can be confusing, especially when we are considering the reception of Plato in monotheistic cultures where spirit is considered divine. The Quran distinguishes soul/*nafs* from spirit/*rūḥ*,[40] although the two are sometimes used interchangeably in later literature. Let us hear from Lane: "Some of the lexicologists assert the *nafs* and the *rūḥ* to be one and the same… others say, that the latter is that whereby is life; and the former, *that whereby is intellect, or reason*" [emphasis in the original]. *Nafs* is used as a reflexive pronoun for God also (and to that extent God has an at least grammatical self), but as Lane also remarked, God breathed of His *rūḥ* into Adam, not of His *nafs*:[41] "I have proportioned him [Adam] and breathed into him of My spirit…" (15:29, 32:9, 38:72).

Although both *nafs* and *rūḥ* have root meanings of breath, during the Translation Movement, *rūḥ* was the translation of *pneuma*, not *psuché*,[42] which as we have seen, was rendered as *nafs*. The *nafsu al-ʻaql* ("the soul of the intellect=intellect soul") and *al-nafsu al-nāṭiqa* ("the speaking soul," i.e. *logistikon*), which Lane also mentioned,[43] are later terms that do not occur in the Quran, but we see that intellect has been understood as a part of the soul. However, it has also been identified with spirit. Inconsistency is due partly to successive eras of terminological expansion; new terms absorbed from other languages or coined through translation[44] are more or less precisely integrated into older paradigms.

Spirit/*rūḥ* is most often understood as non-composite, and soul/*nafs* as arising from the combination of spirit and body. Spirit does not have parts. In the Quran, spirit belongs to God and is at His disposal, while soul is proper to His creatures (except the self-reflexive pronoun as used for God). God "supports" with His spirit those he chooses, prominently Mary and Jesus. In their case, spirit is the "Sacred Spirit" (*rūḥ al-qudus*, 2:87, 2:253, 16:102), commonly identified with the angel Gabriel, who in a reference to his "bringing down" the Quran (16:102) is called "the Trusted Spirit" (*al-rūḥ al-ʻamīn* 26:193). There are also spirits, plural, although of human beings only Jesus is called "a spirit from Him" (4:171). The Quran may seem to deliberately keep spirit

40 A. Arslan's (2006) Turkish translation *irade* ("will") may be a better approximation for the *thumoeides*, but the choice of *rūḥ* for *psuché*, widespread in modern Turkish usage, is problematic also, as we shall see.

41 Lane (1968: vol 8, 2827).

42 *Glossarium Graeco-Arabicum*, https://glossga.bbaw.de/results.php?gr_lexeme=πνεῦμα&ar_lexeme=&ar_root_1=&ar_root_2=&ar_root_3=&ar_root_4=&ar_root_5=&submit-button=, retrieved 17.01.2024.

43 Lane, (1968: vol 8, 2827).

44 i.e. Persian *jān*, or *al-nafsu al-nāṭiqa*, likely coined to render λόγος in Aristotle.

undefined: "They ask you, [O Muhammad], about the spirit. Say, 'The spirit is from the command of my Lord'" (17:85). But "the command" (*'amr*) in this sense has been interpreted as one name for an ontologically defined realm of spirits,[45] and "spirits" have been interpreted as synonymous with "intellects" (*'uqūl*, sing. *'aql*) as well as "angels."[46]

The messenger function of angels may have an ancestor in one of the activities of Greek *daimones*. But spirits, intellects, and angels are grouped together as non-composite, intelligible, created entities. In Islamicate cosmology, they populate the realm more often called *Jabarūt*, which in some ways may be thought of as a realm of abstractions recalling *dianoia*.[47] We find them, like the incorporeal but composite images in the Imaginal Realm, as objects of *theôria* in *al-sayr al-sulūk* literature. Bearing in mind Bénatouïl & Bonazzi's requirements, we can say that the entities in *Jabarūt* and the Imaginal Realm are ontologically and cosmologically defined, superior objects of knowledge. They may in some ways be models to be imitated, and are often imprecisely described as "divine" because of their relative nearness to God, but they are not gods. The status of intellect varies in later works; it may be separable and immortal, or accorded a variously defined unique status.

It seems to be the hasty commanding soul that can "whisper" (50:16; also said of Satan) and entice (12:18, 12:83, 20:96). One can conceal what is in one's soul (2:235, 2:284, 3:154, 5:52, 33:37), lie about one's soul (6:24), and transgress against it (39:53). The soul may be confused and deceived (2:9, 2:187, 4:107); may desire or not desire (2:87, 5:70, 21:102, 41:31, 43:71, 53:23), be envious (2:109), misleading (3:69, 4:113), arrogant and insolent (2:87 25:21), stingy and greedy (4:128, 64:16), and generally do wrong. Cain's *nafs* "permitted" him to murder his brother (5:30).

How did Joseph resist the advances of Potiphar's wife? She tells her friends that she tried "to lure him away from his soul, but he held back" (12:32). "Holding back" seems to be the role of the blaming soul. The soul must be made "patient" (18:27), and "restrained from caprice" (79:40). The soul can feel anxiety and fear (20:67), be terribly constricted (9:118), and be guided (32:13). It can bear witness against itself, in this world and on Judgment Day (4:135, 6:130, 7:37, 9:17, 75:14), so we know that it can assign blame in the Hereafter as well.

Judgment is attributed to the soul. Joseph discriminates between alternative courses of action and makes the wiser choice. The soul can give up some things willingly (4:4), can "wait" (2:28), and have faith (10:100). This seems to

45 WC. Chittick (1998: 276).
46 WC. Chittick (1982: 113).
47 VR. Holbrook (forthcoming 2025a).

be the role of the tranquil soul, to handle the other two parts, like the charioteer *logistikon* in Plato's *Phaedrus* who drives his two horses. The master metaphor of the soul in *Republic* is the city, and there may be an echo in the Quran's use of the adjective "tranquil" only twice; once to describe a soul, and once to describe a city (16:112).

But does the Quran not say anything more specific about intellect?

5 Intellect and Heart

Let us return to the passage where reason is attributed to the heart (22:46.2):

> Have they not traveled through the land, having hearts to reason and ears to hear? Indeed it is not the eyes that are blind, but the hearts in breasts are blind.

Here, "to reason" (ya'qilūna) translates a verb sharing the root of the noun "intellect" ('aql), and literally means "to intellect." The noun would be used during the Translation Movement to render both logos and nous,[48] but does not occur in the Quran. Those who fail "to intellect" are said to be dumb (2.171, 8.22) as well as blind and deaf, associating speech as well as sight and hearing with intellection. The verb occurs in Quranic phrases so numerous as to be refrains, such as, "Will you not (use) intellect?" and, "This is an example for those who (use) intellect." Those in Hell are said to lament, "Had we listened or used intellect, we would not now be companions of the blaze" (67:10). Interestingly, in every instance, intellection guides aright.

Is the role assigned to intellect purely ethical – in the practical sense (if how to avoid Hell is practical knowledge) – as opposed to theoretical? It is by intellection that we interpret the signs of God: "God makes clear to you His signs that you might use intellect" (2:73, 2:242, 57:17). What are customarily called "verses" of the Quran are in fact termed "signs" (*'ayāt*), and the things God has created are also His signs, pointing to their Unseen meanings via "sight." In later literature, "meaning" (*ma'nā*), quiddity" (*māhiyya*), and "reality" (*ḥaqīqa*) are terms used interchangeably for "the fixed (or immutable) entities" (a*'yān thābita*) which are, like the Names and Attributes, present at the highest degree of intelligibility in the cosmos, the realm of God's know-

48 Glossarium Graeco-Arabicum,https://glossga.bbaw.de/results.php?gr_lexeme=&ar_lexeme=ψυχη&ar_root_1=&ar_root_2=&ar_root_3=&ar_root_4=&ar_root_5=&submit-button=, retrieved 17.01.2024.

ledge.[49] All are objects of a *theôria* that is superior to other activities,[50] and the principle that God is loved for His own sake, not out of desire for Paradise or fear of Hell, is established early on.[51]

Still, in the Quran, only the heart is specifically said "to intellect." There are two verb-noun groups for "heart" in the Quran: *qalb* and *fū'ād*. *Qalb* often occurs as the verb "to turn" or "return." While the soul has contesting parts, the heart is changeable, unstable, turning this way and that.[52] In one instance, *qalb* refers to the physical organ (33:4), but *fū'ād*, often translated as "mind," does not. The less frequently mentioned *fū'ād* is characterized by hearing and insight (*baṣira*).

Qalb: God may "intervene between a man and his heart" (8.24); put emotions or inclinations into hearts – regret (3:156), and when the offense is severe, "terror" (3:151, 8:12, 33:26, 59:2), but also "compassion and mercy" (57:27). He may "test hearts for righteousness" (49:3) make hearts "hard" (5:13, 10:88,), or purify (3:154), and guide them (64:11). Hearts may be "diseased" (2:10, 5:52, 33:12, 33:32, 33:60, 57:16, 47:29), "locked" (47:24), harbor "hypocrisy" (9:77, 47:20, 74:31), or be "reassured" (3:126, 5.113) or "sound" (26:89, 37:84). Many instances stress intent; God is repeatedly said to know what is in people's hearts. Unlike soul, the heart is not characterized by perception, but as we know, may be blind. *Ayn*, "eye," can be a metaphor meaning "eye of the soul/heart," as in, "you will see it with the eye of certainty" (102:7). This is very common in later literature, as are the terms *sirr* and *ḍamīr*, meaning "inner heart" or "consciousness," but these usages do not occur in the Quran.

God may "cover" or "set a seal" upon (*ṭaba'a, khatama*) hearts (7:100, 7:101, 9:87, 9.93, 10:74, 42:24) and their "hearing, and a cover over their vision" (2:7, 16:108; 30:59, 40:35, 45:23; 47:16, 63:3) – the familiar connection between heart, hearing, and sight. A "cover" (usually *ghashwa, 'akinna*) is customarily translated as "veil," but this can be misleading. In English, a veil is an article of clothing worn by women. In Arabic, "covered" is one end of the continuum closed/open. In Muhammed's climate, people covered their faces against sandstorms, and there are Arabic, Persian, and Turkish idioms such as "to cover the door," meaning to shut it, or "to cover the lamp," meaning to snuff out a candle or oil wick. In a reference to merchants who cheat customers, the Quran likens the heart to a surface that can rust: "Their profits are rust upon their hearts"

49 Chittick (1982: 113).

50 S. Murata quotes an arch remark al-Ghazali made in this vein: "The food and felicity of the beasts lie in eating, sleeping, and mating. If you are a beast, strive day and night to keep the business of your stomach and private parts in order" (1992: 231).

51 Cornell (2019: 145 et passim).

52 For a discussion of the heart as "turning," see S Murata (1992: 289–320).

(83:14). One of the most common tropes in Islamicate literature is heart as mirror; if it is to reflect higher realities, the heart must be "polished" until is it free of "rust." A large body of writing instructing the reader how to purify the soul (the *tazkiyat al-nafs* genre; in today's Turkish, *nefis tezkiyesi*) arose based on Quran 91:7–10.4: "By the soul, and the proportion He gave it. And inspired it as to its wrong and its right. Indeed, he who purifies it succeeds, and he who buries it fails." "Buries" (*dassā*, from the chapter, "The Sun"), also translated as "veiled," has the sense of obscuring a bright thing, as the sun is obscured on a cloudy day. Another word used is *ghulf*, a covering like the sheath for a sword or quiver for arrows (2:88, 4:155, where it likely means "full," unable/unwilling to contain more).

Much of what is said of the heart is said of the breast (*ṣudūr*) as well. But expansion and contraction is specific to breast (6:125, 94:1). Connection between breast and thought and knowledge emerges from such passages as the following, in answer to those who say resurrection is impossible: "Whether you be stone or iron, or a creature of what in your breasts [thoughts] is hardest [to resurrect]…" (17:50–51); and "The Quran is clear signs within the breasts of those who have been given knowledge (29:49).

Given that soul and heart share faculties, it is logical to suppose they are contiguous. My studies leave me with the impression that the Quran places intellect within the soul, and locates the soul within the heart, so that the clarity of intellect depends upon the clarity of the soul, which depends upon the clarity of the heart. Spirit is pure. Again, for hearts and breasts, as for soul and intellection, the emphasis is upon ethics that prepare one to see higher realities.

6 The Mi'raj as *Theôria* and Imagination

Muhammed's Mi'raj is an exemplar of *theôria* and served as the paradigm for *al-sayr al-sulūk* literature.[53] Referred to in the Quran and expounded in hadith,[54] "Mir'aj" is customarily translated as "Muhammed's ascension to God." His descent was just as important. As Plato's philosopher returns to the cave, Muhammed came back down to relate to his city what he had seen and apply it to the building of his community, and to the first unification of Arabia, under his rule.

---

53 SJ. McMichael (2011: 295).

54 See FS. Colby (2008). Hadith is an extensive literature of reports, verified and classified according to reliability, narrating what Muhammed and his companions said and did.

Quranic cosmography has seven heavens, as well as seven earths, one on top of the other in layers (not spherically, not yet Galenic).[55] The angel Gabriel traveled down the seven heavens as an ambassador from God to Muhammed, and Muhammed traveled up the seven heavens as an ambassador from humankind to God. Gabriel was unable to accompany him all the way, and in this context is often identified with intellect. Intellect can go only so far; there must be the leap beyond.

The Quran was delivered as an oral recitation, as ancient Greek oracle and wisdom was. But Muhammed not only hears Gabriel, he *sees* him. On his journey back down through the heavens, Muhammed also sees human beings long deceased, not ghosts, but visible, and they speak with him. This moved philosophers to investigate imagery in distinctively new ways. How did persons long dead, and Gabriel – a non-corporeal being – appear to Muhammed as if in corruptible bodily form?

The philosophers al-Kindi's and al-Farabi investigated prophecy and resurrection as the spectating of higher realities. In what sense did Muhammed meet God? Are the forms of our resurrected bodies imaginal, without sensible body? Is the pleasure and pain of Paradise and Hell something felt by the soul only? Their work was taken further with innovations in Aristotelian psychology that Avicenna elaborated, after the Brethren of Purity and al-Farabi, introducing five internal senses beyond the five external senses,[56] making the relationship between imagination and theoretical intellect explicit and systematic.

The Islamicate cosmographical schema "the Degrees of Being" (Ar. *maratib al-wujûd*, Tr. *varlık dereceleri*) appears to be a modification of Plato's Divided Line.[57] It is also a map for the individual *theôria* journey. The schema does not treat images as copies of sensible things, but as incorporeal forms of higher realities. *Eikasia*, at the bottom of Plato's Line, is dropped, and the Imaginal Realm is inserted in between *pistis* and *dianoia*. The possibility that Near Eastern Christian iconography is relevant to this reversal of the relationship between sensibles and images remains to be investigated.

In this context, imagination is not "making things up." The order is ontological: all things begin as entities in God's knowledge, then acquire non-

55 AT. Karamustafa (1987: 71–72). It is important to mention that a genre of Mi'raj narrative literature emerged – called Mi'rājnāma and, as in Turkish, the Miraciye preface to a longer work – which in early versions includes a katabasis down through the seven layers of earth; Asin Palacios's assertion that this genre was a source for Dante is well known to specialists of Italian literature (RP. Buckley [2013: 249–252]).

56 HA. Wolfson (1935).

57 VR. Holbrook (forthcoming 2025a).

composite form (spirits, intellects, angels), then composite but incorporeal form (images), and finally bodily form. All forms simultaneously abide in their respective realms, accessible. After death, we return the way we came. But the soul may also visit the Imaginal Realm while the body is alive, and go further, as Muhammed did on the Mi'raj; the deceased persons he saw on his way down were present to him in their imaginal forms. Every image we see in sleep or *theôria* is independent of our minds, although our apperception of it varies in clarity and truth according to the state of our hearts. Such journeys are narrated in the relatively well-known works of the philosopher Suhrawardi (1154–1191), known as "the son of Plato," and Ibn Arabi (1165–1240), Mevlana Rumi (1207–1273), Yunus Emre (1238–1328), and many, many others.

This brings us to the golden age of *al-sayr al-sulūk* literature. The Degrees of Being schema became a model by which older, relatively naturalistic love tales were rewritten as allegories of *theôria*, while old and new genres of philosophical writing more strictly defined continued on. A later example available in translation is the work of Mulla Sadra, a 16th-17th-century Persian philosopher of imagination who is now so popular among film and artificial intelligence theorists that Slavoj Zizek talks about him on Youtube.[58]

7 Conclusion

I hope I have convinced you that the four elements I have reviewed display a Platonic heritage – the analogy between the oracle/poetry pilgrimages of Muhammed's Arabia and the direct apprehension of higher realities; the Quran's ethical justification of this with the example of *sayr* contemplation, the ontology and epistemology of its tripartite soul and Muhammed's Mi'raj. These elements were combined with later extensive development of the psychology of *theôria* in *al-sayr al-sulūk* practice and literature throughout Islamicate cultures that fully articulated the role of the faculty of intellect. I think the Arabic terminology of the Translation Movement must have relied upon and fit with these earlier elements. Corroborating evidence in the hadith corpus remains to be picked out. All of these matters need further research and book-length treatment, but I think we can already see that traces of Platonic *theôria* as cognition can be found in the Arabic of the Quran and flourished in the *al-sayr al-sulūk* genre.

The literature of Islamicate Platonism is as large, perhaps larger, than the Latin/European, and very different. The qualities of *to kalon* as understood by

58 https://www.youtube.com/watch?v=MD-Kx1WUU1Q, 14.07.2024.

Plato were separated into the aesthetic and the moral in Latin and onwards, yet remained united in Arabic, Persian and Turkish usage. What are the consequences? There is much at stake. I also find it likely that German Romantics, who acquired a knowledge of Islamicate literatures more diverse than that of Latin translators and writers, re-introduced some of Islamicate Platonism – although neither did they recognize it as such – to Enlightenment Europe and, also through French Orientalists, to 20th-century continental philosophy.[59]

Bibliography

Primary Sources

Ali, AY., trans. 1937. *The Holy Qur-ān: Arabic Text with English Translation and Commentary*. Lahore.
Asad, M., trans. and comm. 1980. *The Message of the Qur'ān*. London.
Augustine, trans. RS. Pine-Coffin. 1984. *Confessions*. Hammondsworth.
Galip, Şeyh. 2005a. Trans. VR. Holbrook, *Beauty and Love*. New York.
Galip, Şeyh. 2005b. Ed. VR. Holbrook, *Hüsn ü Aşk*. New York.
Hüdayi, M. 2005. *Seyr u Sülûk*. Istanbul.
Nasr, SH., et al., eds. and trans. 2015. *The Study Quran*. San Francisco.
al-Qushayri, Abu'l-Qasim. Trans. AD. Knysh. 2007. *Treatise on Sufism*. Reading.
Ṭūsī, Naṣīr al-Dīn. Trans. GM. Wickens. 1964. *The Nasirean Ethics*. London.
Ṭūsī, Naṣīr al-Dīn. Trans. SJ. Badakhchani. 1999. *Contemplation and Action: The Spiritual Autobiography of a Muslim Scholar*. London.

Other Works Cited

Adamson, P. 2020. "al-Kindi," in *The Stanford Encyclopedia of Philosophy*, https://plato.stanford.edu/archives/win2019/entries/al-kindi/, retrieved January 15th 2024.
Ali, MH. 2022. *Philosophical Sufism: An Introduction to the School of Ibn al-Arabi*. London.
Alon, I. 1991. *Socrates in Medieval Arabic Literature*. Leiden.
Arslan, A. 2006. *İlkçağ Felsefe Tarihi 2: Sofistlerden Platon'a*. Istanbul.
Arzhanov, Y. 2019. "Plato in Syriac Literature," *Le Muséon* 132 (1–2): 1–36.
Bénatouïl, T., and Bonazzi, M., eds. 2012. *Theôria, Praxis and the Contemplative Life in Plato and Aristotle*. Leiden.
Buckley, RP. 2013. *The Night Journey and Ascension in Islam: The Reception of Religious Narrative in Sunni, Shia and Western Culture*. London.

59 VR. Holbrook 2025.

Chittick, WC. 1982. "The Five Divine Presences: From al-Qūnawī to al-Qayṣarī." *Muslim World* 72 (2): 107–128.

Chittick, WC. 1989. *The Sufi Path of Knowledge: Ibn al-'Arabī's Metaphysics of Imagination*. Albany.

Chittick, WC. 1998. *The Self-Disclosure of God: Principles of Ibn al-'Arabī's Cosmology*. Albany.

Colby, FS. 2008. *Narrating Muhammad's Night Journey*. Albany.

Clayton, J., and Rothstein, E. 1991. "Figures in the Corpus: Theories of Influence and Intertextuality." In *Influence and Intertextuality in Literary History*, edited by J. Clayton and E. Rothstein, 3–36. Madison.

Cornell, RE. 2019. *Rabi'a From Narrative to Myth*. London.

Dukes, Kais. 2009–2017. *The Qur'anic Arabic Corpus*. https://corpus.quran.com, retrieved January 31st 2024.

Fahd, T. 2012. "Kāhin." In *Encyclopaedia of Islam, Second Edition*, edited by P. Bearman et al., http://dx.doi.org/10.1163/1573-3912_islam_SIM_3784, retrieved February 10th 2024.

Fischer, A. 2012. "Kāhin." In *Encyclopaedia of Islam, First Edition (1913–1936)*, edited by M. Th. Houtsma et al., http://dx.doi.org/10.1163/2214-871X_ei1_SIM_3801, retrieved February 10th 2024.

Glossarium Graeco-Arabicum. 1994–. Ruhr-Universität Bochum, Seminar für Orientalistik und Islamwissenschaft, https://glossga.bbaw.de, retrieved February 15th 2024.

Grunebaum, GE. von. 2012. "Iʿdjāz." In *Encyclopaedia of Islam, Second Edition*, edited by P. Bearman et al., http://dx.doi.org/10.1163/1573-3912_islam_SIM_3484, retrieved January 19th 2024.

Gutas, D. 1998. *Greek Thought, Arabic Culture: The Graeco-Arabic Translation Movement in Baghdad and Early 'Abbasid Society (2nd–4th/8th–10th centuries)*. London.

Holbrook, VR. Forthcoming 2026. "Divided Line and Degrees of Being: Plato and Islamicate Cosmology." In *The Reception of Plato From Late Antiquity to the Middle Ages*, edited by Eva Anagnostou-Laoutides et al. Berlin.

Holbrook, VR. 2025. "The Separation of Goodness and Beauty: Plato, Galip, Lacan." In *Love in Early Modern Ottoman Poetry*, edited by Christiane Czygan. Berlin.

Karamustafa, AT. 1987. "Cosmographical Diagrams." In *The History of Cartography 2.1, Cartography in the Traditional Islamic and South Asian Societies*, edited by JB. Harley and D. Woodward, 71–89.

Lane, EW. 1968. *An Arabic-English Lexicon in Eight Parts*. Beirut.

Lorenz, H. 2009. "Ancient Theories of the Soul." In *The Stanford Encyclopedia of Philosophy*, https://plato.stanford.edu/entries/ancient-soul, retrieved February 1st 2024.

McMichael, SJ. 2011. "The Night Journey (al-isrāʾ) and Ascent (al-miʿrāj) of Muhammad in Medieval Muslim and Christian Perspectives." *Islam and Christian–Muslim Relations* 22 (3): 293–309.

Moseley, GJ. 2017. *Plato Arabus: On the Arabic Transmission of Plato's Dialogues. Texts and Studies.* Unpublished PhD thesis, Yale University.

Murata, K. 2018. *Beauty in Sufism: The Teachings of Rūzbihān Baqlī.* Albany.

Murata, S. 1992. *The Tao of Islam: A Sourcebook on Gender Relationships in Islamic Thought.* Albany.

Nightingale, AW. 2001. "On Wandering and Wondering: '*Theôria*' in Greek Philosophy and Culture." *Arion: A Journal of Humanities and the Classics*, 3rd Series, Vol 9, No 2 (Fall): 23–58.

Shahîd, I. 2012. "ʿUkāẓ." In *Encyclopaedia of Islam, Second Edition*, edited by P. Bearman et al., http://dx.doi.org/10.1163/1573-3912_islam_SIM_7689, retrieved January 12th 2024.

Tinti, I. 2022. "Ancient Armenian Platonic Translations." In *Armenia through the Lens of Time: Multidisciplinary Studies in Honour of Theo Maarten van Lint*, edited by F. Alpi et al., 213–233. Leiden.

Trieger, A. 2012. *Inspired Knowledge in Islamic Thought: al-Ghazali's Theory of Mystical Cognition and Its Avicennian Foundation.* London.

Weir, TH. 2012. "Dj̲āhilīya." In *Encyclopaedia of Islam, First Edition (1913–1936)*, edited by M.Th. Houtsma et al., http://dx.doi.org/10.1163/2214-871X_ei1_SIM_1984, retrieved January 13th 2024.

Wolfson, HA. 1935. "The Internal Senses in Latin, Arabic, and Hebrew Philosophic Texts." *The Harvard Theological Review* 28 (2): 69–133.

Index

anamnêsis 64, 69*n*18, 89, 164, 173, 194 (-tically)

contemplation 7–8, 15–16, 16*n*8–9, 19, 53*n*21, 56, 56*n*26, 58, 61–62, 70, 72, 76, 76*n*48, 116–123, 127–128, 133, 135, 163, 166–169, 172, 180, 183, 183*n* 7, 184, 192, 195, 218, 248–251, 259, 261, 265, 275
contemplative 13–14, 20, 23, 163, 193, 259

dialectic(s)/dialectical/dialectician 9, 27*n*5, 34*n*18, 35*n*19, 43–46, 50–57, 75, 87, 107, 112, 120, 123–126, 129, 130–131, 134
doxa 78–79, 82, 82*n*12, 117*n*3, 118*n*6, 132, 244 (en-)

epistêmê 78–86, 117, 117*n* 3–4, 118*n*6, 119, 125*n* 18, 127*n*21, 132–133*n*30, 135

geometry 25–37 (passim), 45, 120, 127, 135, 147
(idea/form of) the good 14, 16, 23, 27, 34, 43–44, 46, 52–54, 57, 62, 74, 76, 89, 116–135, 172*n*26, 242, 244, 250

intelligible (passim)

knowledge (passim). See also *epistêmê*

logistikê 25–37 (passim)

mathematics 2, 25–37 (passim), 43–58 (passim), 90, 103
mysteries/mysticism/mystic/mystical 118, 119*n*9, 120, 123, 126, 129, 146*n*13, 170*n*18, 175*n*33, 179, 184–201 (passim), 207, 214–215, 218, 218*n*109, 240–241, 249*n* 29, 250, 250*n*30, 253*n*41, 257

noêsis/noetic 34–36, 48, 61, 75–76, 82, 90–91, 119–120, 130–131, 142, 147, 155–156, 158–159, 177*n*40, 194–195, 199*n*89, 248, 248*n*26, 249–251

perception (passim)

reason (in Plato) 8–19, 22–23, 74–75, 88–89, 223, 264, 271
reasoning 10, 28*n*6, 54, 80*n*4, 89*n*22, 93, 118, 131, 175, 148, 249, 264
recollection 60–64, 69–69, 72, 74, 76, 79, 84–89, 96– 98, 105–106, 113, 164, 167–168, 173–175, 178*n*43, 180, 191, 191*n* 52, 195, 201, 245*n* 16

sensible (passim)
synoptic (vision) 44, 53–57, 167

tripartition (of the soul) 8–10, 10*n*5, 11, 20, 22*n*29, 247

www.ingramcontent.com/pod-product-compliance
Ingram Content Group UK Ltd.
Pitfield, Milton Keynes, MK11 3LW, UK
UKHW020846290926
15470UKWH00004B/1584

9 789004 747111